AF352940

The Health Benefits of Fruits and Vegetables

The Health Benefits of Fruits and Vegetables

Special Issue Editors

Mercedes Del Río Celestino
Rafael Font Villa

MDPI • Basel • Beijing • Wuhan • Barcelona • Belgrade • Manchester • Tokyo • Cluj • Tianjin

Special Issue Editors

Mercedes Del Río Celestino
Agri-food Laboratory of Córdoba
CAGPDS, Junta de Andalucía
Córdoba, Spain

Rafael Font Villa
Agri-food Laboratory of Córdoba
CAGPDS, Junta de Andalucía
Córdoba, Spain

Editorial Office
MDPI
St. Alban-Anlage 66
4052 Basel, Switzerland

This is a reprint of articles from the Special Issue published online in the open access journal *Foods* (ISSN 2304-8158) (available at: https://www.mdpi.com/journal/foods/special_issues/Fruits_phytochemicals_Epidemiological).

For citation purposes, cite each article independently as indicated on the article page online and as indicated below:

LastName, A.A.; LastName, B.B.; LastName, C.C. Article Title. *Journal Name* **Year**, *Article Number*, Page Range.

ISBN 978-3-03928-829-8 (Hbk)
ISBN 978-3-03928-830-4 (PDF)

Cover image courtesy of Mercedes del Río Celestino.

© 2020 by the authors. Articles in this book are Open Access and distributed under the Creative Commons Attribution (CC BY) license, which allows users to download, copy and build upon published articles, as long as the author and publisher are properly credited, which ensures maximum dissemination and a wider impact of our publications.

The book as a whole is distributed by MDPI under the terms and conditions of the Creative Commons license CC BY-NC-ND.

Contents

About the Special Issue Editors

Mercedes Del Río Celestino, Ph.D. in Biological Sciences (2000), has been working in different institutes and research centers in Spain (IAS-CSIC, Córdoba) and Belgium (Université Libre de Bruxelles), and has been awarded a contract from the "Ramon y Cajal" Spanish post-doc program. She was a permanent researcher in the Department of Plant Breeding and Biotechnology at the IFAPA (Almería, Spain) from 2009 to 2018. She is currently a research scientist at the Agri-Food Laboratory of Córdoba (Spain). She has published over 70 peer-reviewed scientific papers and various book chapters in the area of Plant Breeding. For several years, Dr. Del Río Celestino has been studying the genetic control of the fatty acids of the Ethiopian mustard seed and how to increase the added value of the fruit of zucchini (Cucurbita pepo subsp. pepo) through nutritional quality. Her achievements include obtaining materials from Ethiopian mustard with different profiles of the fatty acid composition of the seed adapted to the semi-arid conditions of Southern Spain and the obtaining of the first TILLING platform in zucchini fruit. Later, after the toxic waste spill of the Aznalcóllar mine in 1999, she was part of the CSIC expert Group and a member of the Bioremediation Network for monitoring and recovering the ecosystem altered by metalloids. She has been a pioneer in genotoxicity and cytotoxicity studies in complex biological matrices (horticultural products) for quickly and economically determining its toxicity. Her research also focuses on the developments of chemometric and spectroscopic methods for determining quality components in horticultural products.

Rafael Font Villa, Ph.D. in Biological Sciences (2003) has been working in Spain (IAS-CSIC, Córdoba) and has obtained a pre-doctoral mobility fellowship for short stays in different research centers in Reino Unido (SAK, Aberdeen), Italy (University of Genoa), and Belgium (Université Libre de Bruxelles). He was a permanent researcher in the Department of Food Science and Health, IFAPA Center La Mojonera (Almería, Spain) from 2009 to 2018. He is currently a research scientist at the Agri-Food Laboratory of Córdoba (Spain). He has published over 70 peer-reviewed scientific papers and various book chapters in the Plant Breeding and Postharvest areas. For several years, Dr. Font has been studying the genetic control of quality components (glucosinolates, fiber, fatty acids) of the Brassica seed and how to increase the added value of the fruit of zucchini (Cucurbita pepo subsp. pepo) through nutritional quality. After the toxic waste spill of the Aznalcóllar mine in 1999, he was a pioneer in using chemometric and spectroscopic methods for determining metalloids in different matrices (soil, plant, animals) and quality components in horticultural products. He has worked in genotoxicity and cytotoxicity studies in complex biological matrices (horticultural products) for quickly and economically determining its toxicity. His research also focuses on postharvest technologies such as controlled ripening, edible coating, temperature management, and chemical treatment methods that are potential tools to reduce fruit and vegetable postharvest losses.

 *foods*

Editorial

The Health Benefits of Fruits and Vegetables

Mercedes del Río-Celestino *and Rafael Font

Agri-Food Laboratory, Council of Agriculture, Fisheries and Rural Development of Andalusia (CAPDER), 14004 Córdoba, Spain; rafaelm.font@juntadeandalucia.es
* Correspondence: mercedes.rio.celestino@juntadeandalucia.es; Tel.: +34-671-532-238

Received: 16 March 2020; Accepted: 17 March 2020; Published: 23 March 2020

Abstract: We edited this Special Issue with the objective of bringing forth new data on the phytochemicals from vegetables and fruits, which are recommended for their health-promoting properties. Epidemiological, toxicological and nutritional studies suggested an association between fruit and vegetable consumption and lower incidence of chronic diseases, such as coronary heart problems, cancer, diabetes, and Alzheimer's disease. In this Special Issue, the protective roles (antioxidant and others bioactivities), new sustainable approaches to determine the quality, and the processing techniques that can modify the initial nutritional and antioxidant content of fruits, vegetables and additives have been addressed.

Keywords: fruits; vegetables; biological studies; processing techniques

---

Qualitative and quantitative evaluations of the health-beneficial properties of fruits, vegetables and additives have been addressed in this Special Issue, using highly sensitive techniques as well as in vivo and in vitro models.

Several biological activities have already been reported for kiwifruit (*Actinidia macrosperma*) cultivars, such as antioxidant, anticancer, anti-inflammatory, and antimicrobial activities. This Special Issue contains the first report on a study supporting the potential anti-hypertensive activities of kiwifruits [1]. The results of this study clearly indicate that the flavonoid-rich extract from *A. macrosperma* shows potential as a food or nutraceutical source of anti-hypertensive agents. Further investigations using experimental animal models and human clinical trials are required to explore the anti-hypertensive properties of *A. macrosperma*.

Drosophila is a reliable model to evaluate the toxicity, genotoxicity and other degenerative processes of food or chemical structures [2]. The results obtained in this eukaryote organism are considered translational and highly specific, as more than 80% of genes related to human disease are homologous in *Drosophila* [3]. Additionally, the proapoptotic capacities against cancer processes have been evaluated through the determination of the cytotoxic, clastogenic, and DNA epigenetic modulator activity against in an in vitro human cancer model (HL60 cell line).

The results from biological activity showed that onion and garlic induced DNA damage in HL60 by necrosis—in concordance with the cytotoxic and DNA-fragmentation results [4]. The chemo-preventive activity of garlic could be associated with its distinctive organosulfur diallyl disulphide compounds (DADS). Supplementary studies are needed to clarify the cell death pathway against garlic and DADS.

Important information is added to the agrifood industry as the new data provided in this Special Issue [5] suggest that short-aged fermented black garlic (13 days) has higher biological activities than the longer-fermented ones, and even more than raw white garlic. This could have important industrial and economic consequences. Taking both the physicochemical and biological data, black garlic aged for 1 day was shown to have the best nutraceutical properties. These findings are relevant for black-garlic-processing agrifood companies, as the cost and processing time are significantly reduced to 13 days aging.

Other studies suggest that freeze-dried Czech beers have no severe potential adverse effects [6]. Moreover, all the substances were able to inhibit tumor cell growth and induce DNA damage in the HL-60 cells at different levels (proapoptotic, single/double strand breaks and methylation status). However, further investigations are needed to clarify the effects of beer to other diets, as well as its important role in the prevention of chronic diseases, which mainly are related to the intake of antioxidants. Despite the promising results obtained for the different freeze-dried beers and their materials, their consumption must be moderated due to the known negative effects induced by alcohol.

New scientific data have been added in this Special Issue in relation to the biological and nutritional effects that food additives (Riboflavin, Tartrazine, Carminic Acid, Erythrosine, Indigotine, and Brilliant Blue) have on time-related degenerative processes. The overall results support the idea that a high chronic intake of food coloring throughout an entire life is not advisable [7], since the in vitro results in HL-60 cells showed that the tested food colorings increased tumor cell growth but did not induce any DNA damage or modifications in the DNA methylation status at their acceptable daily intake concentrations. More research on the biological effects that different concentrations of food colorings could have in model systems is warranted.

Caramel (caramel color E150d-class IV: CAR) is one of the most worldwide consumed additives and is produced by heating carbohydrates from vegetable sources (glucose, sucrose, invert sugar, etc.) in the presence of caramelization promoters (ammonia or ammonium in class III and IV, respectively). The results reported that CAR was neither toxic nor genotoxic and showed antigenotoxic effects in *Drosophila* [8]. Moreover, caramel showed chemopreventive activity and modified the methylation status of HL-60 cell line. Nevertheless, much more information about the mechanisms of gene therapies related to epigenetic modulation by food is necessary.

In this Special Issue, we present a systematic, broad-scale metabolomic investigation of 11 species of dried and fresh edible and medicinal mushrooms [9]. The nutritional component analysis of these selected 11 species suggested that mushrooms contained a wide range of proteins, carbohydrates, amino acids, vitamins, and small molecules. The results showing the chemical components of the selected mushrooms provide fundamental data for the development of functional foods from mushrooms.

Besides the approaches in improving the scientific work to back-up the results, there is a need and clear evolution in the methodologies too in terms of respect to the environment, with more and more conscious labs using greener alternatives to implement sustainable practices from the field to the lab. Standard wet chemistry analytical techniques currently used to determine plant fiber constituents (as those described above) are costly, time-consuming and destructive. Calibration equations based on Near Infrared Reflectance Spectroscopy confirm that this technology could be very useful for the rapid evaluation of acid detergent fiber content in turnip greens and turnip tops (*Brassica rapa* L. subsp. *rapa*) [10].

In addition, the germmplasm of *Brassica napus* and *Brassica rapa* evaluated in this Special Issue displayed variability in the fatty acid composition of its seed oil [11]. Further research will be needed for some accessions having seeds with reduced or increased values of erucic acid content, in order to select valuable genotypes that could be used for both nutritional and industrial applications.

The processing techniques at the industrial scale like pasteurization, concentration, and freezing could also modify the initial nutritional and antioxidant content of citrus juices. In this sense, it has been shown that the juice extraction processes employed have influenced the chemical composition and functional properties of bergamot juice (*Citrus bergamia Risso et Poit., Rutaceae*). Results from this study suggest that extracting juice under the screw press extractor process increased the amount of phytochemical content and total antioxidant activity, more so than using an in-line extractor and hand-squeezing juicing process [12].

Some recent publications have described the beneficial effects of black garlic in the prevention or improvement of cardiovascular diseases, diabetes, obesity, or cancerigenous processes, among others. Black garlic is obtained from raw garlic through a multi-step heating process at a controlled temperature and humidity during a variable period of time. Toledano-Medina et al. [13] have pointed out that an excessive duration in the heating process is detrimental to the final product. The product's antioxidant capacity diminishes after reaching a prior maximum value when the process is extended, although the polyphenol content goes on increasing.

Food and nutrition education, food product development, and marketing efforts are called upon to improve adolescent food choices and make less-processed snack food options more appealing and accessible to diverse consumers. Examples of processing levels of snack food items which have the ability to influence adolescent taste preferences are included in this Special Issue [14]. Ultra-processed and processed foods have a large appeal for adolescents, potentially leading to over consumption and unhealthy snacking decisions. Unprocessed and minimally processed food options are not chosen as frequently as processed and ultra-processed foods when all four processing options are made available to an audience of adolescent children.

The analyses conducted in this Special Issue have showed differences in the expression levels of carotenoid biosynthetic genes and carotenoid content between different *Cucurbita melo* cultivars [15]. These findings will contribute to a foundation for the elucidation of carotenoid biosynthesis in *C. melo*. In addition, further investigations regarding molecular genetics and enzyme activities may help to identify key genes for improving the carotenoid accumulation in melon fruit.

In summary, as most of the authors have stated, further research is required in relation to each and every one of the presented papers in the "The Health Benefits of Fruits and Vegetables"—this assures an exciting time for the researchers in this field and for the general public interested in the relationship between vegetables and health.

Author Contributions: M.d.R.-C. and R.F. conceived and wrote this Editorial. All authors have read and agreed to the published version of the manuscript.

Conflicts of Interest: The authors declare no conflict of interest.

References

1. Hettihewa, S.K.; Hemar, Y.; Rupasinghe, H.P.V. Flavonoid-Rich Extract of *Actinidia macrosperma* (A Wild Kiwifruit) Inhibits Angiotensin-Converting Enzyme In Vitro. *Foods* **2018**, *7*, 146. [CrossRef] [PubMed]

2. Graf, U.; Wurgler, F.E.; Katz, A.J.; Frei, H.; Juon, H.; Hall, C.B.; Kale, P.G. Somatic mutation and recombination test in *Drosophila melanogaster*. *Environ. Mutagen.* **1984**, *6*, 153–188. [CrossRef] [PubMed]

3. Reiter, L.T.; Potocki, L.; Chien, S.; Gribskov, M.; Bier, E. A systematic analysis of human disease-associated gene sequences in *Drosophila melanogaster*. *Genome Res.* **2001**, *11*, 1114–1125. [CrossRef] [PubMed]

4. Fernández-Bedmar, Z.; Demyda-Peyrás, S.; Merinas-Amo, T.; del Río-Celestino, M. Nutraceutic Potential of Two *Allium* Species and Their Distinctive Organosulfur Compounds: A Multi-Assay Evaluation. *Foods* **2019**, *8*, 222. [CrossRef] [PubMed]

5. Toledano Medina, M.Á.; Merinas-Amo, T.; Fernández-Bedmar, Z.; Font, R.; del Río-Celestino, M.; Pérez-Aparicio, J.; Moreno-Ortega, A.; Alonso-Moraga, Á.; Moreno-Rojas, R. Physicochemical Characterization and Biological Activities of Black and White Garlic: *In Vivo* and *In Vitro* Assays. *Foods* **2019**, *8*, 220. [CrossRef] [PubMed]

6. Merinas-Amo, T.; Merinas-Amo, R.; García-Zorrilla, V.; Velasco-Ruiz, A.; Chladek, L.; Plachy, V.; del Río-Celestino, M.; Font, R.; Kokoska, L.; Alonso-Moraga, Á. Toxicological Studies of Czech Beers and Their Constituents. *Foods* **2019**, *8*, 328. [CrossRef] [PubMed]

7. Merinas-Amo, R.; Martínez-Jurado, M.; Jurado-Güeto, S.; Alonso-Moraga, Á.; Merinas-Amo, T. Biological Effects of Food Coloring in *In Vivo* and *In Vitro* Model Systems. *Foods* **2019**, *8*, 176. [CrossRef] [PubMed]

8. Mateo-Fernández, M.; Alves-Martínez, P.; Del Río-Celestino, M.; Font, R.; Merinas-Amo, T.; Alonso-Moraga, Á. Food Safety and Nutraceutical Potential of Caramel Colour Class IV Using *In Vivo* and *In Vitro* Assays. *Foods* **2019**, *8*, 392. [CrossRef] [PubMed]

9. Liu, D.; Chen, Y.-Q.; Xiao, X.-W.; Zhong, R.-T.; Yang, C.-F.; Liu, B.; Zhao, C. Nutrient Properties and Nuclear Magnetic Resonance-Based Metabonomic Analysis of Macrofungi. *Foods* **2019**, *8*, 397. [CrossRef] [PubMed]

10. Obregón-Cano, S.; Moreno-Rojas, R.; Jurado-Millán, A.M.; Cartea-González, M.E.; De Haro-Bailón, A. Analysis of the Acid Detergent Fibre Content in Turnip Greens and Turnip Tops (*Brassica rapa* L. Subsp. rapa) by Means of Near-Infrared Reflectance. *Foods* **2019**, *8*, 364.

11. Cartea, E.; De Haro-Bailón, A.; Padilla, G.; Obregón-Cano, S.; del Río-Celestino, M.; Ordás, A. Seed Oil Quality of *Brassica napus* and *Brassica rapa* Germplasm from Northwestern Spain. *Foods* **2019**, *8*, 292. [CrossRef]

12. Cautela, D.; Vella, F.M.; Laratta, B. The Effect of Processing Methods on Phytochemical Composition in Bergamot Juice. *Foods* **2019**, *8*, 474. [CrossRef]

13. Toledano Medina, M.Á.; Pérez-Aparicio, J.; Moreno-Ortega, A.; Moreno-Rojas, R. Influence of Variety and Storage Time of Fresh Garlic on the Physicochemical and Antioxidant Properties of Black Garlic. *Foods* **2019**, *8*, 314. [CrossRef] [PubMed]

14. Svisco, E.; Byker Shanks, C.; Ahmed, S.; Bark, K. Variation of Adolescent Snack Food Choices and Preferences along a Continuum of Processing Levels: The Case of Apples. *Foods* **2019**, *8*, 50. [CrossRef] [PubMed]

15. Tuan, P.A.; Lee, J.; Park, C.H.; Kim, J.K.; Noh, Y.-H.; Kim, Y.B.; Kim, H.; Park, S.U. Carotenoid Biosynthesis in Oriental Melon (*Cucumis melo* L. var. makuwa). *Foods* **2019**, *8*, 77. [CrossRef] [PubMed]

© 2020 by the authors. Licensee MDPI, Basel, Switzerland. This article is an open access article distributed under the terms and conditions of the Creative Commons Attribution (CC BY) license (http://creativecommons.org/licenses/by/4.0/).

Article

Flavonoid-Rich Extract of *Actinidia macrosperma* (A Wild Kiwifruit) Inhibits Angiotensin-Converting Enzyme In Vitro

Sujeewa K. Hettihewa [1], Yacine Hemar [1] and H. P. Vasantha Rupasinghe [2,*]

[1] School of Chemical Sciences, University of Auckland, 23 Symonds Street, Auckland 1142, New Zealand; krishanthi2001@yahoo.com (S.K.H.); y.hemar@auckland.ac.nz (Y.H.)

[2] Department of Plant, Food, and Environmental Sciences, Faculty of Agriculture, Dalhousie University, Truro, NS B2N 5E3, Canada

* Correspondence: vrupasinghe@dal.ca; Tel.: +1-902-893-6623; Fax: +1-902-893-1404

Received: 6 August 2018; Accepted: 3 September 2018; Published: 5 September 2018

Abstract: Increasing interest in flavonoids in kiwifruit is due to the health-promoting properties of these bioactives. Inhibition of the angiotensin-converting enzyme (ACE) is one of the main therapeutic targets in controlling hypertension. The present study investigated the ACE inhibitory activity of flavonoid-rich extracts obtained from different kiwifruit genotypes. The flavonoid-rich extracts were prepared from fruits of *Actinidia macrosperma*, *Actinidia deliciosa* cv Hayward (Green kiwifruit), and *Actinidia chinensis* cv Hort 16A (Gold kiwifruit) by steeping the lyophilized fruit samples in 70% aqueous acetone, followed by partitioning the crude extracts with hexane. The composition of each extract was analyzed using ultrahigh-performance liquid chromatography-mass spectrometry (UPLC-MS/MS). The ACE inhibitory activity of the fruit extracts was performed using a fluorescence-based biochemical assay. The subclass flavonol was the most abundant group of flavonoids detected in all the extracts tested from three different kiwifruit cultivars. Quercetin-3-*O*-galactoside, quercetin-3-*O*-glucoside, quercetin-3-*O*-rhamnoside, quercetin-3-*O*-rutinoside, quercetin-3-*O*-arabinoglucoside, catechin, epigallocatechin gallate, epigallocatechin, chlorogenic, ferulic, isoferulic, and caffeic acid were prominent phenolics found in *A. macrosperma* kiwifruit. Overall, the flavonoid-rich extract from *A. macrosperma* showed a significantly ($p < 0.05$) high percentage of inhibition (IC_{50} = 0.49 mg/mL), and enzyme kinetic studies suggested that it inhibits ACE activity in vitro. The kiwifruit extracts tested were found to be moderately effective as ACE inhibitors in vitro when compared to the other plant extracts reported in the literature. Further studies should be carried out to identify the active compounds from *A. macrosperma* and to validate the findings using experimental animal models of hypertension.

Keywords: ACE; *Actinidia macrosperma*; flavonoids; polyphenols; hypertension; kiwifruit

---

1. Introduction

Hypertension has become a common risk factor for cardiovascular disease around the world, which affects all ages, from children to adults [1]. It is reported that the angiotensin-converting enzyme (ACE) plays an important role in the renin–angiotensin aldosterone system (RAAS) by cleaving angiotensin I to angiotensin II, which is responsible for increasing blood pressure [2,3]. Thus, inhibition of ACE has been identified as a major therapeutic target for controlling over-activation of RAAS. Prescription drugs such as captopril, ramipril, lisinopril, and enalaprilare are synthetic ACE inhibitors, which have been widely used in the treatment of hypertension [4]. Because these drugs are often reported to have undesirable side effects, interest in searching natural sources of

ACE inhibitors has increased. Most studies have shown that plant extracts, which are rich in specific peptides and flavonoids are found to be effective as natural ACE inhibitors [5–16]. Flavonoid-rich fruits and vegetables and their products have now gained attention for their capability to manage blood pressure [17–20].

Kiwifruit has a reputation for being particularly nutritious and medicinally important [21–25]. The most commonly consumed kiwifruits in the world are Green (*Actinidia deliciosa* (A. Chev.) 'Hayward') and Zespri® Gold kiwifruit (*A. sinensis* Planch. 'Hort16A') [21]. A few other varieties (e.g., baby kiwifruit) are grown commercially, and a number of other varieties are currently being assessed for future commercialization. *Actinidia macrosperma* is a noncommercial type of kiwifruit which is orange and small-sized with large seeds, and has relatively thick, highly-colored, hairless skin. It is reported that the roots and stems of these plants have been extensively employed to treat various ailments, such as leprosy, abscess, rheumatism, arthritis inflammation, jaundice, and abnormal leucorrhoea, in Chinese traditional medicine [26].

The present study investigated the in vitro ACE inhibitory activity of flavonoid-rich extracts obtained from kiwifruit cultivars, namely *A. macrosperma*, *A. deliciosa* (Hayward), and *A. chinensis* (Hort 16A), using a fluorescence-based biochemical assay, followed by determination of the kinetic parameters of the inhibition.

2. Methodology

2.1. Chemicals

The ACE extracted from rabbit lung, histidine-L–hippuryl-L–leucine–chloride (HHL), histidine leucine (His–Leu), NaOH, HCl, ethanol anhydrous, captopril, *O*-phaldialdehyde, dimethyl sulfoxide, and HPLC grade methanol were purchased from Sigma–Aldrich Canada Ltd., Oakville, ON, Canada. Borate saline buffer (100 mM boric acid, 1.5 M sodium chloride, sterile, pH adjusted to 8.3) was purchased from Teknova, Hollister, CA, USA. All other reagents and consumables were purchased from Fisher Scientific, Ottawa, ON, Canada.

2.2. Plant Materials

The fruits of *A. macrosperma* were collected between April and August 2010 at the research orchard of Plant and Food Ltd. in Te Puke Bay, New Zealand. *A. deliciosa* and *A. chinensis* were purchased at a market in Auckland, New Zealand.

2.3. Preparation of Flavonoid-Rich Kiwifruit Extracts

The lyophilized ground fruit sample from *A. macrosperma* kiwifruit (5 g) was steeped in 70% aqueous acetone (100 mL) in a Scott Duran bottle (250 mL) for 6 hours in the dark, with nitrogen gas purging at 30 °C. The extract was filtered through a glass filter and the filtrate was collected in an ice bath. The residue was subjected to re-extraction, and then filtrates were combined together and concentrated on a rotary evaporator (Buchi, New Zealand) below 35 °C under a vacuum. The crude extracts from *A. deliciosa* and *A. chinensis* kiwifruits were also prepared using the same extraction procedure described above. The flavonoid-rich extracts were obtained by partitioning the crude extracts with hexane. The composition of each extract was analyzed using ultra-high performance liquid chromatography-mass spectrometry (UPLC-MS/MS).

2.4. UPLC-MS/MS Analysis of Phenolics in the Kiwifruit Extracts

Analyses of major individual phenolic compounds present in the flavonoid-rich kiwifruit extracts were performed at the Department of Plant, Food, and Environmental Sciences, Faculty of Agriculture, Dalhousie University, Truro, Nova Scotia, Canada, according to the procedure reported by Rupasinghe et al. [27]. All analyses were performed using a Waters H-class UPLC separations module (Waters, Milford, MA, USA), coupled with a Quattro *micro* API MS/MS system and controlled with Masslynx V4.0 data analysis system (Micromass, Cary, NC, USA). The column used was an Aquity BEH C_{18} (100 mm $\times$ 2.1 mm $\times$ 1.7 μm) (Waters, Milford, MA, USA). For the separation of the flavonol, flavan-3-ol, phenolic acid, and dihydrochalcone compounds, a gradient elution was carried out, with 0.1% formic acid in water (solvent A) and 0.1% formic acid in acetonitrile (solvent B) at a flow rate of 0.3 mL/min. A linear gradient profile was used, with the following proportions of solvent A applied at time t (min) (t, A%): (0, 94%), (2, 83.5%), (2.61, 83%), (2.17, 82.5%), (3.63, 82.5%), (4.08, 81.5%), (4.76, 80%), (6.75, 20%), (8.75, 94%), (12, 94%).

Electrospray ionization in negative ion mode (ESI-) was used for the analysis of the flavonol, flavan-3-ol, phenolic acid, and dihydrochalcone compounds. The following conditions were used: Capillary voltage -3000 V, and nebulizer gas (N_2) temperature 375 °C at a flow rate of 0.3 mL/min. The cone voltage (25 to 50 V) was optimized for each compound. Multiple reactions–monitoring (MRM) mode using specific precursor/product ion transitions was employed for quantification in comparison with standards: m/z 301→105 for quercetin (Q), m/z 609→301 for Q-3-O-rutinoside, m/z 463→301 for Q-3-O-glucoside and Q-3-O-galactoside, m/z 448→301 for Q-3-O-rhamnotoside, m/z 595→301 for Q-3-O-peltoside, m/z 273→167 for phloritin, m/z 435→273 for phloridzin, m/z 353→191 for chlorogenic acid, m/z 179→135 for caffeic acid, m/z 193→134 for ferulic acid and isoferulic acid, m/z 289→109 for catechin and epicatechin, and m/z 305→125 for epigallocatechin. The quantification of each analyte was performed using calibration curves created using the external standards.

2.5. Assay for ACE Inhibitory Activity

The in vitro ACE inhibitory activity of flavonoid-rich extracts prepared was performed according to the methods published by Cinq-Mars et al. [9] and Balasuriya and Rupasinghe [15], with slight modifications. The ACE enzyme inhibition assay was carried out with the presence of 2.5 mU ACE in buffer (30 μL), 0.78 mM HHL in buffer (150 μL), sodium borate buffer (pH 8.3) (9 μL), and different concentrations of test compounds (21 μL) in the Eppendorf tubes. All the experimental units, including testing units, negative control (without ACE and inhibitors), positive control (with ACE but no inhibitors), and the standard solution made of 10 mg/L captopril in 10% DMSO in buffer (with ACE), were run in triplicates for each experiment. All the experimental units were incubated at 37 °C using a shaker oven (Model: HP 50, Apollo Instrumentation for Molecular Biology, San Diego, CA, USA) for 1 h, followed by adding 0.35 M NaOH (150 μL) to stop the enzyme activity in the experimental unit. The formation of His–Leu by the cleavage of HHL in the presence of ACE was quantified through a spectrophotometric method based on fluorescence (excitation at 360 nm and emission at 500 nm). The mean fluorescence values of the samples were obtained in triplicates, and the percentage of inhibition of the enzyme was expressed in comparison with the positive control (Equation (1)):

$$\text{Percent enzyme inhibition (\%)} = (1 - (F_{\text{sample}} - F_{\text{sample blank}})/F_{\text{positive control}}) \times 100 \qquad (1)$$

where F = fluorescence.

Dose-responsive enzyme inhibition was determined using different concentrations of each extract. The concentration of the tested extracts which could inhibit 50% of enzyme activity (IC_{50}) was calculated using linear regression analysis plot of % ACE inhibition vs. concentrations of tested extract.

2.6. Determination of Kinetic Parameters of ACE Inhibition

Enzyme kinetic analysis for ACE activity was performed by following the method published by Balasuriya and Rupasinghe [15]. Briefly, each experimental unit consisted of 2.5 mU ACE in buffer (30 μL), the relevant concentration (0.125, 0.25, 0.5, 1, 2, 4, 8 mM) of HHL in buffer (150 μL), and sodium borate buffer (pH 8.3) (30 μL) in an Eppendorf tube. All the experimental units were incubated at 37 °C using a shaker oven (Model: HP 50, Appolo Instrumentation for Molecular Biology, CA, USA) for 1 h, followed by adding 0.35 M NaOH (150 μL) to stop the enzyme activity in the experimental unit. A known concentration of extracts obtained in 70% aqueous acetone from *A. macrosperma*, *A. deliciosa*, and *A. chinensis* kiwifruit was subjected to enzyme kinetic analysis for the ACE activity of inhibitors. Kinetic parameters were calculated by adjusting curves to the Michaelis–Menten kinetic equation (Equation (2)):

$$V_0 = V_{max} \, (S)/(K_m + (S)) \tag{2}$$

where V_0 is the initial reaction rate, V_{max} is the maximum reaction rate, K_m is the Michaelis–Menten constant, and S is substrate concentration. The reaction rate of formation of His–Leu was plotted against the different substrate concentrations to obtain the saturation curves to derive Lineweaver–Burk double reciprocal plots to determine the type of the inhibition (Equation (3)):

$$1/V = (K_m/V_{max}) \, (1/S) + 1/V_{max} \tag{3}$$

K_i (dissociating constant) was determined using the following equation:

$$m_i = m \, (1 + (I)/K_i) \tag{4}$$

where m_i: slope from linear plot from the inhibited reaction, m: slope from linear plot from the noninhibited reaction, [I]: concentration of inhibitor, K_i: dissociating constant of the inhibitor (inhibitory constant).

2.7. Statistical Analysis

All measurements were conducted in triplicate, and the results were expressed as mean $\pm$ SD. The effect of kiwifruit cultivar on the percentage of inhibition of ACE was analyzed through analysis of variance (ANOVA), using Originpro8 software (Origin Lab, Northampton, MA, USA). Pair wise multiple comparisons were evaluated based on Tukey's significance difference test used in origin. Differences at $p < 0.05$ were considered significant.

3. Results and Discussion

3.1. ACE Inhibition

Kiwifruits, namely *A. deliciosa* (Hayward), *A. chinensis* (Hort 16A), and *A. macrosperma*, have been evaluated for their many pharmacological applications towards the management and treatment of human diseases, including antimicrobial activity [28], antioxidant activity [21–23,29] immune modulatory activity [30], and anticancer activity [24]. As far as we are aware, there are no reports on ACE inhibitory activity found in kiwifruits. Therefore, the potential anti-hypertensive activity of fruit extracts from different kiwifruit genotypes grown in New Zealand was evaluated through the inhibition of ACE, a key regulatory enzyme of RAAS [3]. All tested extracts inhibited ACE activity in a dose-dependent manner (Figure 1), with different IC_{50} ranging from 0.49 to 69.54 mg/mL of flavonoid-rich extracts (Table 1). The potential anti-hypertensive activities of kiwifruit cultivars showed that *A. macrosperma* possesses the lowest IC_{50} compared to the other two commercially grown cultivars (Table 1). This, in theory, indicates that the flavonoid-rich extract obtained from *A. macrosperma* possesses strong anti-hypertensive agents. These observations were supported by the significantly higher total phenolic and total flavonoid contents determined by LC-MS/MS (Table 2). The subclass flavonol was the most abundant group of flavonoids detected in all extracts tested, which were obtained from three different kiwifruit cultivars. Quercetin, quercetin-3-galactoside, quercetin-3-glucoside, quercetin-3-rhamnoside, quercetin-3-rutinoside, quercetin-arabinoglucoside, catechin, epigallocatechin gallate, epigallocatechin, chlorogenic, ferulic, isoferulic, and caffeic acids were prominent phenolics found in *A. macrosperma* kiwifruit. Some of these phenolics might be responsible for the strong ACE inhibition activity determined in the extract of *A. macrosperma* fruit. This observation is well supported by the literature, indicating that quercetin, quercetin sugar derivatives, and flavan-3-ols such as catechins, epigallocatechin gallate, and epigallocatechin show ACE inhibitory activity [15,31].

However, there are various reports that demonstrated flavonoids and flavonoid-rich plant extracts inhibit ACE activity. Loizzo et al. [10] reported hypertensive in vitro activities of the MeOH extracts and some flavonoids, namely apigenin, luteolin, kaempferol-3-*O*-α-arabinopyranoside, kaempferol-3-*O*-β-galactopyranoside, and quercetin-3-*O*-α-arabinopyranoside, isolated from *Ailanthus excelsa* (Roxb). The ACE inhibitory properties of flavonoid-rich apple peel extracts and selected quercetin derivatives have been reported [15]. The research findings carried out by Persson et al. [32] showed a dose-dependent ACE inhibition of major flavan-3-ols, namely, catechins, (−)-epicatechins, (−)-epigallocatechin, (−)-epicatechin gallate, and (−)-epigallocatechin gallate, isolated from green and black tea using a human umbilical vein endothelial cells (HUVEC) culture model. It has been reported that aqueous extracts of *Ginkgo biloba*, which is rich in quercetin derivatives as the major flavonoids, has a higher ACE inhibitory activity than that of ethanol extracts [33]. Oh et al. reported that fractions of stonecrop (*Sedum sarmentosum*) and five purified flavonols had ACE inhibitory properties [5]. However, one of the limitations of in vitro assays of ACE inhibitory activity of flavonoids-rich extracts is that most of the flavonoids exist as their metabolites in the central circulation.

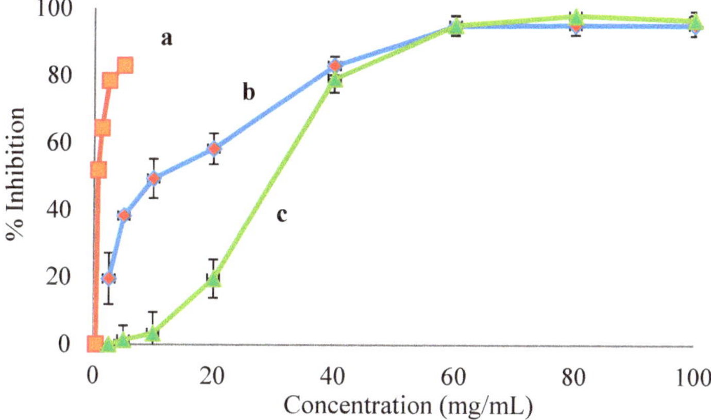

Figure 1. The dose–response curves of the percentage of angiotensin-converting enzyme (ACE) activity inhibition by the extracts of different kiwifruit genotypes: (**a**) *A. macrosperma* (wild kiwifruit); (**b**) *A. chinensis* cv Hort 16A (gold kiwifruit); and (**c**) *A. deliciosa* cv Hayward (green kiwifruit).

Table 1. IC_{50} of ACE inhibitory activity of extracts from different kiwifruit genotypes.

Kiwifruit Genotypes	IC_{50} (mg/mL)
A. macrosperma	0.49
A. chinesis cv Hort 16A	12.81
A. deliciosa cv Hayward	30.49

ACE is a zinc-containing peptidyldipeptide hydrolase where the active site is known to have three parts. Therefore, it is well reported that the ACE inhibitory in vitro activity of flavonoids may be due to the formation of chelate complexes with the zinc atom within the active center of zinc-dependent metallopeptidases, or possibly due to the formation of hydrogen bridges between the inhibitor and phenolics near at the active site [34]. Thus, the presence of phenolic and flavonoid content in the extract would have contributed towards ACE inhibition.

Table 2. Polyphenols content of three kiwifruit genotypes measured using UPLC-MS/MS.

Group/Name of the Flavonoid	Concentration of Phenolics (µg/g DW)		
	A. macrosperma	*A. chinensis*	*A. deliciosa*
Flavonol			
Quercetin-3-*O*-Galactoside	470.9	205.19	441.39
Quercetin-3-*O*-Glucoside	4.16	0.45	0.14
Quercetin Arabinoglucoside	2.53	0.06	0.21
Quercetin-3-*O*-Rhamnoside	2.99	0.61	0.31
Quercetin	2.56	nd	0.17
Quercetin-3-*O*-Rutinoside	1.96	0.29	0.55
Flavanol			
Epigallocatechin	1.55	0.61	0.46
Catechin	54.31	0.75	0.30
Epicatechin	0.91	5.15	0.74
Epigallocatechingallate	0.75	nd	0.40
Dihydrochalcones			
Phloridzin	3.12	2.03	5.08
Phloritin	0.14	0.21	0.14
Phenolic acids			
Chlorogenic acid	1.97	0.39	0.28
Caffeic acid	1.64	0.04	0.08
Ferulic acid	4.70	0.42	0.60
Isoferulic acid	32.71	15.12	28.17
Total Phenolics	586.9	231.32	479.02

DW: dry weight of the fruit; nd: not detected. UPLC-MS/MS: ultrahigh-performance liquid chromatography-mass spectrometry.

3.2. Determination of Kinetic Parameters of ACE Inhibition

To study the type of inhibition of the ACE activity, enzyme kinetic studies were performed. Table 3 shows the kinetics of ACE activity without an inhibitor and in the presence of a known concentration of *A. macrosperma* (2.64 mg/L), *A chinensiscv* Hort 16 A (13 mg/mL), and *A. deliciosa* cv Hayward (31 mg/mL). ACE activity showed a Michaelis–Menten mechanism. The kinetic parameters obtained from these curves are shown in Table 3. The maximum rates of substrate hydrolysis (V_{max}) and Michaelis–Menten constant (K_m) were determined to characterize the kind of inhibition exerted by the extracts. The V_{max} was not significantly altered by each inhibitor, which suggests that it was a noncompetitive inhibitor for ACE. These findings are in agreement with the studies on flavonoid-rich apple peel extracts [15], flavan-3-ols, and anthocyanins-rich food extracts [6,34].

Table 3. Enzyme kinetic parameters of ACE inhibition by the extracts of three different kiwifruit genotypes.

Extract	Concentration Tested (mg/mL)	K_m (mM)	V_{max} (mM/min)	K_i (mg/mL)
No inhibitor	0	0.074	0.024	
A. chinesis cv Hort 16A	13.00	2.036	0.033	44.516
A. deliciosa cv Hayward	31.00	4.849	0.033	64.041
A. macrosperma	2.60	7.258	0.069	78.312

4. Conclusions

Several biological activities have already been reported for kiwifruit cultivars, such as antioxidant, anticancer, anti-inflammatory, and antimicrobial activities. This is the first report on a study supporting the potential anti-hypertensive activities of kiwi fruits. The results of this study clearly indicate that the flavonoid-rich extract from *A. macrosperma* shows potential as a food or nutraceutical source of anti-hypertensive agents. Overall, the flavonoid-rich extract from *A. macrosperma* showed a significantly ($p < 0.05$) high percentage of inhibition (IC_{50} = 0.49 mg/mL) in vitro. Kinetic determinations suggested that the flavonoids-rich extract obtained from *A. macrosperma* kiwifruit possibly inhibits enzyme activity either through nonspecific binding to the enzyme or by competing with the substrate for the active side. The subclass flavonol was the most abundant group of flavonoids detected in the tested extracts of three different kiwifruit genotypes. Quercetin-3-galactoside, quercetin-3-glucoside, quercetin-3-rhamnoside, quercetin-3-rutinoside, quercetin-arabinoglucoside, catechin, epigallocatechin gallate, epigallocatechin, chlorogenic, ferulic, isoferulic, and caffeic acid were prominent phenolic compounds found in *A. macrosperma* kiwifruit. Further investigations using experimental animal models and human clinical trials are required to explore the anti-hypertensive properties of *A. macrosperma*.

Author Contributions: Conceptualization, Y.H.; Data curation, S.K.H.; Funding acquisition, Y.H. and H.P.V.R.; Investigation, S.K.H.; Supervision, Y.H. and H.P.V.R.; Writing—original draft, S.K.H.; Writing—review & editing, H.P.V.R.

Funding: This research was funded by the ZESPRI kiwifruit industry in New Zealand and the Canada Research Chair program of Canada.

Conflicts of Interest: The authors declare no conflicts of interest.

References

1. Dojki, F.K.; Bakris, G.L. Blood Pressure Control and Cardiovascular/Renal Outcomes. *Endocrinol. Metab. Clin. N. Am.* **2018**, *47*, 175–184. [CrossRef] [PubMed]
2. Atlas, S.A. The renin-angiotensin aldosterone system: Pathophysiological role and pharmacologic inhibition. *J. Manag. Care Pharm.* **2007**, *13*, S9–S20. [CrossRef] [PubMed]
3. Balasuriya, B.W.N.; Rupasinghe, H.P.V. Plant flavonoids as angiotensin-converting enzyme inhibitors in regulation of hypertension. *Funct. Foods Health Dis.* **2011**, *5*, 172–188.
4. Quan, A. Fetopathy associated with exposure to angiotensin-converting enzyme inhibitors and angiotensin receptor antagonists. *Early Hum. Dev.* **2006**, *82*, 23–28. [CrossRef] [PubMed]
5. Oh, H.; Kang, D.G.; Kwon, J.W.; Kwon, T.O.; Lee, S.Y.; Lee, D.B.; Lee, H.S. Isolation of angiotensin-converting enzyme (ACE) inhibitory flavonoids from Sedum sarmentosum. *Biol. Pharm. Bull.* **2004**, *27*, 2035–2037. [CrossRef] [PubMed]
6. Actis-Goretta, L.; Ottaviani, J.I.; Fraga, C.G. Inhibition of angiotensin-converting enzyme activity by flavanol-rich foods. *J. Agric. Food Chem.* **2006**, *54*, 229–234. [CrossRef] [PubMed]
7. Ottaviani, J.I.; Actis-Goretta, L.; Villordo, J.J.; Fraga, C.G. Procyanidin structure defines the extent and specificity of angiotensin I converting enzyme inhibition. *Biochimie* **2006**, *88*, 359–365. [CrossRef] [PubMed]
8. Braga, F.C.; Serra, C.P.; Viana Júnior, N.S.; Oliveira, A.B.; Côrtes, S.F.; Lombardi, J.A. Angiotensin-converting enzyme inhibition by Brazilian plants. *Fitoterapia* **2007**, *78*, 353–358. [CrossRef] [PubMed]
9. Cinq-Mars, C.D.; Li-Chan, E.C.Y. Optimizing angiotensin I-converting enzyme inhibitory activity of Pacific hake (*Merluccius productus*) fillet hydrolysate using response surface methodology and ultrafiltration. *J. Agric. Food Chem.* **2007**, *55*, 9380–9388. [CrossRef] [PubMed]
10. Loizzo, M.R.; Said, A.; Tundis, R.; Rashed, K.; Statti, G.A.; Hufner, A.; Menichini, F. Inhibition of angiotensin-converting enzyme (ACE) by flavonoids isolated from *Ailanthus excelsa* (Roxb) (simaroubaceae). *Phytother. Res.* **2007**, *21*, 32–36. [CrossRef] [PubMed]
11. Farzamirad, V.; Aluko, R.E. Angiotensin-converting enzyme inhibition and free-radical scavenging properties of cationic peptides derived from soybean protein hydrolysates. *Int. J. Food Sci. Nutr.* **2008**, *59*, 428–437. [CrossRef] [PubMed]
12. Hong, F.; Ming, L.; Yi, S.; Zhanxia, L.; Yongquan, W.; Chi, L. The antihypertensive effect of peptides: A novel alternative to drugs? *Peptides* **2008**, *29*, 1062–1071. [CrossRef] [PubMed]
13. Guang, C.; Phillips, R.D. Plant food-derived angiotensin I converting enzyme inhibitory peptides. *J. Agric. Food Chem.* **2009**, *57*, 5113–5120. [CrossRef] [PubMed]
14. Udenigwe, C.C.; Lin, Y.S.; Hou, W.C.; Aluko, R.E. Kinetics of the inhibition of renin and angiotensin I-converting enzyme by flaxseed protein hydrolysate fractions. *J. Funct. Foods* **2009**, *1*, 199–207. [CrossRef]
15. Balasuriya, N.; Rupasinghe, H.P.V. Antihypertensive properties of flavonoid-rich apple peel extract. *Food Chem.* **2012**, *135*, 2320–2325. [CrossRef] [PubMed]

16. Aluko, R.E. Structure and function of plant protein-derived antihypertensive peptides. *Curr. Opin. Food Sci.* **2015**, *4*, 44–50. [CrossRef]

17. Hodgson, J.M.; Devine, A.; Puddey, I.B.; Chan, S.Y.; Beilin, L.J.; Prince, R.L. Tea intake is inversely related to blood pressure in older women. *J. Nutr.* **2003**, *133*, 2883–2886. [CrossRef] [PubMed]

18. Ahmed, F.; Siddesha, J.M.; Urooj, A.; Vishwanath, B.S. Radical scavenging and angiotensin-converting enzyme inhibitory activities of standardized extracts of Ficus racemosa stem bark. *Phytother. Res.* **2010**, *24*, 1839–1843. [CrossRef] [PubMed]

19. Lecour, S.; Lamont, K.T. Natural polyphenols and cardioprotection. *Mini-Rev. Med. Chem.* **2011**, *11*, 1191–1199. [PubMed]

20. Balasuriya, N.; Rupasinghe, H.P.V.; Sweeney, M.; McCarron, S.; Gottschall-Pass, K. Antihypertensive effects of apple peel extract on spontaneously hypertensive rats. *Pharmacologia* **2015**, *6*, 371–376.

21. Ferguson, A.R. The need for characterization and evaluation of germplasm: Kiwifruit as an example. *Euphytica* **2007**, *154*, 371–382. [CrossRef]

22. Nishiyama, I. Fruits of the Actinidia Genus. *Adv. Food Nutr. Res.* **2007**, *52*, 293–324. [PubMed]

23. Fiorentino, A.; D'Abrosca, B.; Pacifico, S.; Mastellone, C.; Scognamiglio, M.; Monaco, P. Identification and assessment of antioxidant capacity of phytochemicals from kiwi fruits. *J. Agric. Food Chem.* **2009**, *57*, 4148–4155. [CrossRef] [PubMed]

24. Hunter, D.C.; Greenwood, J.; Zhang, J.; Skinner, M.A. Antioxidant and 'natural protective' properties of kiwifruit. *Curr. Top. Med. Chem.* **2011**, *11*, 1811–1820. [CrossRef] [PubMed]

25. Stonehouse, W.; Gammon, C.S.; Beck, K.L.; Conlon, C.A.; von Hurst, P.R.; Kruger, R. Kiwifruit: Our daily prescription for health. *Can. J. Physiol. Pharmacol.* **2012**, *91*, 442–447. [CrossRef] [PubMed]

26. Lu, Y.; Zhao, Y.; Fu, C. Biological activities of extracts from a naturally wild kiwifruit, *Actinidia macrosperma*. *Afr. J. Agric. Res.* **2011**, *6*, 2231–2234.

27. Rupasinghe, H.P.V.; Wang, L.; Huber, G.M.; Pitts, N.L. Effect of baking on dietary fibre and phenolics of muffins incorporated with apple skin powder. *Food Chem.* **2008**, *107*, 1217–1224. [CrossRef]

28. Lu, Y.; Zhao, Y.; Fu, C. Preliminary evaluation of antimicrobial and cytotoxic activities of extracts from *Actinidia macrosperma*. *Adv. Mater. Res.* **2012**, *455*, 1200–1203. [CrossRef]

29. Latocha, P.; Krupa, T.; Wołosiak, R.; Worobiej, E.; Wilczak, J. Antioxidant activity and chemical difference in fruit of different *Actinidia* sp. *Int. J. Food Sci. Nutr.* **2010**, *61*, 381–394. [CrossRef] [PubMed]

30. Lu, Y.; Fan, J.; Zhao, Y.; Chen, S.; Zheng, X.; Yin, Y.; Fu, C. Immunomodulatory activity of aqueous extract of *Actinidia macrosperma*. *Asia Pac. J. Clin. Nutr.* **2007**, *16* (Suppl. 1), 261–265.

31. Actis-Goretta, L.; Ottaviani, J.I.; Keen, C.L.; Fraga, C.G. Inhibition of angiotensin-converting enzyme (ACE) activity by flavan-3-ols and procyanidins. *FEBS Lett.* **2003**, *555*, 597–600. [CrossRef]

32. Persson, I.A.L.; Josefsson, M.; Persson, K.; Andersson, R.G. Tea flavanols inhibit angiotensin-converting enzyme activity and increase nitric oxide production in human endothelial cells. *J. Pharm. Pharmacol.* **2006**, *58*, 1139–1144. [CrossRef] [PubMed]

33. Pinto, M.D.S.; Kwon, Y.I.; Apostolidis, E.; Lajolo, F.M.; Genovese, M.I.; Shetty, K. Potential of *Ginkgo biloba* L. leaves in the management of hyperglycemia and hypertension using in vitro models. *Bioresour. Technol.* **2009**, *100*, 6599–6609. [CrossRef] [PubMed]

34. Ojeda, D.; Jiménez-Ferrer, E.; Zamilpa, A.; Herrera-Arellano, A.; Tortoriello, J.; Alvarez, L. Inhibition of angiotensin converting enzyme (ACE) activity by the anthocyanins delphinidin- and cyanidin-3-*O*-sambubiosides from Hibiscus sabdariffa. *J. Ethnopharmacol.* **2010**, *127*, 7–10. [CrossRef] [PubMed]

© 2018 by the authors. Licensee MDPI, Basel, Switzerland. This article is an open access article distributed under the terms and conditions of the Creative Commons Attribution (CC BY) license (http://creativecommons.org/licenses/by/4.0/).

Article

Nutraceutic Potential of Two *Allium* Species and Their Distinctive Organosulfur Compounds: A Multi-Assay Evaluation

Zahira Fernández-Bedmar [1,*], Sebastián Demyda-Peyrás [2], Tania Merinas-Amo [1] and Mercedes del Río-Celestino [3]

[1] Department of Genetics, University of Córdoba, Campus Rabanales, Gregor Mendel Building, 14071 Córdoba, Spain; tania.meram@gmail.com

[2] Institute of Veterinary Genetics (IGEVET), Facultad de Ciencias Veterinarias, UNLP-CONICET, Universidad Nacional de La Plata, La Plata 1900, Argentina; sdemyda@igevet.gob.ar

[3] Agrifood Laboratory, CAPDER, Avda. Menéndez Pidal s/n, 14004 Córdoba, Spain; mercedes.rio.celestino@juntadeandalucia.es

[*] Correspondence: b12febez@uco.es or zanoferbed@gmail.com; Tel.: +34-957-218-674

Received: 23 May 2019; Accepted: 18 June 2019; Published: 21 June 2019

Abstract: This study aimed to evaluate the biological activities of two *Allium* species (garlic and onion) as well as diallyl disulphide (DADS) and dipropyl disulphide (DPDS) as their representative bioactive compounds in a multi-assay experimental design. The genotoxic, antigenotoxic, and lifespan effects of garlic, onion, DADS, and DPDS were checked in *Drosophila melanogaster* and their cytotoxic, pro-apoptotic, and DNA-clastogenic activities were analyzed using HL60 tumoral cells. All compounds were non-genotoxic and antigenotoxic against H_2O_2-induced DNA damage with a positive dose-response effect and different inhibition percentages (the highest value: 95% for DADS) at all tested concentrations. Daily intake of *Allium* vegetables, DADS, or DPDS had no positive effects on flies' lifespan and health span. Garlic and DADS exerted the highest cytotoxic effects in a positive dose-dependent manner. Garlic and DADS exerted a DNA-internucleosomal fragmentation as an index of induced proapoptotic activity on HL60 cells. *Allium* vegetables and DADS were able to induce clastogenic strand breaks in the DNA of HL60 cells. This study showed the genomic safety of the assayed substances and their protective genetic effects against the hydrogen peroxide genotoxine. Long-term treatments during the whole life of the *Drosophila* genetic model were beneficial only at low-median concentrations. The chemo-preventive activity of garlic could be associated with its distinctive organosulfur DADS. We suggest that supplementary studies are needed to clarify the cell death pathway against garlic and DADS.

Keywords: garlic; onion; antigenotoxicity; longevity; cytotoxicity; comet assay

1. Introduction

The Mediterranean diet is one of the best nutritional patterns for humans due to its demonstrated beneficial effects on health. This diet, which is based on the high consumption of fruit, vegetables, wine, olive oil, and fish as the main animal protein contribution, is a type of a healthy and well-balanced food intake [1]. Today, most of the studies asserting these well-being effects agree to point the increased antioxidant and phenolic contents as the cause of its properties [2]. Diet-derived antioxidants are implicated in maintaining a balanced homeostasis and scavenging reactive oxygen species (ROS) as a major part of a highly efficient defensive biological network, which neutralizes the oxidative stress and complements the endogenous defense enzymes [3].

Garlic (*Allium sativum*) and onion (*Allium cepa*) are two native vegetables from Asia and are widely used in different gastronomic cultures and traditional medicines for centuries [4]. According to the Food Administration Organization (FAO), these vegetables are two of the most important crops worldwide with a production of 20,000 tons of garlic and 100,000 tons of onion, respectively, in 2015, which shows a trend toward an increased consumption in the recent years due to the expansion of the Mediterranean and Asian cuisine. Both volatile and non-volatile compounds are found in Allium species. Non-volatile compounds named sapogenins, saponins, and flavonoids whose contents are differentially distributed in garlic and onions. The distinctive flavonoids present in onions are different than in garlic (quercetin, kaempferol, and luteolin in onion and myricetin, apigenin and quercetin in garlic) [5]. Quercetin, which is the major flavonoid present in onions, helps prevent glycation of collagens, which is a leading causative factor for the development of cardiovascular complication in diabetic patients. Moreover, quercetin and kaempferol from onions also possess anticarcinogenic properties [6,7]. With regard to the phenolic acids present in the matrix of garlic and onions, gallic acid is one of them and it has several reported bioactivities such as antineoplastic, bacteriostatic, antioxidant, and anticancer. Protocatechuic acid is found in these vegetables as well. This molecule has been found to have an antihepatotoxic, anti-inflammatory, free radical scavenger, including chemopreventive and apoptotic bioactivities among others [8,9]. Nevertheless, despite the above described differences between garlic and onion non-volatile content, these two species contain a unique and distinctive group of volatile organosulfur compounds.

These vegetables have been linked to preventive effects against several diseases such as cancer, obesity, diabetes type-2, coronary heart disease, and hypertension, among others [5,10–12]. These pleiotropic effects were associated with the high content of thiosulfinates, which is a group of volatile organosulfur compounds that originated from the decomposition of the allicin. These are also responsible for their typical pungent aroma and taste [7,13–15]. However, both vegetables showed a high variability with respect to the thiosulfinate profiles among strains including diallyl sulfide (DAS), diallyl disulfide (DADS), and diallyl trisulfide (DATS) normally higher in garlics and dipropyl sulfide (DPS) and dipropyl disulfide (DPDS) higher in onions [16,17]. Garlic oils and extracts were associated with several health-benefit activities, such as a protective capacity against DNA damage induced by oxidative stress, increased hydrogen peroxide (H_2O_2) scavenging activity, and ability to reduce the bioactivity of carcinogens and tumor cell proliferation [18–21]. These capacities were directly linked to DADS, one of their major and most garlic distinctive constituents, which was widely studied and characterized as non-genotoxic, antigenotoxic, inhibitor of cell proliferation and pro-apoptotic in different cancer cell lines like leukemia, colon, prostate, lung, bladder, and skin [22–29].

On the other hand, onions are more versatile vegetables that can also be consumed as fresh and processed products. Both forms also showed a high oxy-radical scavenging capacity [30] as well as an antigenotoxic effect [31]. In addition, garlic ethanolic extracts and oils showed antimutagenic activity [32] and also decreased the viability and increased the apoptosis in several cancer cell lines like HL60, MDA-MB-231, A549, and B16F10 [33–36]. In this case, their pro-healthy properties were widely related to DPDS, which is one of its most representative organosulfur compounds. This molecule was previously associated with strong anticarcinogenic activity [37] and a protective effect against a DNA strand break and oxidative damage [38,39]. Nevertheless, this compound had no anti-tumor effects in mice [40], which means it did not decrease tumor cell growth and did not induce DNA-internucleosomal fragments on cancer cell lines by acting alone [29,40–42].

Then, we performed a qualitative and quantitative evaluation of the health-beneficial properties of garlic, onion, and their representative organosulfur compounds (DADS and DPDS) in a multi-assay experimental design using in vivo and in vitro models. We assessed their genotoxic, antigenotoxic, and lifespan effects in *Drosophila melanogaster* flies, which is a widely used experimental model closely related to humans. Additionally, we evaluated their proapoptotic capacities against cancer processes through the determination of their cytotoxic, clastogenic, and DNA epigenetic modulator activity against in an in vitro human cancer model (HL60 cell line).

2. Materials and Methods

2.1. Allium Vegetables and Single Compounds

Two *Allium* species and two of its most distinctive organosulfur compounds were assayed. Garlic (*Allium sativum*, purple variety) and onion (*Allium cepa*, Victoria variety) were purchased in a local market of Cordoba (Spain). Thiosulfinates, DADS from garlic, and DPDS from onions, which had 80% and $\geq$97% of purity, respectively, were purchased from Sigma (St. Louis, MI, USA, Cat numbers 317691 and 43550, respectively) and were used without further purification.

2.2. Preparation of the Samples

Garlic samples and onions were washed twice with distilled water, cut in slim slices, and freeze-dried at −80 °C. After that, both samples were lyophilized, pulverized with a mortar pestle, sieved, and stored at 25 °C in the dark until use.

2.3. In Vivo Assays

2.3.1. Somatic Mutation and Recombination Test (SMART)

Two *Drosophila melanogaster* strains carrying visible wing genetic markers were used in our experimental design: the flare (flr) strain flr^3/ln (3LR) TM3, Bd^s and the multiple wing-hair (mwh) strain mwh/mwh. The multiple wing hairs (mwh, 3_0.3) marker is a recessive viable mutation in homozygous flies, which produces multiple-hairs trichomes in the fly adult body [43]. The flare (flr^3, 3_38.3) marker is a homozygous recessive lethal mutation, which produces malformed individual wing hairs in somatic cells of larvae. The flr^3 allele is retained in a balancer chromosome carrying multiple inversions and a homozygous lethal dominant visible marker expressed in the edge wing [44].

Genotoxicity was determined using the SMART test as described by Graf and Wurgler [45] including a negative control of pure water. The antigenotoxic activity was also determined using a modified SMART test following our standard protocols [46]. Optimally virgin flr^3/ln (3LR) TM3, ri p^p sep bx^{34e} e^s Bd^S (flare) females were crossed with mwh/mwh strain males, obtaining 72 h transheterozygous F1 larvae after an 8-hour egg-laying on fresh yeast. Larvae were fed with *Drosophila* Instant Medium (Formula 4-24, Carolina Biological Supply, Burlington, NC, USA) in 4 mL vials. Genotoxicity assays consisted of eight experimental groups by supplementing the base larvae food (0.85 g) with different concentrations of onion (0.625 and 5 mg/mL), garlic (0.625 and 5 mg/mL), DADS (4 mM and 34 mM), and DPDS (4 mM and 33 mM). The concentration ranges of single compounds were selected to mimic those described in the fresh *Allium* sp. and they cover the lower and higher estimated content values [47]. Negative (distilled water) and positive (0.12 M H_2O_2) concurrent controls were included. Antigenotoxicity experimental design was similar to the genotoxicity assays by concurrently treating the larvae with the tested substances supplemented with H_2O_2 (0.12 M) as a positive geno-toxicant control. The emerged adults in each group were stored in 70% ethanol until analysis.

Forty wings of heterozygous flies (mwh/flr^3) treated with each compound and concentration were removed and mounted on slides with Faure's solution (Arabic gum 50 g (Sigma, Cat Number G9752), glycerol of 20 mL (Sigma, Cat Number G5516), chloral hydrate of 50 g (Sigma, Cat Number C8383), and distilled water of 50 mL). Both dorsal and ventral surfaces were screened under a bright light microscope at 400× magnification to detect small single spots (1–2 mwh or flr^3 cells), large single spots (three or more cells), and twin spots (adjacent mwh and flr^3 cells). Single spots are produced by gene mutation, somatic recombination, and deletion between the two markers. Twin spots are produced uniquely by recombination between the flr^3 marker and the centromere.

In order to evaluate the possible genotoxic effect, the frequencies of total spots per wing of each series were statistically compared with the total spots of the negative control with the non-parametric U-test of Mann, Whitney, and Wilcoxon [48]. Antigenotoxicity was determined as the inhibition percentage (IP) using the total spots per wing determined at each concentration with the following formula [49].

$$IP = ((a - b)/a) \times 100, \tag{1}$$

where a represents the frequency of total spots induced by the treatment with genotoxine alone, and b represents the frequency of total spots obtained with genotoxine plus substance tested in the different combined treatments.

2.3.2. Longevity Assays

All the longevity experiments were performed following our standard procedures [50]. Transheterozygous larvae from a 12-h egg-laying with the same genetic background described above were used in the life and health-span trials. Health span is the healthy adult period of unimpaired life that precedes functional decline [51]. It is important to consider the quality of a prolonged life and, for this reason, health span is a new focus in aging research. Synchronized larvae of 72 ± 12 h were clustered in groups of 100 individuals in glass vials with 0.85 g of *Drosophila* Instant Medium in 4 mL of water solutions of the different experimental concentrations assayed (0.625, 1.25, 2.5, and 5 mg/mL for *Allium* vegetables, 4, 8, 16, and 33 mM for DPDS, and 4, 8, 17, and 34 mM for DADS). The emerged flies were anesthetized under CO_2, separated into 10 single-sex groups, transferred to longevity vials and fed with the same treated medium during the whole experimental design. A concurrent treatment was also included using distilled water as a negative control. The survivors were counted and the medium was renewed twice a week until all individuals die. Survival curves were plotted as estimated by the Kaplan-Meier method and the statistical significance of curves were assessed using the Log-Rank (Mantel-Cox) method using the SPSS 15.0 statistics software (SPSS Inc. Headquarters, Chicago, IL, USA).

2.4. In Vitro Assays

2.4.1. Cell Line Cultures and Cytotoxicity Assay

In vitro assays were performed using the promyelocytic leukemia HL60 cell line. Some of the genetic characteristics of this tumor cell line are the following: karyotypic abnormalities (monosomy, trisomy, and tetrasomy), and different chromosomal translocations. On the molecular genetic level, the HL60 cell line has deletions in the p53 gene on chromosome 17pl3 and one allele of the GM-CSF gene on chromosome 5q21–q23 is rearranged and partly deleted as well [52].

Cells were cultured at 2.5×10^5 cells/mL following our standard protocol [53] in complete RPMI 1640 medium (BioWhittaker, Basel, Switzerland; BE12-167F) containing 10% heat-inactivated fetal bovine serum (BioWhittaker, de14-801F), L-glutamine 200 mM (Sigma, G7513), and antibiotic-antimycotic solution (Sigma, A5955) at 37 °C in a humidified atmosphere of 5% CO_2. Two passes per week were performed and the experiments were carried with cells with no more than 20 passes. Cell viability was evaluated by the Trypan blue exclusion assay. To ensure the proper behavior of the cell line, proliferation was followed at 0, 4, 24, 48, and 72 h checkpoints. Control cells doubled every 24 four-hour exponentially ($y = 100446e^{0.0345x}$), which reached the maximum at 72 h. Cells (1×10^5 cells/mL) were seeded and incubated for 72 h in 96 well plates supplemented with six different concentrations *Allium species* (ranging from 0.002 mg/mL to 0.06 mg/mL) and 6 different concentrations of thiosulfinates (ranging from 0.012 mM to 0.4 mM). A concurrent negative control (base medium without supplementation) was also run. After incubation, Trypan blue was added to the cell suspension (1:1 ratio) and cells were counted in a Neubauer chamber under an inverted microscope at 100× magnification. Cell viability was expressed as a percentage of survival with respect to control after a 72-h period. IC_{50} values (concentration of tested molecule causing 50% of cell growth inhibition) and EC_{50} values (concentration of a tested substance that complements a system and gives a half-maximal growth response) were estimated for each treatment. Viability curves were plotted as mean viability ± standard deviation of three independent replicas in each substance and concentration.

2.4.2. Inter-Nucleosomal DNA Fragmentation Assay

HL60 cells (1.5×10^6 cells/mL) were incubated with the same compounds and concentrations as in cytotoxicity assays for 5 h in 12-well plates. Thereafter, cells were harvested, centrifuged at 2500 rpm. for 5 min, and washed with phosphate buffer saline (PBS). Total DNA was extracted using a commercial DNA-extraction kit (Blood Genomic DNA Extraction Mini Spin Kit, Canvax Biotech, Cordoba, Spain), according to the manufacturer's instructions and subsequently treated with RNase overnight in order to eliminate a false positive. DNA yielding was quantified in a Nanodrop™ (Thermo Scientific, Madrid, Spain). A total of 1.5 µg of DNA per sample was electrophoresed in a 2% agarose gel, stained with ethidium bromide, and run by 120 m at 60 V. Internucleosomal DNA fragmentation was determined by the presence of ladder band patterns with 200 bp multiple fragments.

2.4.3. Evaluation of DNA Breakage Ability: Comet Assay

DNA strand break ability of the compounds was determined by the alkaline comet assay, as described Olive and Banáth [54] with minor modifications. HL60 cells (5×10^5 cells) were plated in 1.5 mL of culture medium supplemented with different concentrations of onion (0.004, 0.016, and 0.06 mg/mL), garlic (0.002, 0.004, and 0.008 mg/mL), DPDS (0.025, 0.1, and 0.4 mM) and DADS (0.01, 0.025, and 0.05 mM) and incubated for 5 h. After treatment, cells were washed and adjusted to 6.25×10^4 cells/mL in PBS. Then, cells (1.6×10^4) were suspended in a 75 µL pre-warmed low melting point agarose (A4018, Sigma) and 50 µL of the suspension were rapidly spread on microscope slides and covered with coverslips. After gelling for 30 min at RT, the coverslips was gently removed and the slides were put in a tank filled with lysis solution (2.5M NaCl (S3014, Sigma), 100mM Na-EDTA (1.09992, Sigma), 10mM Tris (T4661, Sigma), 250mM NaOH (S8045, Sigma), 10% DMSO (D8418, Sigma), and 1% Triton X-100 (T8787, Sigma), pH = 13 at 4 °C for 1 h. Next, slides were removed from the lysis solution and incubated in alkaline electrophoresis buffer (300 mM NaOH (S8045, Sigma) and 1 mM Na-EDTA (1.09992, Sigma), pH = 13 at 4 °C for 20 to 30 min. Electrophoresis was then carried out in a fresh-made electrophoresis buffer for 15 min at 20 V and 400 mA in dark conditions. After electrophoresis, slices were gently washed in cold fresh-made neutralization buffer (0.4 M Tris-HCl buffer, pH 7.5) for 10 min and allowed to dry overnight at RT in dark conditions. Lastly, gels were stained with 7 µL propidium iodide (S7109, Sigma), covered with a coverslip, and photographed at 400× magnification in a Leica DM2500 epifluorescence microscope with a microscope. At least 50 cells were assessed for each treatment. Data were analyzed using the Open CometTM software [55]. The statistical ANOVA-Tukey test was applied [56] using the SPSS 15.0 statistics software (SPSS Inc. Headquarters, Chicago, IL, USA) in order to compare the results obtained for the different treatments and the negative control.

2.4.4. Epigenetic Analysis of Repetitive Sequences on DNA of HL60 Cells

HL60 cells were plated and treated with two concentrations of Allium species (0.002 and 0.06 mg/mL) and two concentrations of thiosulfinates (0.012 and 0.4 mM) for 5 h. Genomic DNA from HL60 cells was isolated in the same way as described in the DNA fragmentation section. After that, Bisulphite-modified DNA from food coloring treatments (EZ DNA Methylation Gold™Kit) was used as a template for fluorescence-based real-time quantitative Methylation-Specific PCR (qMSP). A qMSP were carried out according to the protocol described by Merinas-Amo et al. [57] in 48 well plates in the MiniOpticon Real-Time PCR System (MJ Mini Personal Thermal Cycler, Bio-Rad) and was analyzed by Bio-Rad CFX Manager 3.1 Software. The final reaction mixture (V = 10 µL) consisted of: 1 µL of bisulfite converted genomic DNA, 2 µL of milliQ water, 5 µM of each forward and reverse primer, 2 µL of iTaq™ Universal SYBR® GreenSupermix (Bio-Rad, which contained antibody-mediated hot-start iTaqDNA polymerase, dNTPs, MgCl2, SYBR® Green I dye, enhancers, stabilizers, and a blend of passive reference dyes including ROX and fluorescein).

qMSP conditions included initial denaturalization at 95 °C for 3 min and amplification, which consisted of 45 cycles at 95 °C for 10 s, 60 °C for 15 s, and 72 °C for 15 s, taking a picture at the end of each elongation cycle. After that, the melting curve was determined by increasing 0.5 °C each 0.05 s from 60 °C to 95 °C and taking pictures.

Repetitive elements were selected in order to analyze a wide range of human genomic DNA. While Alu and LINE sequences are interspersed throughout the genome, satellites are confined to the centromere areas [58–61]. Alu M1, LINE-1, and Sat-α sequences were used and the housekeeping Alu-C4 was used as a reference to correct for total DNA input. All primers were obtained from Isogen Life Science and their sequences are as follows: Alu-C4 (forward: 5′-GGTTAGGTATAGTGGTTTATATTTGTAATTTTAGTA-3′; reverse: 5′-ATTAACTAAACTAATCTTA AACTCCTAACCTCA-3′), Alu-M1 (forward: 5′-ATTATGTTAGTTAGGATGGTTTCGATTTT-3′; reverse: 5′-CAATCGACCGAACGCGA-3′); LINE-1 (forward: 5′-GGACGTATTTGGAAAATCGGG-3′; reverse: 5′-AATCTCGCGATACGCCGTT-3′); Sat-α (forward: 5′-TGATGGAGTATTTTTAAAATATAC GTTTTGTAGT-3′. For detailed information on the primers, see Weisenber et al. [62].

The relative yielded results were normalized with the housekeeping sequence Alu C4 using the Nikoliaidis et al. [63] and the Liloglou et al. [64] comparative C_T method.

- C_T of the target gene was normalized with respect to the referent gene (ΔC_T).
- ΔC_T of each experimental sample or reference ($\Delta C_{T,r}$) were compared with ΔC_T of the calibrator sample ($\Delta C_{T,cb}$): $\Delta\Delta C_T$.
- The relative value of each sample is defined by the formula below.

$$2^{-(\Delta C_{T,r} - \Delta C_{T,cb})} = 2^{-\Delta\Delta C_T}$$

Each sample was analyzed in triplicate. One-way ANOVA and post hoc Tukey's tests were used to evaluate the differences among the tested compound, repetitive elements, and concentrations.

3. Results

3.1. SMART Test

The results of genotoxicity and antigenotoxicity are shown in Table 1. All the assayed compounds were non-genotoxic in the flies at all tested concentrations. Both *Allium* vegetables showed no differences compared with water control in single and total spots. Validation of the experimental design was assessed by the results of the positive control (H_2O_2, 0.37 total spots/wing), which agreed with our previous results [50,65]. The antigenotoxic potency of *Allium* sp. vegetables, DPDS and DADS against H_2O_2 exhibited a clear positive dose-response effect even though the lowest concentration of garlic was not statistically different with respect to the positive control (Figure 1), which shows the DADS the highest IP value (95%).

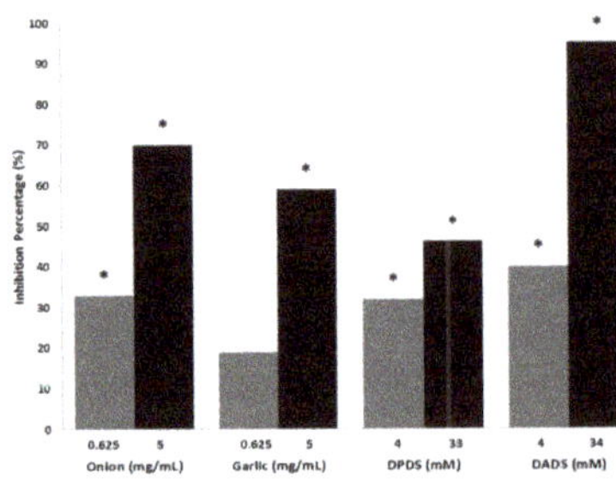

Figure 1. Mutagenicity inhibition percentages produced by onion, garlic, dipropyl disulphide (DPDS), and diallyl disulphide (DADS) against H_2O_2 – DNA induced damage (*Drosophila melanogaster* model). *: Statitiscal significance compared with the positive control using the Kastenbaum-Bowman binomial test with significance levels $\alpha = \beta = 0.05$.

Table 1. Genotoxicity and antigenotoxicity results obtained in the SMART test when flies were fed with different concentrations of onions, garlic, and organosulfur DPDS and DADS in single and combined treatments.

Compounds	N	Clones Per Wings (Number of Spots) [1]				
		Small Spots (1–2 cells)	Large Spots (>2 cells)	Twin Spots	Total Spots	Mann-Whitney Test [3]
Controls						
H_2O	40	0.10 (4) [2]	0	0	0.10 (4)	
H_2O_2 (0.12M)	40	0.30 (12)	0.05 (2)	0.02 (1)	0.37 (15) +	Ω
Onion (mg/mL)						
0.625	40	0.17 (7)	0	0	0.17 (7) i	Δ
5	38	0.08 (3)	0.03 (1)	0	0.10 (4) i	Δ
$0.625 + H_2O_2$	40	0.20 (8)	0	0.05 (2)	0.25 (10) λ	Ω
$5 + H_2O_2$	38	0.08 (3)	0	0.03 (1)	0.11 (4) β	
Garlic (mg/mL)						
0.625	40	0.07 (3)	0	0	0.07 (3) i	Δ
5	40	0.05 (2)	0	0	0.05 (2) i	Δ
$0.625 + H_2O_2$	40	0.27 (11)	0	0.02 (1)	0.30 (12) λ	Δ
$5 + H_2O_2$	40	0.15 (6)	0	0	0.15 (6) β	
DPDS (mM)						
4	40	0.22 (9)	0.02 (1)	0.02 (1)	0.27 (11) i	Δ
33	40	0.07 (3)	0.07 (3)	0	0.15 (6) i	Δ
$4 + H_2O_2$	40	0.20 (8)	0.05 (2)	0	0.25 (10) λ	Ω
$33 + H_2O_2$	40	0.17 (7)	0.02 (1)	0	0.20 (8) λ	Ω
DADS (mM)						
4	40	0.15 (6)	0	0	0.15 (6) i	Δ
34	26	0.04 (1)	0	0	0.04 (1) i	Δ
$4 + H_2O_2$	40	0.20 (8)	0.02 (1)	0	0.22 (9) λ	Ω
$34 + H_2O_2$	40	0.02 (1)	0	0	0.02 (1) β	

[1] Statistical diagnosis according to Frei and Würgler [48]. + (positive) and i (inconclusive) versus negative control. β (significantly different) and λ (inconclusive) versus positive control. m: multiplication factor. Kastenbaum-Bowman Test without Bonferroni correction and probability levels $\alpha = \beta = 0.05$. [2] Number of spots or clones in parentheses. [3] Inconclusive and positive results were resolved using the Mann-Whitney U-test. Delta marker (Δ) means no differences between the treatments and the concurrent control. Ohm marker (Ω) means differences between the treatments and the concurrent control.

3.2. Longevity Assays

Flies' survival curves for all treatments are plotted in Figure 2. In general, all treatments induce lifespan maintenance. As shown in Table 2, only DPDS significantly decreased the lifespan at two supplementation levels (8 and 16 mM). DPDS and DADS significantly decreased the mean health span by 17% and 14%, respectively, only at the highest concentrations. It is noteworthy that there is an agreement between lifespan and health span significances of DPDS at 8 and 16 mM.

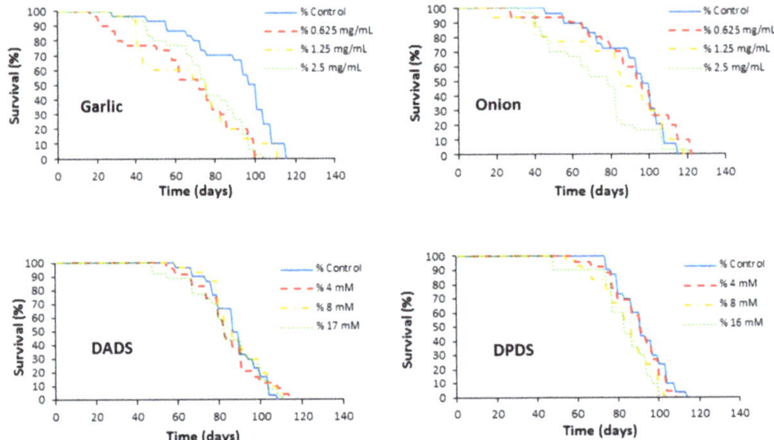

Figure 2. Effects of garlic, onion, DADS, and DPDS supplementation on the lifespan of *Drosophila melanogaster*.

Table 2. Effects of the tested compounds at different concentrations on the *Drosophila melanogaster* mean lifespan and health span.

	Mean Lifespan (Days)	Mean Lifespan Difference (%) [a]	Health-Span (75th Percentile) (Days)	Health-Span Difference (%) [a]
Onion (mg/mL)				
Control	92.24 ± 3.58	0	76.00 ± 12.63	0
0.625	95.77 ± 3.45	4	83.00 ± 5.04	9
1.25	92.83 ± 3.36	1	83.00 ± 5.08	9
2.5	81.92 ± 4.98	−11	65.00 ± 13.34	−11
Garlic (mg/mL)				
Control	81.25 ± 4.57	0	51.14 ± 4.31	0
0.625	79.51 ± 3.30	−2	58.83 ± 1.76	15
1.25	76.21 ± 4.46	−6	47.29 ± 3.73	−7
2.5	77.68 ± 3.35	−4	53.57 ± 3.48	5
DPDS (mM)				
Control	91.63 ± 2.06	0	77.37 ± 1.05	0
4	89.58 ± 2.39	−2	73.00 ± 2.80	−6
8	84.23 ± 2.38*	−8	67.12 ± 2.69 **	−13
16	82.25 ± 3.23*	−10	64.20 ± 6.63 *	−17
DADS (mM)				
Control	88.00 ± 2.26	0	73.11 ± 2.40	0
4	84.64 ± 3.01	−4	68.00 ± 3.34	−7
8	88.83 ± 2.36	1	75.22 ± 2.50	3
17	86.21 ± 3.27	−2	62.87 ± 3.96 *	−14

[a] Difference between treated flies and the concurrent negative control (water) in percentage. Positive results indicate that lifespan was increased and negative results indicate that lifespan was decreased. Statistical significance: * = $p \leq 0.05$, ** = $p \leq 0.01$ (log-Mantel-Cox test).

3.3. Cytotoxicity and Proapoptotic Assays in Leukemia Cells

The cytotoxic effects of *Allium* vegetables and their distinctive compounds (DADS and DPDS) on the survival of HL60 cells are shown in Figure 3. Garlic and DADS exerted a cytotoxic effect on cell growth in a positive dose-dependent manner after 72 h of incubation, with EC_{50} of 0.003 mg/mL in the case of garlic and IC_{50} of 0.06 mM in the case of DADS. On the contrary, the effect observed in DPDS was smaller, with a high IC_{50} of 0.25 mM and it was absent in onion treatments in which the cytotoxic effect resulted only in a growth inhibition of 30% at the higher tested concentrations.

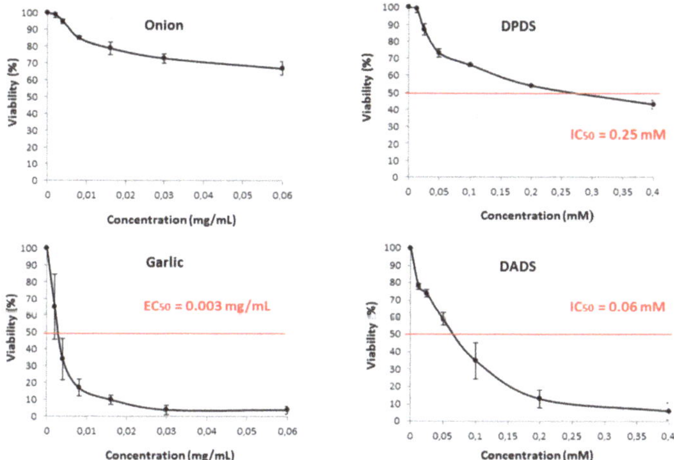

Figure 3. Viability of HL60 cells treated during 72 h with different concentrations of onion, garlic, and their respective organosulfur compound, DPDS, and DADS. Curves are plotted as mean percentages with respect to the control (three independent replicates). IC_{50}: Inhibition concentration 50 for the tested organosulfur. EC_{50}: effective concentration 50 for the tested extracts.

The results of proapoptotic effects of different concentrations of garlic, onion, DADS, and DPDS in HL60 cells measured as internucleosomic programmed fragmentation [66] are shown in Figure 4. DNA fragmentation was observed at high concentrations of garlic (0.03 and 0.06 mg/mL) and DADS (0.1, 0.2, and 0.4 mM). Nevertheless, any DNA inter-nucleosomal fragments were induced neither by onions nor by DPDS at the assayed concentrations.

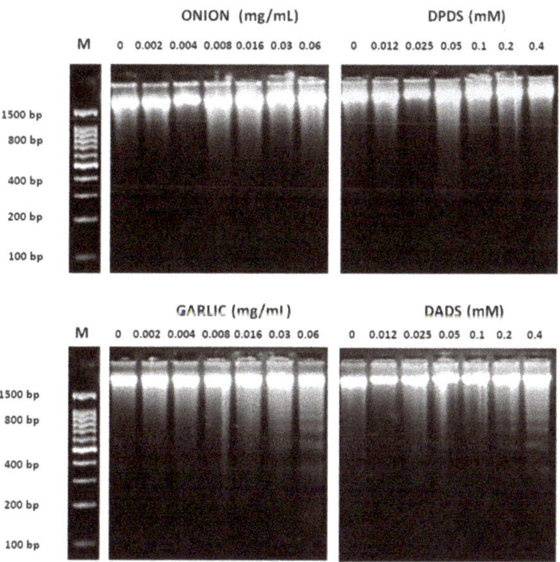

Figure 4. Inter-nucleosomal DNA fragmentation. HL-60 cells were exposed to various concentrations of onion, garlic, and their distinctive organo-sulfurs for 5 h. DNA was extracted from cells and was subject to 2% agarose gel electrophoresis at 50 V for 90 min. M: DNA size marker.

3.4. DNA Single Strand Breaks

Both vegetables induced a significant ($p \leq 0.001$) increase in the tail moment (TM) at all tested concentrations. On the contrary, only DADS (garlic with organosulfur) was able to induce a significant ($p \leq 0.01$) increase of this parameter at 28 and 56 µM (Figure 5).

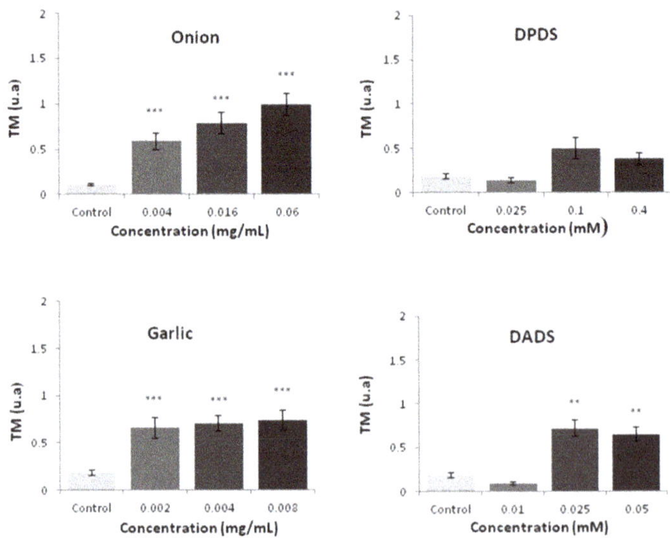

Figure 5. HL60 DNA integrity measured by the comet assay after 5 h of treatment with different concentrations of the tested compounds. Data are expressed as a TM parameter [54]. Statitiscal significance compared with a negative control: *** $p \leq 0.000$ and ** $p \leq 0.01$ for mean values of three independent replicates.

3.5. Methylation Status

The relative normalized expression of three repetitive sequences (Alu M1, LINE-1, and Sat-α) studied in HL-60 cells treated with different concentrations of *Allium* sp. vegetables, DPDS, and DADS is shown in Figure 6. After one-way ANOVA and post hoc Tukey's test, statistical results showed a significant hypermethylation level at LINE-1 and Sat- α repetitive sequences at the highest concentration tested of onion and DPDS. Moreover, garlic exhibited a significant hypermethylation status at the highest concentration tested in LINE-1 and at the lowest concentration tested in Sat-α sequences. Contrarily, a significant hypomethylation level of both assayed concentrations of DADS and garlic is shown in the Alu M1 repetitive sequences. The rest of the concentrations showed a similar methylation level to that of the normalized control.

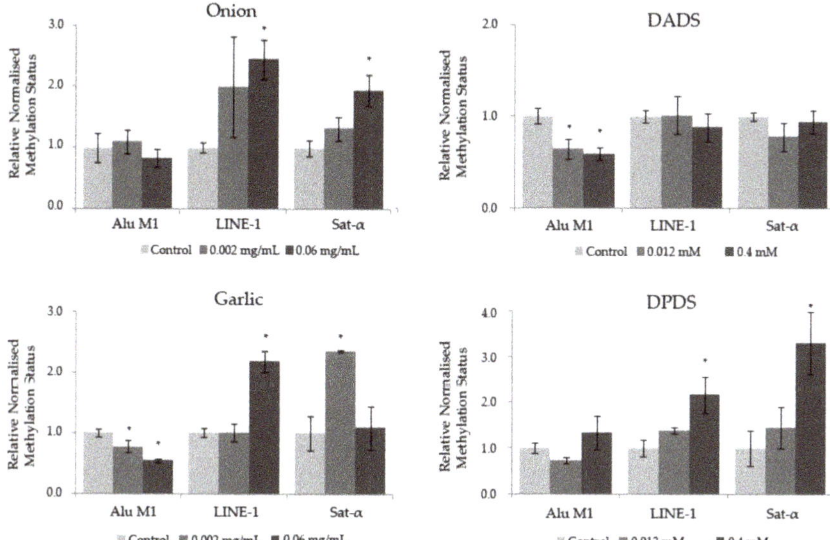

Figure 6. Methylation status of *Allium* sp. vegetables, DADS, and DPDS in HL-60 cells. Relative normalized expression data of each repetitive element (Alu M1, LINE-1, and Sat-α). Values represent the mean ± SE from three independent experiments. * $p \leq 0.05$.

4. Discussion

4.1. In Vivo Assessment of the Safety, Protection, and Lifespan Modulation

Garlic samples and onions have traditionally been used as food sources around the world across centuries likely due to their demonstrated particular flavor but also due to the health benefits, such as the prevention of cardiovascular diseases, cancer, and even aging [7]. Despite their popularity, the number of systematic, integrated, and multifocal studies assessing the genotoxic, antigenotoxic, and health span effects are scarce, and even less for assessing their distinctive organosulfur compounds (DADS and DPDS).

Our in vivo DNA stability studies (genotoxicity, antigenotoxicity, and longevity) were carried out using *D. melanogaster* flies. These organisms are widely used as a genetic animal model due to their homology with several mammal models in biological, physiological, and neurological traits [67,68]. It was demonstrated that more than 70% of human disease-causing genes have a functional homolog in this fly model [69]. Additionally, this particular model was also largely used to evaluate the genotoxicity of different biological compounds and molecules due to its accuracy, robustness, and reproducibility [70–72].

Carcinogen molecules and mutagenic properties should be taken into account and carefully evaluated in every complex mixture to be proposed for food. For this reason, genotoxic screening assays are considered as the first mandatory step, with the *Drosophila* wing spot test one of the most reliable methodologies to be employed as an ideal assay to evaluate biological products aimed to use in human and animal diets. To our knowledge, this is the first study to characterize the genotoxic effect of garlic, onion, and their two major and distinctive organosulfur constitutive molecules (DADS and DPDS, respectively) using the *D. melanogaster* animal model. Previous studies determined the lack of mutagenicity of these vegetables in a *Salmonella typhimurium* and in yeast models [73,74]. It has also been demonstrated that aqueous garlic extracts (5% *v/v*), fine garlic powder supplementation (7.5, 5, and 2.5 g/kg body weight), and fresh garlic bulb extracts (3, 6, and 12 mg/culture) were safe in vitro (cell lines) and non-animal models [74–77].

Our results for onion supplementation in the *Drosophila* model demonstrated a lack of genotoxicity, which validates previous reports obtained by Kulkarni et al. [78] in several *Salmonella* strains. In the same way, DPDS and DADS, which are the active principles in garlic and onion, were also non-genotoxic in our SMART trials. Despite the fact that onions are widely employed in the human diet, the number of genotoxicity studies carried out in DPDS are scarce [37,79]. Our study was the first to test the safety and protective effects of this compound using in vivo models. Nevertheless, previous reports assessing these particular molecules are controversial. For instance, Musk et al. demonstrated that DADS induced both chromosome aberrations and sister chromatid exchanges, characterized as genotoxic effects, at lower concentrations (below 0.07 mM) in a Chinese hamster ovary cell line [80]. However, this controversy could partially be explained due to methodological differences (in vivo vs. in vitro models) and the concentrations were tested. Controversial results are commonly found for a single molecule when it is tested in different assays and in vivo carcinogenic trials are needed.

One of the strategies for coping with the food and environmental genotoxic compounds is to identify effective antimutagens and anticarcinogens in order to increase man's exposure to them as a way to decrease the cancer incidence [81]. This is the second step in the search of real nutraceutical substances. In our case, antigenotoxicity assays were conducted using hydrogen peroxide as a positive geno-toxicant model since this compound is able to induce somatic mutation and mitotic recombination in *D. melanogaster* [65], which affects the DNA integrity and stability.

Similar results to ours were reported on the desmutagenic activity of onions. Ethanolic extracts showed a strong inhibitory effect against NDBA in prokaryotes [32] and Welsh onion juice suppressed the mutagenic activity of benzo[a]pyrene (BaP) and 4-nitroquinoline 1-oxide (4QNO) and reduced the number of 2,4-dimethoxybenzaldehyde (DMBA)-induced chromosome aberrations in rats [82] while onion supplementation protected *D. melanogaster* against urethane-induced DNA damage [31]. All those reports validate our findings since onion supplementation reduced the mutagenic effects of H_2O_2 by as much as 65% in a dose-dependent manner. In the same way, DPDS showed des-mutagenic properties when it was tested as an individual molecule despite being at a lower extent when compared with the effect on onions. In this sense, DPDS strongly increased dimethyl nitrosamine (DMN) mutagenicity in *S. typhimurium* [37] and reduced NPYR/NDMA-induced oxidative DNA damage in HepG2 cells at 5 μM [38]. However, our study is the first one demonstrating that *Allium* vegetables have a protective role against H_2O_2 induced damage using the *D. melanogaster* model, which is a more adequate model widely used to extrapolate to mammals. This effect could be due to its well-known scavenging potential against free-radicals of their respective organosulfur compounds [18,83,84] since similar results were observed in the vegetables and simple molecule assessments.

The desmutagenic activity of garlic and different types of garlic extracts were previously described in several induced mutagenesis models. It was demonstrated that garlic and garlic water extracts protected against gamma-radiation and cyclophosphamide in mice [75,77,85]. In the same way, methanolic and ethanolic garlic extracts, even prepared by different processing methods (raw, grilled, and pickled), showed inhibitory activities on H_2O_2-induced DNA damage in human leukocytes [86] and reduced the chromosomal aberrations induced by DMBA in mice bone marrow [87]. In the same way, raw garlic methanolic extracts reduced the urethane mutagenicity in standard and high bioactivated *D. melanogaster* crosses [31]. In our experimental design, garlic clearly behaves as an anti-genotoxin, which could potentially be explained by the fact that concurrent experiments using DADS as simple molecule also inhibited the 95% of the H_2O_2-induced DNA damage. This desmutagenic property of DADS was previously proposed in several reports using different mutagenic substances such as (+)-anti-7β,8α-dihydroxy-9α,10α-oxy-7,8,9,10-tetrahydrobenzo[a]pyrene (BPDE), styrene oxide (SO), 4-NQO, aflatoxin B1 (AFB1), *N*-nitrosodimethylamine (NDMA), and 1-nitrosopyrrolidine (NPYR) [38,79,88].

Longevity assays are one of the most simple and efficient methodological approaches to evaluate the aging and anti-aging effects of simple compounds and complex mixtures on higher organisms. *D. melanogaster* is considered a very useful genetic model on aging research since its similarities with human metabolic pathways controlling nutrient uptake, storage, and metabolism [89,90]. In addition, this model has a short lifespan compared with similar in vivo models, which reduces the experimental periods.

To our knowledge, this is the first assessment on the effect of onions, DADS, and DPDS on the *D. melanogaster* lifespan and one of the few available assessing this effect in garlic samples [91]. These results support the hypothesis that individual organosulfur compounds can reduce longevity to some extent. These compounds could primarily be responsible for the apparent reduced viability observed in some cohort groups of flies. A similarity between the complete food and their distinctive compounds in the lifespan behavior is observed, although, in the case of vegetables, the lifespan is not significantly reduced when compared to the concurrent negative controls. Being onion and garlic complex mixtures of many individual molecules, the final outcome of such a complex trait longevity appears to be an additive combination of positive and negative synergic effects of the molecular components of vegetables with many of them phenolic and organosulfur, which are not included in the present study. Previous reports showed beneficial effects of garlic extracts on animal lifespan, including *D. melanogaster* and *C. elegans* [21]. Those differences could be due to the different tested samples being raw garlic in our study and garlic extracts in previous reports. In this sense, Prowse et al. demonstrated the insecticidal activity of garlic juices across several life stages of flies at a wide range of concentrations (0.25%–5%) in two dipteran pests (*Delia radicum* and *Musca domestica*) [92]. These results agree with the fact that similar but not significant effects on lifespan were caused by garlic, onion, DADS, and DPDS in our *D. melanogaster* experiments. It is noticeable that high doses were used for medicinal purposes in human acute treatments [93]. Thus, high dosages of garlic would not be advisable to be used in long-term chronic treatments due to the adverse effects that could be associated, even though nutraceuticals or dietary supplements include the bioactive compounds at higher doses than those used in this study.

4.2. In Vitro Assessment of the Cytotoxic, Clastogenic Activities and Methylation Status

Our results showed that only garlic and DADS have a strong cytotoxic effect and induce a clear DNA pro-apoptotic inter-nucleosomal fragmentation against HL60 cells. Previous reports demonstrated that garlic and DADS exerted a chemo-preventive effect through different pathways: (i) by increasing apoptosis and *Bcl-2* expression and decreasing p53 protein and *Bax* expression in lung cancer cells (NCI-H1299) [94], (ii) by increasing intracellular ROS in A549 cells [22], (iii) by inhibiting cell proliferation in CaCo-2 and HT-29 cells repressing histone deacetylase activity and histone hyperacetylation and increasing the p21(waf1/cip1) expression [95], and (iv) by inducing apoptosis by activating caspase-3 expression in HL60 cells [96]. In addition, Yang et al. observed that DADS supplementation (0.5, 10, and 25 µM) had a pro-apoptotic effect in COLO 205 cell line by inducing reactive oxygen species and caspase cascade [23]. On the contrary, the cytotoxic effect exerted by the onion and DPDS is relatively weak and their molecular mechanism is less clear. As an example, Sundaram and Milner [97] demonstrated that DPDS (100 µM) was an inefficient molecule to inhibit the cell growth and to induce programmed cell death in tumor cells (HCT-15). However, Wu et al. suggested that onion oil induces cell cycle arrest and apoptosis through ROS production in A549 cells [35]. It was also proposed that the carcinogenic inhibition mechanism of DADS is mediated through a modulation of the P450 cytochrome–dependent monooxygenases and/or the acceleration of carcinogen detoxification through phase II-enzymes upregulation [98,99]. In our case, the chemo-preventive properties of raw onion samples and DADS were weak despite the type of sample employed.

DNA inter-nucleosomal fragmentation was defined as one of the hallmarks of cellular apoptosis, even though it cannot be considered as a single criterion to assess the apoptotic cell death [100]. In order to determine the ability of our tested substances to induce DNA breaks in HL60 cells, we employed a single cell gel electrophoresis (comet) assay, which it is widely used to detect the apoptotic capability of mixtures and single compounds to induce DNA damage [101,102]. Currently, this procedure is being widely employed to evaluate the DNA stability in normal and carcinogenic cell lines against different substances, due to its robustness and reliability [103]. In this methodology, we employed the tail moment (TM) index, which is an accurate parameter to quantify the DNA migration and, thus, the DNA fragmentation status [54]. With this parameter, we differentiated apoptosis-induced from necrosis-induced DNA damage as follows: a TM > 30 is considered to be an indicator of apoptosis and a TM between 5 and 30 a.u. (arbitrary units) is considered to be a necrotic process [104].

In this study, we determined for the first time the DNA-damage exerted by garlic, onion, DADS, and DPDS through the alkaline "comet assay" in HL60 leukemic cells in order to assess their potential anticarcinogenic effect. Our results (Figure 6) showed that onion and garlic induced DNA damage in HL60 by necrosis (short tails, TM < 2) being in concordance with our cytotoxic and DNA-fragmentation results. Similar results were observed in DADS and DPDS, but in a lower extent, which suggested the total absence of proapoptotic activity in the entire compound tested at the different assayed concentrations.

Our results with DPDS disagree with those obtained by Arranz and Haza [38], who showed that DPDS could act in a positive dose-dependent manner since the higher concentrations tested (>5 µM) caused DNA damage in HepG2 cells (data not shown) by the comet assay. Arranz et al. assaying higher concentrations of DADS (>5 µM), showed DNA damage in HepG2 cells in the alkaline comet assay [38]. However, controversial results were also reported by Belloir et al., which suggests that DADS was not genotoxic at concentrations between 5 to 100 µM in the same in vitro model [105].

Despite a non-significant relationship shown in the methylation status of *Allium* sp. vegetables, DPDS, and DADS in the three repetitive sequences studied, a general tendency to hypermethylate the genomic randomized-distributed sequences of the HL-60 cells (LINE-1 and Sat-α) is shown.

Based on our knowledge, no previous studies about the in vitro effects that *Allium* sp. vegetables, DPDS, and DADS have in the methylation status of three repetitive sequences of treated HL-60 tumor cells. Taking into account that methylation of the repetitive sequences is understood as an important genomic protective mechanism [62,106], high concentrations of *Alliium* sp. vegetables and DPDS could have positive effects on tumor cells, that could be an interesting chemo-preventive effect. On the other hand, negative effects on tumor cells are related to garlic and DADS in the short repetitive element studied (Alu M1).

5. Conclusions

To sum up, our experimental results provide the evidence that (i) garlic, onion, DADS, and DPDS are safe substances, which exert an antigenotoxic effect against oxidative mutagens in a dose-dependent manner. (ii) The decrease of lifespan induced in the *Drosophila* animal model by DPDS at the highest concentrations could be a signal that the long-term consumption of complex mixtures is safe only at low concentrations. (iii) Garlic exerted a clear chemo-preventive effect, with its distinctive organosulfur DADS as the most likely cause of such activities. (iv). The slight cytotoxic effect of onions is probably mediated by a non-apoptotic mechanism. Overall, this study could be a baseline for further supplementary studies to clarify the cell death pathway induced by garlic and DADS. (v) A general increase of the methylation status in LINE-1 and Sat-α repetitive sequences of HL-60 treated cells are shown in onions, garlic, and DPDS, which is related to a genomic protective mechanism.

Author Contributions: M.d.R.-C. designed this study and revised the manuscript. Z.F.-B. performed all in vivo and in vitro assays and statistical analysis as well as the interpretation of the results. T.M.-A. carried out the analysis and the interpretation of the epigenetic results. Z.F.-B. and S.D.-P. wrote the manuscript. All the listed authors have read and approved the submitted manuscript.

Funding: This research received no external funding.

Conflicts of Interest: The authors declare no conflict of interest.

References

1. De Lorgeril, M.; Salen, P. The Mediterranean-style diet for the prevention of cardiovascular diseases. *Public Health Nutr.* **2006**, *9*, 118–123. [CrossRef] [PubMed]
2. Trichopoulou, A.; Vasilopoulou, E. Mediterranean diet and longevity. *Br. J. Nutr.* **2000**, *84*, S205–S209. [CrossRef] [PubMed]
3. Ambrosone, C.B.; Freudenheim, J.L.; Thompson, P.A.; Bowman, E.; Vena, J.E.; Marshall, J.R.; Graham, S.; Laughlin, R.; Nemoto, T.; Shields, P.G. Manganese superoxide dismutase (MnSOD) genetic polymorphisms, dietary antioxidants, and risk of breast cancer. *Cancer Res.* **1999**, *59*, 602–606. [PubMed]
4. Rivlin, R.S. Historical perspective on the use of garlic. *J. Nutr.* **2001**, *131*, 951S–954S. [CrossRef] [PubMed]
5. Lanzotti, V. The analysis of onion and garlic. *J. Chromatogr. A* **2006**, *1112*, 3–22. [CrossRef] [PubMed]
6. Santhosha, S.G.; Jamuna, P.; Prabhavathi, S.N. Bioactive components of garlic and their physiological role in health maintenance: A review. *Food Biosci.* **2013**, *3*, 59–74. [CrossRef]
7. Corzo-Martínez, M.; Corzo, N.; Villamiel, M. Biological propierties of onions and garlic. *Trends Food Sci. Technol.* **2007**, *18*, 609–625. [CrossRef]
8. Khadem, S.; Marles, R.J. Monocyclic phenolic acids; hydroxi-polihidroxibenzoic acids: Ocurrence and recent bioactivity studies. *Molecules* **2010**, *15*, 7985–8005. [CrossRef]
9. Heleno, S.A.; Martins, A.; Queiroz, M.J.R.P.; Ferreira, I.C.F.R. Bioactivities of phenolic acids: Metabolites versus parent compounds: A review. *Food Chem.* **2015**, *173*, 501–513. [CrossRef]
10. Yoshinari, O.; Shiojima, Y.; Igarashi, K. Anti-obesity effects of onion extract in Zucker diabetic fatty rats. *Nutrients* **2012**, *4*, 1518–1526. [CrossRef]
11. Yamamoto, Y.; Aoyama, S.; Hamaguchi, N.; Rhi, G.S. Antioxidative and antihypertensive effects of Welsh onion on rats fed with a high-fat high-sucrose diet. *Biosci. Biotechnol. Biochem.* **2005**, *69*, 1311–1317. [CrossRef]
12. Sengupta, A.; Ghosh, S.; Bhattacharjee, S. *Allium* vegetables in cancer prevention: An overview. *Asian Pac. J. Cancer Prev.* **2004**, *5*, 237–245.
13. Cerella, C.; Dicato, M.; Jacob, C.; Diederich, M. Chemical properties and mechanisms determining the anti-cancer action of garlic-derived organic sulfur compounds. *Anti-Cancer Agents Med. Chem.* **2011**, *11*, 267–271. [CrossRef]
14. Amagase, H.; Petesch, B.L.; Matsuura, H.; Kasuga, S.; Itakura, Y. Intake of garlic and its bioactive components. *J. Nutr.* **2001**, *131*, 955S–962S. [CrossRef] [PubMed]
15. Brodnitz, M.H.; Pascale, J.V.; Van Derslice, L. Flavor components of garlic extract. *J. Agric. Food Chem.* **1971**, *19*, 273–275. [CrossRef]
16. Shukla, Y.; Kalra, N. Cancer chemoprevention with garlic and its constituents. *Cancer Lett.* **2007**, *247*, 167–181. [CrossRef] [PubMed]
17. Shaath, N.A.; Flores, F.B. Egyptian onion oil. *Dev. Food Sci.* **1998**, *40*, 443–453.
18. Benkeblia, N. Free-radical scavenging capacity and antioxidant properties of some selected onions (*Allium cepa* L.) and garlic (*Allium sativum* L.) extracts. *Braz. Arch. Biol. Technol.* **2005**, *48*, 753–759. [CrossRef]
19. Park, J.H.; Park, Y.K.; Park, E. Antioxidative and antigenotoxic effects of garlic (*Allium sativum* L.) prepared by different processing methods. *Plant Foods Hum. Nutr.* **2009**, *64*, 244–249. [CrossRef]
20. Tanaka, S.; Haruma, K.; Yoshihara, M.; Kajilyama, G.; Kira, K.; Amagase, H.; Chayama, K. Aged garlic extract has potential suppressive effect on colorectal adenomas in humans. *J. Nutr.* **2006**, *136*, 821S–826S. [CrossRef]
21. Huang, C.H.; Hsu, F.Y.; Wu, Y.H.; Zhong, L.; Tseng, M.Y.; Kuo, C.J.; Hsu, A.L.; Liang, S.S.; Chiou, S.H. Analysis of lifespan-promoting effect of garlic extract by an integrated metabolo-proteomics approach. *J. Nutr. Biochem.* **2015**, *26*, 808–817. [CrossRef] [PubMed]
22. Wu, X.-J.; Kassie, F.; Mersch-Sundermann, V. The role of reactive oxygen species (ROS) production on diallyl disulfide (DADS) induced apoptosis and cell cycle arrest in human A549 lung carcinoma cells. *Mutat. Res.* **2005**, *579*, 115–124. [CrossRef] [PubMed]
23. Yang, J.-S.; Chen, G.-W.; Hsia, T.-C.; Ho, H.-C.; Ho, C.-C.; Lin, M.-W.; Lin, S.-S.; Yeh, R.-D.; Ip, S.-W.; Lu, H.-F.; et al. Diallyl disulfide induces apoptosis in human colon cancer cell line (COLO 205) through the induction of reactive oxygen species, endoplasmic reticulum stress, caspases casade and mitochondrial-dependent pathways. *Food Chem. Toxicol.* **2009**, *47*, 171–179. [CrossRef] [PubMed]
24. Jakubikova, J.; Sedlak, J. Garlic-derived organosulfides induce cytotoxicity, apoptosis, cell cycle arrest and oxidative stress in human colon carcinoma cell lines. *Neoplasma* **2005**, *53*, 191–199.

25. Tan, H.; Ling, H.; He, J.; Yi, L.; Zhou, J.; Lin, M.; Su, Q. Inhibition of ERK and activation of p38 are involved in diallyl disulfide induced apoptosis of leukemia HL-60 cells. *Arch. Pharm. Res.* **2009**, *31*, 786–793. [CrossRef] [PubMed]

26. Lu, H.; Sue, C.C.; Yu, C.S.; Chen, S.C.; Chen, G.W.; Chung, J.G. Diallyl disulfide (DADS) induced apoptosis undergo caspase-3 activity in human bladder cancer T24 cells. *Food Chem. Toxicol.* **2008**, *42*, 1543–1552. [CrossRef]

27. Lin, Y.T.; Yang, J.S.; Lin, S.Y.; Tan, T.W.; Ho, C.C.; Hsia, T.C.; Chiu, T.H.; Yu, C.S.; Lu, H.F.; Weng, Y.S.; et al. Diallyl disulfide (DADS) induces apoptosis in human cervical cancer Ca Ski cells via reactive oxygen species and Ca2+-dependent mitochondria-dependent pathway. *Anticancer Res.* **2008**, *28*, 2791–2799.

28. Gayathri, R.; Gunadharini, D.-N.; Arunkumar, A.; Senthilkumar, K.; Krishnamoorthy, G.; Banudevi, S.; Vignesh, R.C.; Arunakaran, J. Effects of diallyl disulfide (DADS) on expression of apoptosis associated proteins in androgen independent human prostate cancer cells (PC-3). *Mol. Cell. Biochem.* **2009**, *320*, 197–203. [CrossRef]

29. Sundaram, S.-G.; Milner, J.-A. Diallyl disulfide inhibits the proliferation of human tumor cells in culture. *BBA-Mol. Basis Dis.* **1996**, *1315*, 15–20. [CrossRef]

30. Yang, J.; Meyers, K.-J.; van der Heide, J.; Liu, R.H. Varietal differences in phenolic content and antioxidant and antiproliferative activities of onions. *J. Agric. Food Chem.* **2004**, *52*, 6787–6793. [CrossRef]

31. Aunanan, A.; Kangsadalampai, K. Effect of Preparation and Temperature treatments on Antimutagenicity against Urethane in Drosophila melanogaster and Antioxidant activity of Three Allium members. *Thai J. Toxicol.* **2010**, *23*, 108–116.

32. Ikken, Y.; Morales, P.; Martinez, A.; Marín, M.L.; Haza, A.I.; Cambero, M.I. Antimutagenic effect of fruit and vegetable ethanolic extracts against N-nitrosamines evaluated by the Ames test. *J. Agric. Food Chem.* **1999**, *47*, 3257–3264. [CrossRef] [PubMed]

33. Seki, T.; Tsuji, K.; Hayato, Y.; Moritomo, T.; Ariga, T. Garlic and onion oils inhibit proliferation and induce differentiation of HL-60 cells. *Cancer Lett.* **2000**, *160*, 29–35. [CrossRef]

34. Wang, Y.; Tian, W.-X.; Ma, X.-F. Inhibitory effects of onion (Allium cepa L.) extract on proliferation of cancer cells and adipocytes via inhibiting fatty acid synthase. *Asian Pac. J. Cancer Prev.* **2012**, *13*, 5573–5579. [CrossRef] [PubMed]

35. Wu, X.-J.; Stahl, T.; Hu, Y.; Kassie, F.; Mersch-Sundermann, V. The production of reactive oxygen species and the mitochondrial membrane potential are modulated during onion oil–induced cell cycle arrest and apoptosis in A549 cells. *J. Nutr.* **2006**, *136*, 608–613. [CrossRef] [PubMed]

36. Shrivastava, S.; Ganesh, N. Tumor inhibition and Cytotoxicity assay by aqueous extract of onion (*Allium cepa*) & Garlic (*Allium sativum*): An in-vitro analysis. *Int. J. Phytomed.* **2010**, *2*, 80–84.

37. Guyonnet, D.; Belloir, C.; Suschetet, M.; Siess, M.-H.; Le Bon, A.-M. Liver subcellular fractions from rats treated by organosulfur compounds from *Allium* modulate mutagen activation. *Mutat. Res.* **2000**, *466*, 17–26. [CrossRef]

38. Arranz, N.; Haza, A.I.; García, A.; Möller, L.; Rafter, J.; Morales, P. Protective effects of organosulfur compounds towards N-nitrosamine-induced DNA damage in the single-cell gel electrophoresis (SCGE)/HepG2 assay. *Food Chem. Toxicol.* **2007**, *45*, 1662–1669. [CrossRef]

39. Arranz, N.; Haza, A.I.; García, A.; Delgado, E.; Rafter, J.; Morales, P. Effects of organosulfurs, isothiocyanates and vitamin C towards hydrogen peroxide-induced oxidative DNA damage (strand breaks and oxidized purines/pyrimidines) in human hepatoma cells. *Chem. Biol. Interact.* **2007**, *169*, 63–71. [CrossRef]

40. Singh, S.V.; Pan, S.S.; Srivastava, S.K.; Xia, H.; Hu, X.; Zaren, H.A.; Orchard, J.L. Differential induction of NAD (P) H: Quinone oxidoreductase by anti-carcinogenic organosulfides from garlic. *Biochem. Biophys. Res. Commun.* **1998**, *244*, 917–920. [CrossRef]

41. Merhi, F.; Auger, J.; Rendu, F.; Bauvois, B. *Allium* compounds, dipropyl and dimethyl thiosulfinates as antiproliferative and differentiating agents of human acute myeloid leukemia cell lines. *Biol. Targets Ther.* **2008**, *2*, 885–895.

42. García, A.; Morales, P.; Arranz, N.; Delgado, M.A.; Rafter, J.; Haza, A.I. Antiapoptotic effects of dietary antioxidants towards N-nitrosopiperidine and N-nitrosodibutylamine-induced apoptosis in HL-60 and HepG2 cells. *J. Appl. Toxicol.* **2009**, *29*, 403–413. [CrossRef] [PubMed]

43. Yan, J.; Huen, D.; Morely, T.; Johnson, G.; Gubb, D.; Roote, J.; Adler, P.N. The multiple-wing-hairs gene encodes a novel GBD–FH3 domain-containing protein that functions both prior to and after wing hair initiation. *Genetics* **2008**, *180*, 219–228. [CrossRef]

44. Ren, N.; Charlton, J.; Adler, P.N. The flare gene, which encodes the AIP1 protein of Drosophila, functions to regulate F-actin disassembly in pupal epidermal cells. *Genetics* **2007**, *176*, 2223–2234. [CrossRef] [PubMed]

45. Graf, U.; Wurgler, F.E.; Katz, A.J.; Frei, H.; Juon, H.; Hall, C.B.; Kale, P.G. Somatic mutation and recombination test in *Drosophila melanogaster*. *Environ. Mutagen.* **1984**, *6*, 153–188. [CrossRef] [PubMed]

46. Anter, J.; Fernández-Bedmar, Z.; Villatoro-Pulido, M.; Demyda-Peyrás, S.; Moreno-Millán, M.; Alonso-Moraga, A.; Muñoz-Serrano, A.; Luque de Castro, M.D. A pilot study on the DNA-protective, cytotoxic, and apoptosis-inducing properties of olive-leaf extracts. *Mutat. Res. Genet. Toxicol. Environ. Mutagen.* **2011**, *723*, 165–170. [CrossRef]

47. Block, E. The Organosulfur Chemistry of the Genus *AZZium*-Implications for the Organic Chemistry of Sulfur. *Angew. Chem. Int. Ed. Engl.* **1992**, *31*, 1135–1178. [CrossRef]

48. Frei, H.; Würgler, F.E. Optimal experimental design and sample size for the statistical evaluation of data from somatic mutation and recombination tests (SMART) in *Drosophila*. *Mutat. Res.* **1995**, *334*, 247–258. [CrossRef]

49. Abraham, S.K. Antigenotoxicity of coffee in the *Drosophila* assay for somatic mutation and recombination. *Mutagenesis* **1994**, *9*, 383–386. [CrossRef]

50. Fernández-Bedmar, Z.; Anter, J.; de la Cruz-Ares, S.; Muñoz-Serrano, A. Role of citrus juices and distinctive components in the modulation of degenerative processes: Genotoxicity, antigenotoxicity, cytotoxicity, and longevity in Drosophila. *J. Toxicol. Environ. Health A* **2011**, *74*, 1052–1066. [CrossRef]

51. Lee, K.S.; Lee, B.S.; Semnani, S.; Avanesian, A.; Um, C.Y.; Jeon, H.J.; Seong, K.M.; Yu, K.; Min, K.J.; Jafari, M. Curcumin extends life span, improves health span, and modulates the expression of age-associated aging genes in *Drosophila melanogaster*. *Rejuvenation Res.* **2010**, *13*, 561–570. [CrossRef] [PubMed]

52. Birnie, G. The HL60 cell line: A model system for studying human myeloid cell differentiation. *Br. J. Cancer* **1988**, *9*, 41–45.

53. Anter, J.; Demyda-Peyrás, S.; Delgado de la Torre, M.P.; Campos-Sánchez, J.; Luque de Castro, M.D.; Munoz-Serrano, A.; Alonso-Moraga, A. Biological and health promoting activity of vinification byproducts produced in Spanish vineyards. *S. Afr. J. Enol. Vitic.* **2015**, *36*, 126–133. [CrossRef]

54. Olive, P.L.; Banáth, J.P. The comet assay: A method to measure DNA damage in individual cells. *Nat. Protoc.* **2006**, *1*, 23–29. [CrossRef]

55. Gyori, B.M.; Venkatachalan, G.; Thiagarajan, P.S.; Hsu, D.; Clement, M.V. OpenComet: An automated tool for comet assay image analysis. *Redox Biol.* **2014**, *2*, 457–465. [CrossRef] [PubMed]

56. Serpeloni, J.M.; Bisarro dos Reis, M.; Rodrigues, J.; Campaner dos Santos, L.; Vilegas, W.; Varanda, E.A.; Dokkedal, A.L.; Cólus, I.M.S. In vivo assessment of DNA damage and protective effects of extracts from *Miconia species* using the comet assay and micronucleus test. *Mutagenesis* **2008**, *23*, 501–507. [CrossRef] [PubMed]

57. Merinas-Amo, T.; Merinas-Amo, R.; Alonso-Moraga, A. A clinical pilot assay of beer consumption: Modulation in the methylation status patterns of repetitive sequences. *Sylwan* **2017**, *161*, 134–156.

58. Deininger, P.L.; Moran, J.V.; Batzer, M.A.; Kazazian, H.H. Mobile elements and mammalian genome evolution. *Curr. Opin. Genet. Dev.* **2003**, *13*, 651–658. [CrossRef] [PubMed]

59. Ehrlich, M. DNA hypomethylation, cancer, the immunodeficiency, centromeric region instability, facial anomalies syndrome and chromosomal rearrangements. *J. Nutr.* **2002**, *132*, 2424S–2429S. [CrossRef] [PubMed]

60. Lee, C.; Wevrick, R.; Fisher, R.; Ferguson-Smith, M.; Lin, C. Human centromeric DNAs. *Hum. Genet.* **1997**, *100*, 291–304. [CrossRef]

61. Weiner, A.M. SINEs and LINEs: The art of biting the hand that feeds you. *Curr. Opin. Cell Biol.* **2002**, *14*, 343–350. [CrossRef]

62. Weisenberger, D.J.; Campan, M.; Long, T.I.; Kim, M.; Woods, C.; Fiala, E.; Ehrlich, M.; Laird, P.W. Analysis of repetitive element DNA methylation by MethyLight. *Nucleic Acids Res.* **2005**, *33*, 6823–6836. [CrossRef]

63. Nikolaidis, G.; Raji, O.Y.; Markopoulou, S.; Gosney, J.R.; Bryan, J.; Warburton, C.; Walshaw, M.; Sheard, J.; Field, J.K.; Liloglou, T. DNA methylation biomarkers offer improved diagnostic efficiency in lung cancer. *Cancer Res.* **2012**, *72*, 5692–5701. [CrossRef]

64. Liloglou, T.; Bediaga, N.G.; Brown, B.R.; Field, J.K.; Davies, M.P. Epigenetic biomarkers in lung cancer. *Cancer Lett.* **2014**, *342*, 200–212. [CrossRef] [PubMed]

65. Romero-Jimenez, M.; Campos-Sanchez, J.; Analla, M.; Muñoz-Serrano, A.; Alonso-Moraga, A. Genotoxicity and anti-genotoxicity of some traditional medicinal herbs. *Mutat. Res. Genet. Toxicol. Environ. Mutagen.* **2005**, *585*, 147–155. [CrossRef]

66. Masuda, Y.; Asada, K.; Satoh, R.; Takada, K.; Kitajima, J. Capillin, a major constituent of *Artemisia capillaris Thunb.* flower essential oil, induces apoptosis through the mitochondrial pathway in human leukemia HL-60 cells. *Phytomedicine* **2015**, *22*, 545–552. [CrossRef] [PubMed]

67. Reiter, L.T.; Potocki, L.; Chien, S.; Gribskow, M.; Bier, E. A systematic analysis of human disease-associated gene sequences in *Drosophila melanogaster*. *Genome Res.* **2001**, *11*, 1114–1125. [CrossRef]

68. Jones, M.A.; Grotewiel, M. *Drosophila* as a model for age-related impairment in locomotor and other behaviors. *Exp. Gerontol.* **2011**, *46*, 320–325. [CrossRef]

69. Pandey, U.B.; Nichols, C.D. Human disease models in *Drosophila melanogaster* and the role of the fly in therapeutic drug discovery. *Pharmacol. Rev.* **2011**, *63*, 411–436. [CrossRef]

70. Mateo-Fernández, M.; Merinas-Amo, T.; Moreno-Millán, M.; Alonso-Moraga, A.; Demyda Peyrás, S. In Vivo and in vitro genotoxic and epigenetic effects of two types of cola beverages and caffeine: A multiassay approach. *BioMed Res. Int.* **2016**, *2016*, 7574843. [CrossRef]

71. Mukhopadhyay, I.; Chowdhuri, D.K.; Bajpayee, M.; Dhawan, A. Evaluation of in vivo genotoxicity of cypermethrin in *Drosophila melanogaster* using the alkaline Comet assay. *Mutagenesis* **2004**, *19*, 85–90. [CrossRef] [PubMed]

72. Siddique, H.R.; Chowdhuri, D.K.; Saxena, D.K.; Dhawan, A. Validation of *Drosophila melanogaster* as an in vivo model for genotoxicity assessment using modified alkaline Comet assay. *Mutagenesis* **2005**, *20*, 285–290. [CrossRef] [PubMed]

73. Martínez, A.; Ikken, I.; Cambero, M.I.; Marín, M.L.; Haza, A.I.; Casas, C.; Morales, P. Mutagenicity and cytotoxicity of fruits and vegetables evaluated by the Ames test and 3-(4,5-dimethylthiazol-2-yl)-2,5-diphenyltetrazolium bromide (MTT) assay. *Food Sci. Technol. Int.* **1999**, *5*, 431–437. [CrossRef]

74. Chughtai, S.R.; Dhmad, M.A.; Khalid, N.; Mohamed, A.S. Genotoxicity testing of some spices in diploid yeast. *Pakisthan J. Bot.* **1998**, *30*, 33–38.

75. Shukla, Y.; Taneja, P. Antimutagenic effects of garlic extract on chromosomal aberrations. *Cancer Lett.* **2002**, *176*, 31–36. [CrossRef]

76. Abraham, S.K.; Kesavan, P. Genotoxicity of garlic, turmeric and asafoetida in mice. *Mutat. Res.* **1984**, *136*, 85–88. [CrossRef]

77. Sowjanya, B.L.; Devi, K.R.; Madhavi, D. Modulatory effects of garlic extract against the cyclophosphamide induced genotoxicity in human lymphocytes in vitro. *J. Environ. Biol.* **2009**, *30*, 663–666.

78. Kulkarni, M.; Ascough, G.D.; Verschaeve, L.; Baeten, K.; Arruda, M.P.; Van Staden, J. Effect of smoke-water and a smoke-isolated butenolide on the growth and genotoxicity of commercial onion. *Sci. Hort.* **2010**, *124*, 434–439. [CrossRef]

79. Guyonnet, D.; Belloir, C.; Suschetet, M.M.; Siess, M.H.; Le Bon, A.M. Antimutagenic activity of organosulfur compounds from *Allium* is associated with phase II enzyme induction. *Mutat. Res.* **2001**, *495*, 135–145. [CrossRef]

80. Musk, S.; Clapham, P.; Johnson, I. Cytotoxicity and genotoxicity of diallyl sulfide and diallyl disulfide towards Chinese hamster ovary cells. *Food Chem. Toxicol.* **1997**, *35*, 379–385. [CrossRef]

81. Ramel, C.; Alekperok, U.K.; Ames, B.N.; Kada, T.; Wattenberg, L.W. Inhibitors of mutagenesis and their relevance to carcinogenesis: Report by ICPEMC expert group on antimutagens and desmutagens. *Mutat. Res.* **1986**, *168*, 47–65. [CrossRef]

82. Ito, Y.; Maeda, S.; Sugiyama, T. Suppression of 7, 12-dimethylbenz [a] anthracene-induced chromosome aberrations in rat bone marrow cells by vegetable juices. *Mutat. Res.* **1986**, *172*, 55–60. [CrossRef]

83. Prakash, D.; Singh, B.N.; Upadhyay, G. Antioxidant and free radical scavenging activities of phenols from onion (*Allium cepa*). *Food Chem.* **2007**, *102*, 1389–1393. [CrossRef]

84. Imai, J.; Ide, N.; Nagae, S.; Moriguchi, T.; Matsuura, H.; Ytakura, Y. Antioxidant and radical scavenging effects of aged garlic extract and its constituents. *Planta Med.* **1994**, *60*, 417–420. [CrossRef]

85. Singh, S.; Abraham, S.; Kesavan, P. Radioprotection of mice following garlic pretreatment. *Br. J. Cancer* **1996**, *27*, S102–S104.

86. Kim, J.M.; Jeon, G.I.; Park, E.J. Protective Effect of Garlic (*Allium sativum* L.) extracts prepared by different processing methods on DNA damage in human leukocytes. *J. Korean Soc. Food Sci. Nutr.* **2010**, *39*, 805–812. [CrossRef]

87. Sengupta, A.; Ghosh, S.; Das, S. Administration of garlic and tomato can protect from carcinogen induced clastogenicity. *Nutr. Res.* **2002**, *22*, 859–866. [CrossRef]

88. Le Bon, A.M.; Roy, C.; Dupont, C.; Suschetet, M. In vivo antigenotoxic effects of dietary allyl sulfides in the rat. *Cancer Lett.* **1997**, *114*, 131–134. [CrossRef]

89. Anh, N.T.T.; Nishitani, M.; Harada, S.; Yamaguchi, M.; Kamei, K.A. *Drosophila* model for the screening of bioavailable NADPH oxidase inhibitors and antioxidants. *Mol. Cell. Biochem.* **2011**, *352*, 91–98. [CrossRef]

90. Zhang, Z.; Han, S.; Wang, H.; Wang, T. Lutein extends the lifespan of *Drosophila melanogaster*. *Arch. Gerontol. Geriatr.* **2014**, *58*, 153–159. [CrossRef]

91. Lei, M.M.; Xu, M.Y.; Zhang, Z.S.; Zhang, M.; Gao, Y.F. The analysis of saccharide in black garlic and its antioxidant activity. *Adv. J. Food Sci. Technol.* **2014**, *6*, 755–760. [CrossRef]

92. Prowse, G.M.; Galloway, T.S.; Foggo, A. Insecticidal activity of garlic juice in two dipteran pests. *Agric. For. Entomol.* **2006**, *8*, 1–6. [CrossRef]

93. Penner, R.; Fedorak, R.N.; Madsen, K.L. Probiotics and nutraceuticals: Non-medicinal treatments of gastrointestinal diseases. *Curr. Opin. Pharmacol.* **2005**, *5*, 596–603. [CrossRef] [PubMed]

94. Hong, Y.S.; Ham, Y.A.; Choi, J.H.; Kim, J. Effects of allyl sulfur compounds and garlic extract on the expressions of Bcl-2, Bax, and p53 in non small cell lung cancer cell lines. *Exp. Mol. Med.* **2000**, *32*, 127–134. [CrossRef] [PubMed]

95. Druesne, N.; Pagniez, A.; Mayeur, C.; Thomas, M.; Cherbuy, C.; Duée, P.H.; Martel, P.; Chaumontet, C. Diallyl disulfide (DADS) increases histone acetylation and p21waf1/cip1 expression in human colon tumor cell lines. *Carcinogenesis* **2004**, *25*, 1227–1236. [CrossRef] [PubMed]

96. Kwon, K.B.; Yoo, S.J.; Ryu, D.G.; Yang, J.Y.; Rho, H.W.; Kim, J.S.; Park, J.W.; Kim, H.R.; Park, B.H. Induction of apoptosis by diallyl disulfide through activation of caspase-3 in human leukemia HL-60 cells. *Biochem. Pharmacol.* **2002**, *63*, 41–47. [CrossRef]

97. Sundaram, S.G.; Milner, J.A. Diallyl disulfide induces apoptosis of human colon tumor cells. *Carcinogenesis* **1996**, *17*, 669–673. [CrossRef] [PubMed]

98. Chun, H.S.; Kim, H.J.; Choi, E.H. Modulation of cytochrome P4501-mediated bioactivation of benzo [a] pyrene by volatile allyl sulfides in human hepatoma cells. *Biosci. Biotechnol. Biochem.* **2001**, *65*, 2205–2212. [CrossRef]

99. Iciek, M.; Kwiecien, I.; Wlodek, L. Biological properties of garlic and garlic-derived organosulfur compounds. *Environ. Mol. Mutagen.* **2009**, *50*, 247–265. [CrossRef]

100. Cohen, G.M.; Sun, X.M.; Snowden, R.T.; Dinsdale, D.; Skilleter, D.N. Key morphological features of apoptosis may occur in the absence of internucleosomal DNA fragmentation. *Biochem. J.* **1992**, *286*, 331–334. [CrossRef]

101. Balasenthil, S.; Rao, K.; Nagini, S. Garlic induces apoptosis during 7, 12-dimethylbenz [a] anthracene-induced hamster buccal pouch carcinogenesis. *Oral Oncol.* **2002**, *38*, 431–436. [CrossRef]

102. Cardile, V.; Scifo, C.; Russo, A.; Falsaperla, M.; Morgia, G.; Motta, M.; Renis, M.; Imbriani, E.; Silvestre, G. Involvement of HSP70 in resveratrol-induced apoptosis of human prostate cancer. *Anticancer Res.* **2002**, *23*, 4921–4926.

103. Zulkifli, S.; Luddin, N.; Thirumulu, K.P. Perspectives for the Comet assay in assessing genotoxicity—A review. *Curr. Pharmacogenomics Person. Med.* **2016**, *14*, 29–35. [CrossRef]

104. Fairbairn, D.W.; O'Neill, K.L. Necrotic DNA degradation mimics apoptotic nucleosomal fragmentation comet tail length. *In Vitro Cell. Dev. Biol. Anim.* **1995**, *31*, 171–173. [CrossRef] [PubMed]

105. Belloir, C.; Singh, V.; Daurat, C.; Siess, M.H.; Le Bon, A.M. Protective effects of garlic sulfur compounds against DNA damage induced by direct-and indirect-acting genotoxic agents in HepG2 cells. *Food Chem. Toxicol.* **2006**, *44*, 827–834. [CrossRef]

106. Roman-Gomez, J.; Jimenez-Velasco, A.; Agirre, X.; Castillejo, J.A.; Navarro, G.; San Jose-Eneriz, E.; Garate, L.; Cordeu, L.; Cervantes, F.; Prosper, F. Repetitive DNA hypomethylation in the advanced phase of chronic myeloid leukemia. *Leuk. Res.* **2008**, *32*, 487–490. [CrossRef] [PubMed]

© 2019 by the authors. Licensee MDPI, Basel, Switzerland. This article is an open access article distributed under the terms and conditions of the Creative Commons Attribution (CC BY) license (http://creativecommons.org/licenses/by/4.0/).

Article

Physicochemical Characterization and Biological Activities of Black and White Garlic: In Vivo and In Vitro Assays [†]

María Ángeles Toledano Medina [1], Tania Merinas-Amo [2], Zahira Fernández-Bedmar [2,*],
Rafael Font [3], Mercedes del Río-Celestino [3], Jesús Pérez-Aparicio [1], Alicia Moreno-Ortega [4,5],
Ángeles Alonso-Moraga [2] and Rafael Moreno-Rojas [4]

[1] Department of Food Science and Health, IFAPA-Palma del Río, Avda. Rodríguez de la Fuente, s/n,
14700 Palma del Río, Córdoba, Spain; mariaa.toledano@juntadeandalucia.es (M.A.T.M.);
jesus.perez.aparicio@juntadeandalucia.es (J.P.-A.)

[2] Department of Genetics, University of Córdoba, Gregor Mendel Building, Campus Rabanales,
14071 Córdoba, Spain; tania.meram@gmail.com (T.M.-A.); ge1almoa@uco.es (A.A.-M.)

[3] Agrifood Laboratory, CAPDER Avda. Menéndez Pidal s/n, 14004 Córdoba, Spain;
rafaelm.font@juntadeandalucia.es (R.F.); mercedes.rio.celestino@juntadeandalucia.es (M.d.R.-C.)

[4] Department of Bromatology and Food Technology, University of Córdoba, Darwin Building, Campus
Rabanales, 14071 Córdoba, Spain; aliciamorenoortega@hotmail.com (A.M.-O.);
rafael.moreno@uco.es (R.M.-R.)

[5] Department of Food Science and Health, IFAPA-Alameda del Obispo, Avda. Menéndez-Pidal, s/n,
14004 Córdoba, Spain

[*] Correspondence: b12febez@uco.es or zanoferbed@gmail.com; Tel.: +34-957-218-674

[†] The work was a part of María Ángeles Toledano Medina's doctoral thesis.

Received: 11 May 2019; Accepted: 12 June 2019; Published: 21 June 2019

Abstract: White and three types of black garlic (13, 32, and 45 days of aging, named 0C1, 1C2, and 2C1, respectively) were selected to study possible differences in their nutraceutic potential. For this purpose, garlic were physicochemically characterized (Brix, pH, aW, L, polyphenol, and antioxidant capacity), and both in vivo and in vitro assays were carried out. Black garlic samples showed higher polyphenol content and antioxidant capacity than the white ones. The biological assays showed that none of the samples (neither raw nor black garlic) produced toxic effects in the *Drosophila melanogaster* animal genetic model, nor exerted protective effects against H_2O_2, with the exception of the 0C1 black garlic. Moreover, only white garlic was genotoxic at the highest concentration. On the other hand, 0C1 black garlic was the most antigenotoxic substance. The in vivo longevity assays showed significant extension of lifespan at some concentrations of white and 0C1and 1C2 black garlic. The in vitro experiments showed that all of the garlic samples induced a decrease in leukemia cell growth. However, no type of garlic was able to induce proapoptotic internucleosomal DNA fragmentation. Taking into account the physicochemical and biological data, black garlic could be considered a potential functional food and used in the preventive treatment of age-related diseases. In addition, our findings could be relevant for black-garlic-processing agrifood companies, as the economical and timing costs can significantly be shortened from 45 to 13 days of aging.

Keywords: black garlic; physicochemical profile; polyphenol content; HL-60 cell line

---

1. Introduction

Garlic (*Allium sativum*) is probably one of the oldest known medicinal plants, used since ancient times to cure different human diseases. Garlic started taking part in humans' daily diet in Ancient Egypt [1]. Several scientific researches and clinical trials have been conducted during the last decade to determine the effects of garlic consumption on human health. Garlic's principal medicinal uses have focused on prevention and treatment of cardiovascular disease by lowering blood pressure and cholesterol, and, more recently, on its antimicrobial properties and as a preventive agent for cancer [2,3].

The physiological effects of garlic are due mainly to the presence of volatile sulfur compounds like thiosulfates, which give it its characteristic pungent aroma. Several recent studies have shown that these organosulfur compounds show anti-cancer, anti-cardiovascular, anti-neurological, and anti-liver disease effects, as well as effects for the prevention of allergies and arthritis [4–7]. This group of compounds, originating from the allicin decomposition, are associated with *Allium* species' pungent aroma and taste as well as their antioxidant activity [4,8]. However, rats fed with fresh garlic at high doses (0.5 g/kg of body weight/day) showed toxicity in the liver [9].

Even though the health benefits of garlic are known, its global consumption is declining. In general, people are reluctant to eat raw garlic due to its pungent taste, smell, and gastrointestinal discomfort. Because of this, researchers are interested in developing aged garlic products to decrease these negative effects [10].

With a growing awareness of the health benefits of garlic, black garlic, an aged garlic product, has emerged as one of the fastest-growing health-oriented food products in world markets [10]. Black garlic is produced through the natural aging of whole ordinary garlic under controlled high temperature (70 °C) and humidity (90%) conditions for several days, without any artificial treatments or additives [11]. Thermal processes are commonly used in food manufacturing to enhance the sensorial quality of foods, their palatability, and to extend the range of colors, tastes, aromas, and textures in food [12]. In addition, heating processes have led to the formation of biological compounds that are not originally present in food [13]. However, the influence of thermal processes on the concentration of single flavonoids and phenolic acids in garlic still remains unknown.

During aging, the cloves of normal garlic change their color from white to brown and finally become black due to the Maillard reaction. At the same time, unstable compounds in raw garlic are transformed into stable soluble compounds with a high antioxidant power [6,14]; the organoleptic characteristics in black garlic are improved due to the conversion of unstable and odorous compounds to stable and odorless compounds such as S-allyl-L-cysteine (SAC), or decomposed to organosulfur compounds such as diallyl sulphide (DAS), diallyl disulphide (DADS), diallyl trisulphide (DATS), dithiins, and ajoene [4,6]. Previous studies on black garlic have reported that the increase in its antioxidant capacity could be due to the increase in polyphenols and S-allyl-cysteine, a compound derived from alliin, during the heat processing [15].

Compared with fresh garlic, black garlic contains a polyphenol content that is three times higher in whole black garlic bulbs and six times higher in peeled black garlic cloves [11], which is directly related to the increase in the antioxidant activity. The amino acids, carbohydrates, and the S-allyl-L-cysteine contents are increased 2.5 times, 28.7%–47.0%, and eight times, respectively [16,17].

Different beneficial health properties of black garlic have been described previously: (i) antioxidant effects using different indicators such as super-oxide dismutase (SOD), 2,2′-azino-bis(3-ethylbenz-thiazoline-6-sulfonic acid) (ABTS), and hydroxy radical scavenging, as well as Fe^{2+}-chelating activities; (ii) in vivo and in vitro chemopreventive effects in different cancers—including ethanol extracts of aged black garlic, which reduce the viability of several human cancer cell lines (i.e., AGS, A549 lung, HepG2 liver, and MCF-7 breast cancer cells), and hexane extracts, which induce caspase-dependent apoptosis in leukemic cells; (iii) anti-inflammatory effects have been shown by inactivation of NF-κB, upregulation of heme oxygenase-1, and inhibition of the COX-2 and 5-lipooxygenase activities, among other effects [10].

The aim of the present study is to perform a qualitative and quantitative evaluation of the health-beneficial activities of white and three types of black garlic using a multi-assay experimental design at the individual, cell, and DNA levels. We assessed the genotoxic, antigenotoxic, and lifespan effects using an in vivo animal model of the common fruit fly (*Drosophila melanogaster*) and their proapoptotic capacities against cancer processes, including cytotoxicity and clastogenic DNA activity, using an in vitro human cancer model (HL-60 cell line).

2. Materials and Methods

2.1. Preparation of Samples

White and black garlic were used in this study. Raw white garlic was purchased in a local market. Black garlic was manufactured at 60 °C and 90% relative humidity (RH). Samples at 0 (White), 13 (0C1), 32 (1C2), and 45 (2C1) days aging were taken during the manufacturing process. After peeling bulbs, samples were crushed and divided into three subsets (to be physicochemically analyzed at specified times). Garlic samples were lyophilized (−20 °C, less than 1% water) before the biological assays and then dissolved in distilled water to obtain the different concentrations tested. The lyophilized extracts were stored at room temperature in a dark and dry atmosphere until use.

The concentrations of garlic for the different bioassays were established taking into account the average daily food intake of *D. melanogaster* (1 mg/day) and the average body weight (1 mg) [18]. The concentration range for all tested substances was calculated in order to make it comparable to the recommended garlic daily intake for humans. Although there is no standard intake for garlic, the German Kommission E monograph proposed that a daily intake of approximately 1–2 garlic cloves (about 4 g) of intact garlic may have health benefits [19]. Unfortunately, this recommendation is not substantiated by any scientific reference.

2.2. Measurement of Soluble Solid Content, pH, a_w, and Browning Intensity

Total soluble solid content (°Brix), pH, water activity (a_w), and browning intensity (L value) values were determined in triplicate for all samples following the method previously described by Toledano–Medina et al. [11]. Garlic soluble solids (°Brix) were measured with an Abbe Refractometer ORT-1 of KERN (Kern & Sohn GmbH, Balingen-Frommern, Germany). Garlic pH was measured with a pH meter Crison Basic 20 (Crison Instruments, Barcelona, Spain). Garlic water activity (a_w) was measured with an Aqualab Series 3/3TE meter with a temperature stabilizer (MeterGroup, München, Germany). Garlic browning intensity was determined with a Konica Minolta CR-410 Croma Meter colorimeter (Konica Minolta, Inc., Tokyo, Japan) as an L value (L = 100, white; L = 0, black), following the method described previously by Toledano–Medina et al. [11].

2.3. Total Polyphenol Content and Antioxidant Capacity

A Perkin Elmer Lambda 20 UV VIS spectrophotometer (Perkin Elmer, Waltham, MA, USA) was used to determine total polyphenol content and antioxidant capacity in raw and heated garlic. A previous extract was prepared to analyze antioxidant properties. Briefly, samples were lyophilized (−20 °C, less than 1% water) and spliced into five extracts per sample. Garlic extract was prepared dissolving 0.3 g of the lyophilized sample in 10 mL of a mixture of 50% (*v/v*) ethanol and distilled water. Next, samples were stirred for one hour and then filtered using a Buchner funnel with Whatman paper (Whatman PLC, Maidstone, UK) into a vacuum flask connected to a vacuum pump filter. The filtered extract was levelled at 25 mL with a 50% (*v/v*) hydroalcoholic solution.

The polyphenol concentration of garlic samples was determined by the Folin–Ciocalteu method [20]. To a volumetric 25 mL flask, 0.5 mL of extract, 10 mL of distilled water, 1 mL of Folin–Ciocalteu reagent, and 3 mL of sodium carbonate 20% (*w/v*) were added and diluted to volume (25 mL) with distilled water. The mixture was heated to 50 °C for 5 min to accelerate the coloration reaction. Subsequently, it was cooled with water, and the reading was carried out in the spectrophotometer (Perkin Elmer) at 765 nm. The reading was compared with a calibration curve prepared with different gallic acid solutions: 75, 100, 200, 250, 300 ppm. Polyphenol content results were expressed considering the dilution of the sample (0.3 g in 25 mL) in grams of gallic acid equivalents per kilogram of lyophilized sample.

Raw and heated garlic antioxidant capacity was determined by the ABTS method [21]. A mix of 2.557 mL of a solution of 7 mM ABTS reagent (Sigma, St. Louis, MI, USA) and 0.333 mL of a solution of 2.25 mM potassium persulfate in distilled water was made. This solution was stored in darkness for 16 h, enough time for radical cation (ABTS$^{\bullet+}$) formation. Then, 0.15 mL of the ABTS^{+} solution was diluted in 15 mL of ethanol. The absorbance value at 734 nm was adjusted near 0.7 (A_0). Next, 0.980 mL of ABTS^{+} solution and 0.02 mL of garlic extract were added. After stirring, the absorbance was read at 734 nm after 7 minutes (A_1). The inhibition percentage was calculated by the following expression:

$$\% \text{ inhibition} = (A_0 - A_1) \times 100/A_0. \tag{1}$$

A calibration curve was built with the following Trolox (6-hydroxy-2, 5, 7, 8-tetramethylchroman-2-carboxylic acid) concentrations: 0.1, 0.5, 1, and 1.5 mM. Considering the sample dilution, results were expressed in mmol Trolox-equivalents per kilogram of lyophilized sample.

2.4. In Vivo Assays

2.4.1. *D. melanogaster* Strains

The following *Drosophila* strains, each carrying a third chromosome hair marker, were used: (i) *mwh/mwh* are homozygous for the recessive multiple wing hairs (*mwh*) mutation that produces multiple tricomas per cell instead of one [22], and (ii) flr^3/In (*3LR*) *TM3, rip^p sep bx^{34e} e^s Bd^S*, where the flr^3 (*flare*) marker is a homozygous recessive lethal mutation that produces deformed tricomas but is viable in homozygous somatic cells once larvae start the development [23]. For detailed information on the mutations, see Lindsley and Zimm [24].

2.4.2. Toxicity and Antitoxicity Assays

Five concentrations (4, 2, 1, 0.5, and 0.25 mg/mL) for each tested garlic, along with negative (H_2O) and positive (0.12 M H_2O_2) controls were assayed after toxicity screening experiments. The toxicity index was calculated as the percentage of individuals born in each treatment with respect to the negative control. The antitoxicity tests consisted of combined treatments using the same concentrations as in the toxicity assays, with the exception of the highest one (4 mg/mL), by adding the toxicant hydrogen peroxide at 0.12 M [25]. The percentage of emerging adults was compared with the positive control.

2.4.3. Genotoxicity and Antigenotoxicity Assays

The genotoxicity assays were carried out following the method described by Graf et al. [26]. Briefly, trans-heterozygous larvae for *mwh* and flr^3 gene markers were obtained by crossing four-day-old virgin flr^3 females with *mwh* males in a 2:1 ratio. Four days after fertilization, females were allowed to lay eggs in fresh yeast medium (25 g yeast and 4 mL sterile distilled water) during 8 h to obtain synchronized larvae. After 72 ± 4 h, the larvae were collected, washed with distilled water to remove the remaining medium, and transferred, in groups of 100 individuals, to the treatment tubes where they were chronically fed with the different compounds. Treatment tubes contained 0.85 g of *Drosophila* Instant Medium (Formula 4-24, Carolina Biological Supply, Burlington, NC, USA) and 4 mL of solutions with different concentrations of garlic (2 mg/mL and 0.25 mg/mL).

The antigenotoxicity trials were carried out following the method described by Graf et al. [27], which consists of combined treatments of genotoxin (0.12 M H_2O_2) (Sigma, cat. number H-1009) and the same concentrations used in genotoxicity assays of lyophilized garlic. For the evaluation of the inhibition potency, negative (H_2O) and positive (0.12 M H_2O_2) (Sigma, cat. number H-1009) concurrent controls were included. After emergence, adult flies were stored in 70% ethanol until the removal and mounting of wings on slides using Faure's solution (30 g Arabic gum, 20 mL glycerol, 50 g chloral hydrate, and 50 mL distilled water) for mutation screening under a photonic microscope (Leica, Wetzlar, Germany) at 400× magnification.

Similar numbers of male and female wings for each treatment and concentration were mounted, and wing hair mutations were scored among a total of 24,400 monotricoma wild-type cells per wing [28].

Wing hair spots were grouped into three different categories: *S*, a small single spot corresponding to one or two cell clones exhibiting the *mwh* phenotype that occurs in the latest stages of the mitotic division; *L*, a large single spot with three or more cell clones showing *mwh* or *flr³* phenotypes that occur in the early stages of larval development; or *T*, a twin spot corresponding to two juxtapositioned clones, one showing the *mwh* phenotype and other the *flr³* phenotype. Small and large spots are caused by somatic point mutations, chromosome aberrations, and somatic recombinations, while twin spots are produced exclusively by somatic recombinations between the *flr³* locus and the centromere.

The total number of clones was also counted and a multiple-decision procedure was applied to determine whether a result was positive, inconclusive, or negative [29,30]. The inhibition percentages (IPs) for the combined treatments were calculated from the total spots per wing statistics with the following formula [31]:

$$IP = ((\text{single genotoxin} - \text{combined treatment})/\text{single genotoxin}) \times 100. \tag{2}$$

2.4.4. Lifespan Assays

In order to compare the genotoxicity and longevity results, flies that underwent the lifespan trials carried the same genotype as in genotoxicity assays. Hence, the F1 progeny from *mwh* and *flr³* parental strains produced by a 24 h egg-laying in fresh yeast medium was used in the longevity experiments. All experiments were carried out at 25 °C and according to the procedure described by Fernández–Bedmar et al. [25]. Briefly, synchronized 72 ± 12 hour-old trans-heterozygous larvae were washed, collected, and transferred in groups of 100 individuals to test vials containing 0.85 g of *Drosophila* Instant Medium (formula 4-24, Carolina BiologicalSupply, Burlington NC, USA) and 4 mL of the different concentrations of the selected compounds.

Sets of 25 emerged individuals of the same sex were selected and placed into sterile vials containing 0.21 g of *Drosophila* Instant Medium (formula 4-24, Carolina BiologicalSupply, Burlington NC, USA) and 1 mL of the different concentrations of solution of the compounds (4 mg/mL–0.25 mg/mL range). Two replicates were followed during the complete life extension for each control and for the concentrations established. Alive animals were counted, and the media was renewed twice a week.

2.5. In Vitro Assays

2.5.1. HL-60 Cell Line Culture Conditions

Cells were grown in RPMI-1640 medium (Sigma, R5886, St. Louis, MI, USA) supplemented with 50 mL heat-inactivated fetal bovine serum (Linus, S01805, Madrid, Spain), L-glutamine at 200 mM (Sigma, G7513), and antibiotic-antimycotic solution with 10,000 units of penicillin, 10 mg of streptomycin, and 25 µg amphotericin B per mL (Sigma, A5955). Cells were incubated at 37 °C in a humidified atmosphere of 5% CO_2 (Shel Lab, Cornelius, OR, USA) [32]. The cultures were plated at 2.5×10^4 cells/mL density in 10 mL culture bottles and passed every 2 days.

2.5.2. Cytotoxicity Assay

HL-60 cells were placed in 96-well culture plates (2×10^4 cells/mL) and treated for 72 h with the lyophilized white and black garlic at different concentrations (4 mg/mL, 2 mg/mL, 1 mg/mL, 0.5 mg/mL, 0.25 mg/mL, 0.12 mg/mL, 0.06 mg/mL, 0.03 mg/mL, and 0.015 mg/mL for white garlic and 4 mg/mL, 2 mg/mL, 1 mg/mL, 0.5 mg/mL, and 0.25 mg/mL for black garlic samples). This wide range of concentrations was intended to estimate the inhibitory concentration 50 (IC_{50}).

Cell viability was determined by the trypan blue dye (Sigma, T8154) exclusion test. Trypan blue solution was added to the cell cultures at a 1:1 volume ratio and 20 µL of cell suspension were immediately loaded into a Neubauer chamber. Cells were counted with an inverted microscope at

100× magnification (AE30/31, Motic, Wetzlar, Germany). Curves were plotted as survival percentages with respect to the control growing at 72 h. At least three independent repetitions were carried out.

2.5.3. Determination of DNA Fragmentation

DNA fragmentation is a hallmark of apoptosis and has been regarded as a critical step in apoptosis. Briefly, HL-60 cells (1×10^6 cells/mL) were treated with different concentrations of lyophilized garlic (4 mg/mL, 2 mg/mL, 1 mg/mL, 0.5 mg/mL, and 0.25 mg/mL, respectively) for 5 h. Treated cells were collected and centrifuged at 3000 rpm for 5 min, and DNA was extracted with lysis, precipitation, and wash steps according to Merinas–Amo et al. [33]. The total extracted DNA was quantified in a spectrophotometer (Nanodrop® ND-1000, Thermo Fisher Scientific, Waltham, MA, USA), and 1200 ng of DNA was loaded into a 2% agarose gel electrophoresis, stained with ethidium bromide, and visualized under UV light.

2.6. Statistical Analysis

The statistical analysis of the solid content, pH, a_w, browning intensity, polyphenol content, antioxidant capacity, and total polyphenol index for each type of garlic was evaluated with the SPSS Statistics 17.0 software SPSS (IBM, Armonk, NY, USA) using one-way ANOVA and Tukey's test (homogeneous subsets) to assess the significance of the subsets.

Significant differences with respect to the concurrent control in toxicity assays were determined using the Chi-square method, and a concentration was considered as a toxic when the Chi-square value was higher than 5.02.

The frequency of each type of mutant clone/wing in the anti/genotoxicity assays was compared with the negative concurrent control, and significance was given at the 5% error level. Inconclusive and positive results were further analyzed with the Mann–Whitney–Wilcoxon ($\alpha = \beta = 0.05$) nonparametric U-test using the SPSS Statistics 17.0 software SPSS.

The statistical treatment of life- and health-span data for each control and concentration was assessed with the SPSS Statistics 17.0 software, using the Kaplan–Meier method. The significance of the curves was determined using the Log-Rank method (Mantel-Cox).

To obtain the tumor growth inhibition curves, the mean of three independent assays of the alive-treated cells for each compound and concentration was used. The standard errors of the three repetitions were calculated, and the Excel-given curve was added. Finally, the inhibitory concentration 50 (IC_{50}) was estimated.

3. Results and Discussion

3.1. Soluble Solids Content, pH, Water Activity, and Browning Intensity

A weight reduction was observed during the garlic manufacturing procedure, with the 0C1 black garlic being the sample with the nearest weight to the white garlic (Table 1). According to similar studies, changes in garlic weight during processing are mainly caused by a reduction of the amount of water [16]. The main organosulfur in black garlic is considered to be the water-soluble S-allyl-L-cysteine (SAC) [34]. Hence, after aging, SAC increased in the processed black garlic matrix, and its precursor garlicγ-glutamyl-S-allyl-L-cysteine decreased [10]. The manufacturing of black garlic in this manner is not a microbe-associated fermentation but a Maillard and Browning reaction because the processing temperature of garlic does not allow bacterial growth to elicit fermentation [16].

Soluble solids content (°Brix), pH, water activity (a_w), and browning intensity (L) are shown in Table 1. During heat treatment, soluble solids content increased in garlic, whereas pH, a_w, and browning intensity decreased. Similar Tukey's test values were obtained in °Brix readings for white and 0C1 black garlic (40.47), meanwhile, significant soluble solids content differences were observed in the 1C2 and 2C1 black garlic (43.17 and 45.67, respectively). The sugar content (°Brix) of black garlic increased with respect to white garlic. This result is in agreement with the data of Choi et al., which show that sugar content (e.g., glucose, fructose, sucrose, and maltose) increased in black garlic compared to fresh and steamed garlic [35]. Furthermore, this increment might be related to its sweeter taste [16]. pH significantly decreased during the manufacturing process. White garlic pH was the highest with a value of 5.94, whereas black garlic pH decreased rapidly starting at 3.69 and reaching 3.49 at 45 days of aging. These results are in agreement with the report by Shin et al., who showed that black garlic pH decreased from 6.40 to 5.29 after 6 days of aging [36]. The same observation has recently been described [11,37]. Water activity (a_w) decreased with aging to a lesser extent than other parameters because the black garlic was manufactured maintaining a high relative humidity. According to Kaanane and Labuza and Labuza and Saltmarch, the rate of the browning reaction is known to reach a maximum at a_w values in the range of 0.5–0.7 [38,39]. However, significant differences between white and black garlic a_w are found (Table 1). The high RH and time required for producing black garlic in the present study might have created a balanced situation between the a_w of the heated garlic sample and the RH inside the chamber where black garlic was produced. This a_w condition is thought to facilitate the browning reaction in heated garlic samples. As Table 1 shows, browning intensity (L) in white and black garlic was significantly different, with more than 28 units of difference between them, although the 1C2 and 2C1 black garlic showed similar luminescence (17.85 and 17.58, respectively). Browning intensity happened earlier at higher temperatures. Several studies have shown a positive relationship between temperature increasing and browning product formation; however, at the initial induction period a decrease is observed [40,41]. The garlic's color eventually changed to dark brown/black, mainly due to the formation of numerous compounds resulting from the non-enzymatic browning reaction (Maillard reaction).

Table 1. Physicochemical characterization of four types of garlic according to the days of aging.

Type of Garlic	White	0C1 Black	1C2 Black	2C1 Black
Aging process (days)	0	13	32	45
Weight of 10 garlic cloves (g)	49.69 ± 0.25 [a,1]	45.83 ± 0.32 [b,c]	37.57 ± 0.38 [b]	19.67 ± 0.19 [d]
Soluble solid content (°Brix)	40.47 ± 0.29 [c]	40.47 ± 0.34 [c]	43.17 ± 0.48 [b]	45.67 ± 0.42 [a]
pH	5.94 ± 0.01 [a]	3.69 ± 0.03 [b]	3.60 ± 0.04 [c]	3.49 ± 0.06 [d]
Water activity (a_w)	0.97 ± 0 [a]	0.93 ± 0 [c]	0.93 ± 0 [c]	0.93 ± 0 [c]
Browning intensity (L)	47.16 ± 0.15 [a]	18.73 ± 0.21 [c]	17.85 ± 0.24 [b]	17.58 ± 0.25 [b]
Polyphenol content (g GAE/kg)	4.30 ± 0.04 [d]	10.94 ± 0.28 [c]	14.67 ± 0.19 [b]	16.17 ± 0.29 [a]
Antioxidant activity (TROLOX equivalents/kg)	10.20 ± 0.27 [d]	67.65 ± 1.26 [b]	57.35 ± 1.74 [c]	78.61 ± 2.41 [a]

Values are means ± standard error (SE) (n = 3). [1] Different letters (a, b, c, d) in the same row show significant values in a one-way ANOVA using the post hoc Tukey's test. GAE, gallic acid equivalents.

3.2. Total Polyphenol Content and Antioxidant Capacity

Total polyphenol (g/kg in Gallic) content and antioxidant capacity (inhibition percentage) are shown in Table 1. During heat treatment, unstable compounds of raw garlic are transformed into stable soluble compounds with a high antioxidant power [6,14]. Previous studies on black garlic reported that this enhancement of the antioxidant capacity could be due to the increase in polyphenols and *S*-allyl-cysteine, the compound derived from alliin [15]. The antioxidant power of polyphenols has been demonstrated, so it seems logical to state that an increase in polyphenol content in black garlic is responsible for the antioxidant properties in this product [42]. It is well known that the higher antioxidant effect of black garlic is due to the presence of *S*-allyl-cysteine, a compound derived from alliin during heat processing [15].

Significant differences among all the samples were found for the total polyphenol content and the antioxidant capacity. Both parameters increased significantly as heat increased. The highest concentration of polyphenol content was obtained in 2C1 black garlic, although all black garlic samples showed increases between 6 and 12 times in relation to the heat treatment (Table 1). Previous studies carried out with whole bulbs of black garlic at 70, 72, 75, and 78 °C have described an increase of 2–3 times in polyphenol content compared to raw garlic [11,43]. Our results on the increase of polyphenol content after heating agree in part with those obtained by other authors who found a threefold increase in content [43].

To clarify the antioxidant properties of black garlic during aging, we focused on the analysis of total polyphenol content. At the end of the heating process, an increase in antioxidant capacity was observed in garlic. Black garlic samples showed an increase rank of 5.7–7.8 times with respect to white garlic (Table 1). Several studies described that aged black garlic exerts stronger antioxidant activity than white garlic, both in vitro and in vivo assays [15,44]. The total polyphenol content of black garlic was not only significantly higher than that of raw garlic, but also increased significantly at the 13th day of aging. Similar results were obtained by Sasaki et al., who showed an antioxidant potency increase in aged black garlic extracts reaching 25-fold compared with fresh garlic [16]. According to Xu and Chang, heat treatment of the phenolic compounds increased the free fraction of phenolic acids, whereas it decreased the ester, glycoside, and ester-bound fractions, leading to an increase in free phenol forms [45]. Gorinstein et al. showed that the garlic processing conditions lead to changes in the content of its bioactive compounds (polyphenols such as flavonoids and anthocyanins), and this is related to the type and duration of treatment [46]. From the results regarding total polyphenols and antioxidant capacity, it is possible to state that the optimum aging period for black garlic in order to maximize antioxidant content may be 13 days.

3.3. Toxicity/Antitoxicity

The toxicity and antitoxicity of the four samples of tested garlic was assessed in the *D. melanogaster* in vivo model. Figure 1A shows the relative percentage of emerging adults after treating larvae with different concentrations of these substances, showing that none of the garlic samples were toxic at the assayed doses in *D. melanogaster*. These results agree with those by which the safety of garlic extracts was well established through general, chronic, acute, and subacute toxicity, teratogenicity, and toxicity tests conducted by the U.S. Food and Drug Administration, and clinical studies as well [47–52].

Figure 1B shows the results of the antitoxicity assays using hydrogen peroxide as a toxicant. The individuals treated with 0.12 M of the oxidative toxin reached an average survival rate of 63.4% with respect to the water negative control. In addition, 0C1 black garlic was the only preventive substance against H_2O_2 at two of the assayed concentrations (0.5 and 1 mg/mL). On the other hand, white and 1C2 and 2C2 black garlic did not exhibit protective effects against the genotoxin at any tested concentrations (0.25–0.5 and 0.5–1 mg/mL, respectively). Lei et al. studied the effects that black 10–15 days-aged garlic extracts had in *D. melanogaster*. The results from this study demonstrated that black garlic extracts possess strong antioxidant capacities in vitro and in vivo [53].

3.4. Genotoxicity/Antigenotoxicity

To assess the genotoxicity/antigenotoxicity of the studied compounds, we used the SMART (Somatic Mutation and Recombination Test) Test in *D. melanogaster* [27]. Increasing concentrations of tested compounds, a negative control corresponding to water used as a solvent, and a positive control (H_2O_2) for periodic validation of the assay were concurrently assayed. Furthermore, antigenotoxicity experiments were carried out using combined treatments consisting of repeating every concentration tested and by adding the same concentration of hydrogen peroxide, which we have demonstrated to be a potent mutagen in the SMART system [54].

Table 2 shows the results of genotoxicity assays in the SMART test for white and the three black garlics. Negative controls showed a frequency of mutations per wing equal to 0.195, which falls into the historical range for the wing spot test [33,55]. The final concentration of H_2O_2 used (0.12 M) has been demonstrated to exert a potent genotoxic effect capable of inducing somatic mutations and mitotic recombination in *D. melanogaster* [56]. The average frequency of total mutations per wing obtained in the treatment with H_2O_2 was 0.425. For each concentration and compound, single small, single large, twin, and total clones were analyzed in the wings of chronically treated animals. The results showed that all garlic showed non-genotoxic activity except for the white one, which significantly increased the frequency of mutations to 0.425 at the highest concentration tested. Similar results were obtained by Abraham and Kesavan and Shukla and Taneja, who demonstrated that aqueous garlic extracts (5% *v/v*) and fine garlic powder (7.5, 5 and 2.5 g/kg body weight) supplementation do not induce chromosomal aberrations nor DNA damage in mouse bone marrow cells [57,58]. Similar results were obtained by Sowjanya et al. at 3, 6, and 12 mg/culture in human lymphocytes [59] and by Chughtai et al. using extracts of fresh garlic bulbs in a yeast model [60].

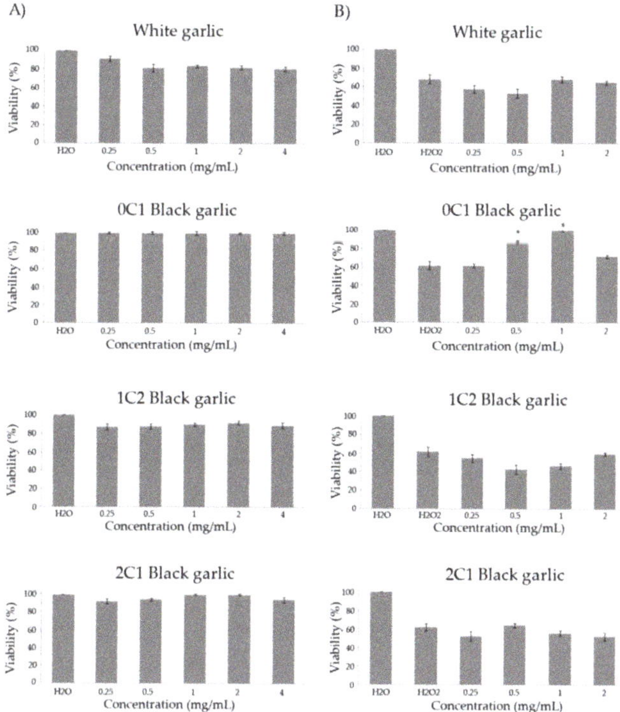

Figure 1. Toxicity (**A**) and antitoxicity (**B**) levels of black and white garlic studied in *D. melanogaster*. (**A**) Percentage of viability of *Drosophila* treated with different concentrations of the assayed garlic. (**B**) Viability of *Drosophila* tested with different concentrations of the tested garlic combined with the genotoxicant hydrogen peroxide at 0.12 M. Values represent the mean ± SE from three independent experiments. *: significant ($p \leq 0.05$), with respect to their concurrent controls. 0C1: black garlic with 13 days aging, 1C2: black garlic with 32 days aging, and 2C1: black garlic with 45 days aging.

Vegetables contain polyphenols and oligoelements with antimutagenic activity [61]. The 0C1 black garlic was the only one able to inhibit the genotoxic activity of hydrogen peroxide in a dose-dependent manner (Table 2). The highest concentration tested for 0C1 black garlic in the combined treatments partially counteracted part of the genotoxic effect of H_2O_2, showing a decrease in the total mutation frequency to 0.266 spots/wing and inhibiting around 37% of the genotoxicity induced by H_2O_2 (without control correction). The rest of the compounds tested did not show significant protective results against DNA damage at the highest concentration and a just a slight inhibition percentage of mutations induced by the genotoxin were observed (24% for white, 18.6% for 1C2 black, and 7.5% for 2C1 black garlic).

In general, garlic has significant antioxidant activity and protective effects against oxidative DNA damage regardless of the processing method [62]. Our antitoxicity and antigenotoxicity results showed that 0C1 black garlic (aged for 13 days) is able to protect from the genomic damage of this genotoxin in a dose-dependent manner. This effect could probably be due to the antioxidative and free-radical scavenging capacity of their respective organosulfur compounds, which agree with previous reports [14,63,64]. Besides the antioxidant activity, our results about the stronger antioxidant activity shown by black garlic, compared with fresh garlic, are in agreement with previous in vivo and in vitro garlic assays [65,66].

Table 2. Genotoxicity and antigenotoxicity of white (0 days aging), 0C1 black (13 days aging), 1C2 black (32 days aging), and 2C1 black (45 days aging) garlic in the *Drosophila* wing spot test.

Compound	Number of Wings	Clones per Wing (n° spots) [1]				Inhibition Percentage (%) [2]
		Small Single Clones (1–2 Cells) $m = 2$	Large Simple Clones (More Than 2 Cells) $m = 5$	Twin Clones $m = 5$	Total Clones $m = 2$	
H_2O	41	0.146 (6)	0.049 (2)	0	0.195 (8)	
H_2O_2	40	0.350 (14)	0.075 (3)	0	0.425 (17) +	
Simple Treatment						
White garlic (mg/mL)						
0.25	40	0.225 (9)	0.025 (1)	0.025 (1)	0.275 (11) −	
2	40	0.375 (15)	0.050 (2)	0.000	0.425 (17) +	
0C1 Black garlic (mg/mL)						
0.25	40	0.175 (7)	0.025 (1)	0.000	0.200 (8) −	
2	41	0.122 (5)	0.000	0.000	0.122 (5) −	
1C2 Black garlic (mg/mL)						
0.25	40	0.200 (8)	0.025 (1)	0.025 (1)	0.250 (10) -	
2	40	0.175 (7)	0.025 (1)	0.000	0.200 (8) -	
2C1 Black garlic (mg/mL)						
0.25	40	0.175 (7)	0.05 (2)	0.000	0.225 (9) −	
2	40	0.250 (10)	0.000	0.000	0.250 (10) −	
Combined Treatment With H_2O_2 (0.12 M)						
White garlic (mg/mL)						
0.25	34	0.235 (8)	0.088 (3)	0.000	0.323 (11) −	24
2	34	0.265 (10)	0.206 (7)	0.000	0.500 (17) +	−17
0C1 Black garlic (mg/mL)						
0.25	30	0.5 (15)	0.033 (1)	0.000	0.533 (16) +	−25.4
2	30	0.233 (7)	0.033 (1)	0.000	0.266 (8) −	37.4
1C2 Black garlic (mg/mL)						
0.25	26	0.307 (8)	0.038 (1)	0.000	0.346 (9) −	18.6
2	38	0.368 (14)	0.053 (2)	0.000	0.421 (16) −	0.17
2C1 Black garlic (mg/mL)						
0.25	28	0.357 (10)	0.036 (1)	0.000	0.393 (11) −	7.5
2	28	0.357 (10)	0.250 (7)	0.000	0.607 (17) +	−42.8

[1] Statistical diagnosis according to Frei and Wurgler [29,30]. +, positive ($p < 0.05$); −, negative. m, multiplication factor. Levels of significance $\alpha = \beta = 0.05$, tail test without Bonferroni correction. Inconclusive results were resolved by Mann–Whitney–Wilcoxon U-test. [2] The inhibition percentages for the combined treatments were calculated from total spots per wing according to Abraham [31].

3.5. Longevity Assays

Drosophila melanogaster is a choice model organism in the study of aging due to its relatively short life expectancy. Moreover, a large number of individuals can be reared in controlled laboratory conditions, and adults show many aspects of the observed cellular senescence events in mammals. Thus, flies have been frequently used to study physiological and pathological processes that affect life expectancy and can help to understand the relationship between nutrient metabolism and the mechanisms of aging [25].

The entire lifespan curves and significances obtained by the Kaplan–Meier method for each substance and concentration are shown in Figure 2 and Table 3, respectively. *Drosophila* had an average lifespan expansion of 60 days in the control treatment. White and 1C2 black garlic significantly increased *Drosophila*'s lifespan at the lowest and the two moderated concentrations tested (0.25, 1, and 2 mg/mL), with an extension with respect to the concurrent control of 10.1, 11.1, and 18.5 days for white garlic and 9.4, 10.1, and 9.8 days for black garlic, respectively (Table 3). Furthermore, every concentration assayed of 0C1 black garlic, except the highest one, induced a lifespan extension of 10 days in *D. melanogaster* compared to the control. On the other hand, 2C1 black garlic did not influence the lifespan extension of *D. melanogaster* at any tested concentration. No previous in vivo studies on longevity activities of black garlic as a food have been reported. However, several authors have reported beneficial effects on animal lifespans using white garlic extracts in *D. melanogaster* at 37.5

and 75 mg/mL, *Caenorhabditis elegans* at 0.05 mg/mL, and senescence-accelerated mice (SAMP8) at 2% (*w/w*) [53,67,68].

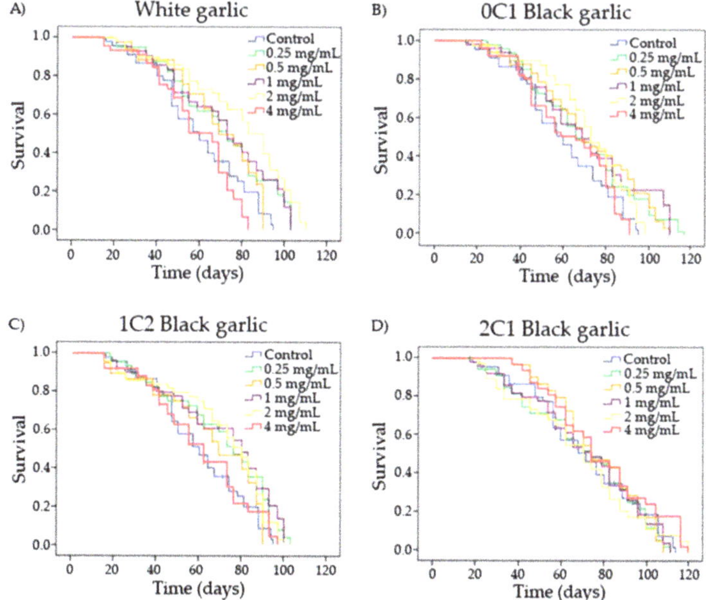

Figure 2. Survival curves of *D. melanogaster* fed with different concentrations of garlic (black and white) over time. **A**) White garlic, **B**) 0C1: black garlic with 13 days aging, **C**) 1C2: black garlic with 32 days aging and **D**) 2C1: black garlic with 45 days aging.

Table 3. Mean and significances of lifespan and healthspan curves for the different garlic treatments assayed in *D. melanogaster*.

Compound Title	Treatment (mg/mL)	Mean Lifespan (Days)		Mean Healthspan (Days)	
Negative Control	0	60.31		32.46	
WG	0.25	70.47	*	38.40	ns
	0.5	68.72	ns	29.40	ns
	1	71.43	**	40.39	*
	2	78.89	***	40.44	*
	4	58.21	ns	31.91	ns
0C1	0.25	70.15	*	37.67	ns
	0.5	71.72	**	38.18	ns
	1	70.73	*	30.85	ns
	2	71.80	*	43.90	*
	4	62.60	ns	33.10	ns
1C2	0.25	69.73	*	29.36	ns
	0.5	65.19	ns	25.15	*
	1	70.49	**	31.10	ns
	2	70.18	*	28.57	ns
	4	60.26	ns	26.20	ns
2C1	0.25	59.06	ns	24.07	ns
	0.5	64.75	ns	41.29	ns
	1	59.82	ns	28.50	ns
	2	57.30	ns	23.46	**
	4	66.38	ns	40.50	ns

Results were calculated by the Kaplan–Meier method, and the significance of the curves was determined by the Log-Rank method (Mantel–Cox). ns: non-significant ($p > 0.05$), *: significant ($p < 0.05$), **: significant ($p < 0.01$), ***: significant ($p < 0.001$). WG: white garlic (0 days aging); 0C1: black garlic (13 days aging); 1C2: black garlic (32 days aging); 2C1: black garlic (45 days aging).

We suggest that the differences found between these results and ours could be due to the different types of sample presentation. We used crude entire garlic material, and all data available elsewhere on lifespan trials come from extracts. In this sense, Prowse et al. demonstrated that garlic juice exerted insecticidal activity across life stages of flies at a wide range of concentrations (0.25–5%) in two dipteran pests (*Delia radicum* and *Musca domestica*) [69]. Lei et al. studied the effects of black 10–15 days-aged garlic extracts on the lifespan of *Drosophila* through the observation of half-life time, and the mean and maximum lifespan of organisms. The results suggested a significant longevity extension in *Drosophila* treated with black garlic extracts in a dose-dependent manner [53].

3.6. Healthspan Assays

In order to know the quality of life of the *Drosophila* treated in the longevity assays, we studied the 25% of individual survival at the top of the lifespan curves obtained in the previous test for each substance and concentration tested. This part of the lifespan is considered the healthspan of a curve and is characterized by low and more or less constant age-specific mortality rate values [70]. The results are shown in Table 3.

Only white and 0C1 black garlic induced a significant increase of healthspan in *D. melanogaster* compared to the control, with an average of 8 and 11.5 days, respectively. In contrast, 1C2 and 2C1 black garlic induced a significant reduction of healthspan in *Drosophila* at moderate concentrations, with a value of 7.3 and 9 days, respectively, with respect to the control. No previous studies about the effects that white and black garlic exert on quality of life have been reported.

3.7. Cytotoxicity

All the substances assayed showed cytotoxic activity against HL-60 tumor cells (Figure 3). White and black garlic showed a dose-dependent response, with an increase in the cytotoxicity level as the concentration of garlic increased. White garlic showed the highest cytotoxic effect against the tumor cells, the inhibitory concentration 50 (IC_{50}) being under 0.03 mg/mL.

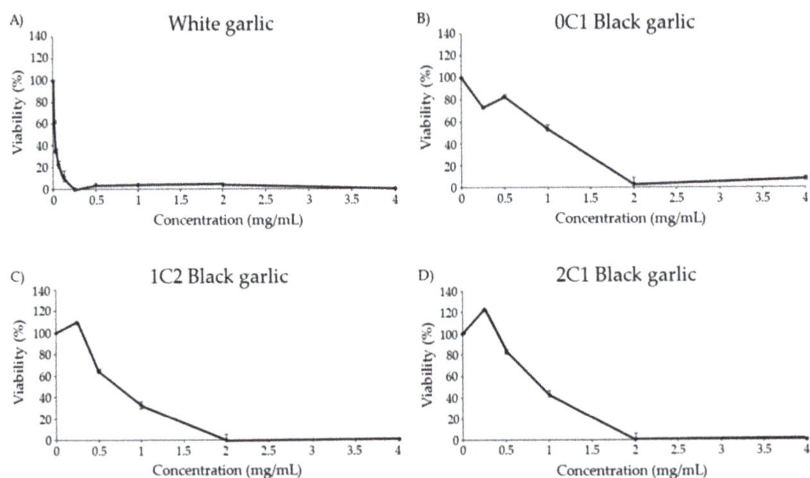

Figure 3. Viability of HL-60 cells treated with different concentrations of black and white garlic for 72 h. (**A**) White garlic, (**B**) 0C1: black garlic with 13 days aging, (**C**) 1C2: black garlic with 32 days aging, and (**D**) 2C1: black garlic with 45 days aging.

The cytotoxicity curve of 0C1 black garlic showed a dose-dependent increase with an IC_{50} value equal to 1 mg/mL. In relation to 1C2 and 2C1 black garlic, no inhibition was observed at the lowest concentration tested, but contrarily, a strong tendency to increase cell growth is observed with an IC_{50} value of 0.7 and 0.9 mg/mL, respectively. Moreover, an eventual cell-growth inhibition was observed in 1C2 and 2C1 black garlic at 2 mg/mL.

A number of studies have demonstrated the chemopreventive activity of garlic by using different garlic preparations, including fresh garlic extract, aged garlic, garlic oil, and a number of organosulfur compounds derived from garlic [71,72]. Such a chemopreventive activity has been attributed to the presence of organosulfur compounds in garlic. Therefore, the consumption of garlic may provide some kind of protection against tumor cell proliferation [73]. Studies on the preventive effects of black garlic extracts also show an induction of in vitro and in vivo inhibition in gastric cancer cell growth, chemopreventive effects in rat colon tumors, and an increase in anti-tumor activity in a mouse model [16,74,75].

3.8. DNA Internucleosomal Fragmentation

The HL 60 cell line belongs to the undifferentiated immortal lines, as they are tumor cells. It is widely investigated as a model for purposes of inducible cell differentiation. This phenomenon might affect the cell's ability to proliferate and therefore their immortality, with the appearance of apoptosis. Compounds capable of inducing differentiation and apoptosis are candidates to act as a chemopreventive agents or cancer chemotherapeutics.

Figure 4 shows the electrophoresis of the genomic DNA of HL-60 cells when treated with different concentrations of white, 0C1, 1C2, and 2C1 black garlic.

DNA internucleosomal fragmentation is represented by a DNA laddering, and it is associated with the activation of the apoptotic way in cancer cells, a hallmark of the genomic integrity [76]. None of the assayed concentrations (4 mg/mL to 0.25 mg/mL) induced internucleosomal fragmentation by the different black garlic treatments, but a slight fragmentation was observed in the lowest assayed concentration of white garlic (0.25 mg/mL). Hence, the cytotoxic activity observed is only induced in a proapoptotic way in the white garlic.

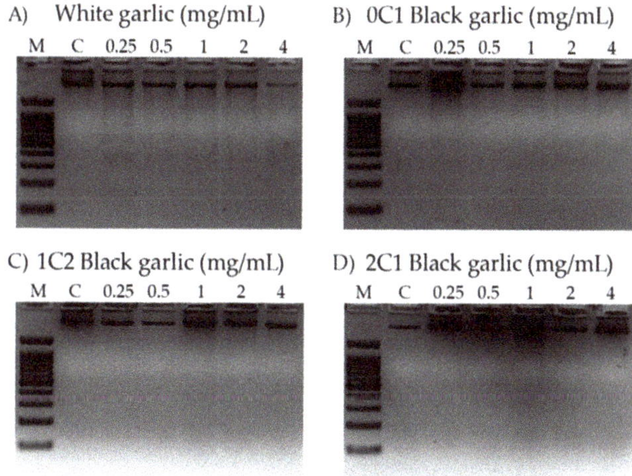

Figure 4. Internucleosomal DNA fragmentation in HL-60 cells treated for 5 h with different concentrations of black and white garlic. DNA fragmentation was detected following electrophoresis in agarose gels and staining with ethidium bromide. M: indicates DNA size marker; C: indicates control (lane 1); 0.25 mg/mL (lane 2); 0.50 mg/mL (lane 3); 1 mg/mL (lane 4); 2 mg/mL (lane 5), and 4 mg/mL (lane 6) of garlic sample. (**A**) White garlic, (**B**) 0C1: black garlic with 13 days aging, (**C**) 1C2: black garlic with 32 days aging, and (**D**) 2C1: black garlic with 45 days aging.

Our results demonstrate that only white garlic has a strong cytotoxic effect and induces slight DNA proapoptotic internucleosomal fragmentation against HL-60 cells. These results agree with several reports demonstrating that garlic exerts a chemopreventive effect by increasing apoptosis in lung cancer cells (NCI-H1299) [77]. On the other hand, our results do not agree with the results obtained by Wang et al., who detected a dose-dependent apoptosis in aged black garlic extract in in vitro studies [74].

4. Conclusions

It is the first time that an investigation of the relationship between the physicochemical characterization and the biological activities of white and black garlic has been carried out. Multifocal studies integrating the toxicity, antitoxicity, genotoxicity, antigenotoxicity, longevity, cytotoxicity, and proapoptotic properties of different types of garlic were followed in order to propose black garlic as a nutraceutical or functional food.

Black garlic aged for thirteen days showed qualitative improved physicochemical characteristics with respect to white garlic and to the other processed black garlic as well. The 0C1 black garlic (13 days aged) showed similar weight and soluble solids content (°Brix) to the raw garlic. All of the black garlic samples had an improved the polyphenol content and inhibition percentage with respect to the white garlic.

All types of garlic were safe, not showing toxicity in the *D. melanogaster* model, except for the white one, although only black garlic aged for 13 days showed slight protection against the oxidative toxicant at the three highest concentrations. Genotoxicity assays revealed that all raw and processed garlic were not genotoxic, with the exception of the higher concentration of white garlic, and exhibit moderate antigenotoxic effects when the imaginal discs are treated with the genotoxin hydrogen peroxide. The longevity assays in *D. melanogaster* yielded a significant extension of lifespan results in several concentrations of white and 0C1 and 1C2 black garlic. Finally, the results achieved in the in vitro experiments for garlic cytotoxicity were hopeful. All studied garlic induced a decrease in leukemia cells growth. However, no type of garlic was able to induce proapoptotic internucleosomal DNA fragmentation.

Important information is added to the agrifood industry as our data suggest that short-aged fermented black garlic (13 days) has higher biological activities than the longer-fermented ones, and even more than white garlic. This could have important industrial and economics consequences. Taking both the physicochemical and biological data, the black garlic aged for 1 days has shown itself to have the best nutraceutical properties. Our findings are relevant for black-garlic-processing agrifood companies as the economical and timing incomes are significantly reduced to 13 days aging.

Author Contributions: M.A.T.M., J.P.-A., and A.M.-O. performed all physicochemical analyses. Z.F.-B., T.M.-A., R.F., and M.d.R.-C. carried out all genotoxicological and antigenotoxicological analyses. R.M.-R. and A.A.-M. designed this study and revised the manuscript. Z.F.-B., T.M.-A., R.F., and M.d.R.-C. wrote this manuscript. T.M.-A. performed the longevity assays. Z.F.-B., T.M.-A., and A.A.-M. performed all in vitro assays. All the listed authors have read and approved the submitted manuscript.

Funding: This research received no external funding.

Conflicts of Interest: The authors declare no conflict of interest.

References

1. Block, E. The chemistry of garlic and onions. *Sci. Am.* **1985**, *252*, 114–121. [CrossRef] [PubMed]
2. Rivlin, R.S. Is garlic alternative medicine? *J. Nutr.* **2006**, *136*, 713S–715S. [CrossRef] [PubMed]
3. Tattelman, E. Health effects of garlic. *Am. Fam. Physician* **2005**, *72*, 103–106. [PubMed]
4. Amagase, H. Clarifying the real bioactive constituents of garlic. *J. Nutr.* **2006**, *136*, 716S–725S. [CrossRef] [PubMed]

5. Cerella, C.; Dicato, M.; Jacob, C.; Diederich, M. Chemical properties and mechanisms determining the anti-cancer action of garlic-derived organic sulfur compounds. *Anti-Cancer Agents Med. Chem.* **2011**, *11*, 267–271. [CrossRef]

6. Corzo-Martínez, M.; Corzo, N.; Villamiel, M. Biological properties of onions and garlic. *Trends Food Sci. Technol.* **2007**, *18*, 609–625. [CrossRef]

7. Chowdhury, R.; Dutta, A.; Chaudhuri, S.R.; Sharma, N.; Giri, A.K.; Chaudhuri, K. In vitro and in vivo reduction of sodium arsenite induced toxicity by aqueous garlic extract. *Food Chem. Toxicol.* **2008**, *46*, 740–751. [CrossRef] [PubMed]

8. Brodnitz, M.H.; Pascale, J.V. Thiopropanal s-oxide: A lachrymatory factor in onions. *J. Agric. Food Chem.* **1971**, *19*, 269–272. [CrossRef]

9. Rana, S.; Pal, R.; Vaiphei, K.; Singh, K. Garlic hepatotoxicity: Safe dose of garlic. *Trop. Gastroenterol.* **2006**, *27*, 26–30.

10. Ryu, J.H.; Kang, D. Physicochemical properties, biological activity, health benefits, and general limitations of aged black garlic: A review. *Molecules* **2017**, *22*, 919. [CrossRef] [PubMed]

11. Toledano-Medina, M.A.; Pérez-Aparicio, J.; Moreno-Rojas, R.; Merinas-Amo, T. Evolution of some physicochemical and antioxidant properties of black garlic whole bulbs and peeled cloves. *Food Chem.* **2016**, *199*, 135–139. [CrossRef] [PubMed]

12. Capuano, E.; Fogliano, V. Acrylamide and 5-hydroxymethylfurfural (HMF): A review on metabolism, toxicity, occurrence in food and mitigation strategies. *LWT—Food Sci. Technol.* **2011**, *44*, 793–810. [CrossRef]

13. Shinkawa, H.; Takemura, S.; Minamiyama, Y.; Kodai, S.; Tsukioka, T.; Osada-Oka, M.; Kubo, S.; Okada, S.; Suehiro, S. S-allylcysteine is effective as a chemopreventive agent against porcine serum-induced hepatic fibrosis in rats. *Osaka City Med. J.* **2009**, *55*, 61–69. [PubMed]

14. Imai, J.; Ide, N.; Nagae, S.; Moriguchi, T.; Matsuura, H.; Itakura, Y. Antioxidant and radical scavenging effects of aged garlic extract and its constituents. *Planta Med.* **1994**, *60*, 417–420. [CrossRef] [PubMed]

15. Lee, Y.-M.; Gweon, O.-C.; Seo, Y.-J.; Im, J.; Kang, M.-J.; Kim, M.-J.; Kim, J.-I. Antioxidant effect of garlic and aged black garlic in animal model of type 2 diabetes mellitus. *Nutr. Res. Pract.* **2009**, *3*, 156–161. [CrossRef]

16. Sasaki, J.-I.; Lu, C.; Machiya, E.; Tanahashi, M.; Hamada, K. Processed black garlic (allium sativum) extracts enhance anti-tumor potency against mouse tumors. *Med. Aromat. Plant Sci. Biotechnol.* **2007**, *227*, 138.

17. Nai-feng, Y.G.-m.X.; Bao-cui, Q.D.-w.Z.; Ya-yu, Z. Study on the technology of black garlic drink fermented by immobilized lactobacillus. *Shandong Food Ferment.* **2012**, *4*, 005.

18. William, W.J.; Carvalho, G.B.; Mak, E.M.; Noelle, N.; Fang, A.Y.; Liong, J.C.; Brummel, T.; Benzer, S. Prandiology of drosophila and the cafe assay. *Proc. Natl. Acad. Sci. USA* **2007**, *104*, 8253–8256.

19. Klein, S.; Rister, R.; Riggins, C. *The Complete German Commission E Monographs: Therapeutic Guide to Herbal Medicines*; American Botanical Council: Austin, TX, USA, 1998; p. 356.

20. Singleton, V.L.; Orthofer, R.; Lamuela-Raventós, R.M. Analysis of total phenols and other oxidation substrates and antioxidants by means of folin-ciocalteu reagent. In *Methods in Enzymology*; Elsevier Ltd.: Amsterdam, The Netherlands, 1999; pp. 152–178.

21. Re, R.; Pellegrini, N.; Proteggente, A.; Pannala, A.; Yang, M.; Rice-Evans, C. Antioxidant activity applying an improved abts radical cation decolorization assay. *Free Radic. Biol. Med.* **1999**, *26*, 1231–1237. [CrossRef]

22. Yan, J.; Huen, D.; Morely, T.; Johnson, G.; Gubb, D.; Roote, J.; Adler, P.N. The multiple-wing-hairs gene encodes a novel GBD-FH3 domain-containing protein that functions both prior to and after wing hair initiation. *Genetics* **2008**, *180*, 219–228. [CrossRef]

23. Ren, N.; Charlton, J.; Adler, P.N. The flare gene, which encodes the aip1 protein of drosophila, functions to regulate f-actin disassembly in pupal epidermal cells. *Genetics* **2007**, *176*, 2223–2234. [CrossRef] [PubMed]

24. Lindsley, D.L.; Zimm, G.G. *The Genome of Drosophila Melanogaster*; Academic Press: Cambridge, MA, USA, 2012.

25. Fernández-Bedmar, Z.; Anter, J.; Alonso Moraga, Á. Anti/genotoxic, longevity inductive, cytotoxic, and clastogenic-related bioactivities of tomato and lycopene. *Environ. Mol. Mutagen.* **2018**, *59*, 427–437. [CrossRef] [PubMed]

26. Graf, U.; Würgler, F.; Katz, A.; Frei, H.; Juon, H.; Hall, C.; Kale, P. Somatic mutation and recombination test in drosophila melanogaster. *Environ. Mutagen.* **1984**, *6*, 153–188. [CrossRef] [PubMed]

27. Graf, U.; Abraham, S.K.; Guzmán-Rincón, J.; Würgler, F.E. Antigenotoxicity studies in drosophila melanogaster. *Mutat. Res.* **1998**, *402*, 203–209. [CrossRef]

28. Moraga, A.A.; Graf, U. Genotoxicity testing of antiparasitic nitrofurans in the drosophila wing somatic mutation and recombination test. *Mutagenesis* **1989**, *4*, 105–110. [CrossRef]
29. Frei, H.; Würgler, F. Statistical methods to decide whether mutagenicity test data from drosophila assays indicate a positive, negative, or inconclusive result. *Mutat. Res.* **1988**, *203*, 297–308. [CrossRef]
30. Frei, H.; Würgler, F.E. Optimal experimental design and sample size for the statistical evaluation of data from somatic mutation and recombination tests (smart) in drosophila. *Mutat. Res.* **1995**, *334*, 247–258. [CrossRef]
31. Abraham, S.K. Antigenotoxicity of coffee in the drosophila assay for somatic mutation and recombination. *Mutagenesis* **1994**, *9*, 383–386. [CrossRef]
32. Gallagher, R.; Collins, S.; Trujillo, J.; McCredie, K.; Ahearn, M.; Tsai, S.; Metzgar, R.; Aulakh, G.; Ting, R.; Ruscetti, F. Characterization of the continuous, differentiating myeloid cell line (hl-60) from a patient with acute promyelocytic leukemia. *Blood* **1979**, *54*, 713–733.
33. Merinas-Amo, T.; Tasset-Cuevas, I.; Díaz-Carretero, A.M.; Alonso-Moraga, Á.; Calahorro, F. In vivo and in vitro studies of the role of lyophilised blond lager beer and some bioactive components in the modulation of degenerative processes. *J. Funct. Foods* **2016**, *27*, 274–294. [CrossRef]
34. Jones, M.; Collin, H.; Tregova, A.; Trueman, L.; Brown, L.; Cosstick, R.; Hughes, J.; Milne, J.; Wilkinson, M.; Tomsett, A. The biochemical and physiological genesis of alliin in garlic. *Med. Aromat. Plant Sci. Biotechnol.* **2007**, *1*, 21–24.
35. Choi, D.-J.; Lee, S.-J.; Kang, M.-J.; Cho, H.-S.; Sung, N.-J.; Shin, J.-H. Physicochemical characteristics of black garlic (*Allium sativum* L.). *J. Korean Soc. Food Sci. Nutr.* **2008**, *37*, 465–471. [CrossRef]
36. Shin, J.-H.; Choi, D.-J.; Lee, S.-J.; Cha, J.-Y.; Kim, J.-G.; Sung, N.-J. Changes of physicochemical components and antioxidant activity of garlic during its processing. *J. Life Sci.* **2008**, *18*, 1123–1131. [CrossRef]
37. Bae, S.E.; Cho, S.Y.; Won, Y.D.; Lee, S.H.; Park, H.J. Changes in s-allyl cysteine contents and physicochemical properties of black garlic during heat treatment. *LWT—Food Sci. Technol.* **2014**, *55*, 397–402. [CrossRef]
38. Kaanane, A.; Labuza, T. The Maillard reaction in foods. In *The Maillard Reaction in Aging, Diabetes and Nutrition*; AR Liss Press, Inc.: New York, NY, USA, 1989.
39. Labuza, T.P.; Saltmarch, M. The nonenzymatic browning reaction as affected by water in foods. In *Water Activity: Influences on Food Quality*; Elsevier: Amsterdam, The Netherlands, 1981; pp. 605–650.
40. Choi, J.-H.; Kim, W.-J.; Yang, J.-W.; Sung, H.-S.; Hong, S.-K. Quality changes in red ginseng extract during high temperature storage. *Appl. Biol. Chem.* **1981**, *24*, 50–58.
41. Kim, M.-H.; Kim, B.-Y. Development of optimum processing conditions in air dried garlics using response surface methodology. *J. Korean Soc. Food Nutr.* **1990**, *19*, 234–238.
42. Kang, Y.-H.; Park, Y.-K.; Lee, G.-D. The nitrite scavenging and electron donating ability of phenolic compounds. *Korean J. Food Sci. Technol.* **1996**, *28*, 232–239.
43. Jang, E.-K.; Seo, J.-H.; Lee, S.-P. Physiological activity and antioxidative effects of aged black garlic (*Allium sativum* L.) extract. *Korean J. Food Sci. Technol.* **2008**, *40*, 443–448.
44. Kakimoto, M.; Suzuki, A.; Nishimoto, I. Iadditives, supplements, antioxidants and phyto-chemicals. *Trends Food Sci. Technol.* **2000**, *11*, 266–267.
45. Xu, B.; Chang, S.K. Total phenolics, phenolic acids, isoflavones, and anthocyanins and antioxidant properties of yellow and black soybeans as affected by thermal processing. *J. Agric. Food Chem.* **2008**, *56*, 7165–7175. [CrossRef]
46. Gorinstein, S.; Leontowicz, M.; Leontowicz, H.; Najman, K.; Namiesnik, J.; Park, Y.-S.; Jung, S.-T.; Kang, S.-G.; Trakhtenberg, S. Supplementation of garlic lowers lipids and increases antioxidant capacity in plasma of rats. *Nutr. Res.* **2006**, *26*, 362–368. [CrossRef]
47. Nakagawa, S.; Masamoto, K.; Sumiyoshi, H.; Harada, H. Acute toxicity test of garlic extract. *J. Toxicol. Sci.* **1984**, *9*, 57–60. [CrossRef] [PubMed]
48. Sumiyoshi, H.; Kanezawa, A.; Masamoto, K.; Harada, H.; Nakagami, S.; Yokota, A.; Nishikawa, M.; Nakagawa, S. Chronic toxicity test of garlic extract in rats. *J. Toxicol. Sci.* **1984**, *9*, 61–75. [CrossRef] [PubMed]
49. Kanezawa, A.; Nakagawa, S.; Sumiyoshi, H.; Masamoto, K.; Harada, H.; Nakagami, S.; Date, S.; Yokota, A.; Nishikawa, M.; Fuwa, T. General toxicity tests of garlic extract preparation (kyoleopin) containing vitamins. *Oyo Yakuri* **1984**, *27*, 909–929.
50. Nakagawa, S.; Masamoto, K.; Sumiyoshi, H.; Kunihiro, K.; FUWA, T. Effect of raw and extracted-aged garlic juice on growth of young rats and their organs after peroral administration. *J. Toxicol. Sci.* **1980**, *5*, 91–112. [CrossRef]

51. Assayed, M.; Khalaf, A.; Salem, H. Protective effects of garlic extract and vitamin C against in vivo cypermethrin-induced teratogenic effects in rat offspring. *Food Chem. Toxicol.* **2010**, *48*, 3153–3158. [CrossRef] [PubMed]

52. Steiner, M.; Khan, A.H.; Holbert, D.; Lin, R. A double-blind crossover study in moderately hypercholesterolemic men that compared the effect of aged garlic extract and placebo administration on blood lipids. *Am. J. Clinic. Nutr.* **1996**, *64*, 866–870. [CrossRef] [PubMed]

53. Lei, M.-M.; Xu, M.-Y.; Zhang, Z.-S.; Zhang, M.; Gao, Y.-F. The analysis of saccharide in black garlic and its antioxidant activity. *Adv. J. Food Sci. Technol.* **2014**, *6*, 755–760. [CrossRef]

54. Fernández-Bedmar, Z.; Alonso-Moraga, A. In vivo and in vitro evaluation for nutraceutical purposes of capsaicin, capsanthin, lutein and four pepper varieties. *Food Chem. Toxicol.* **2016**, *98*, 89–99. [CrossRef]

55. Rojas-Molina, M.; Campos-Sánchez, J.; Analla, M.; Muñoz-Serrano, A.; Alonso-Moraga, Á. Genotoxicity of vegetable cooking oils in the drosophila wing spot test. *Environ. Mol. Mutagen.* **2005**, *45*, 90–95. [CrossRef]

56. Romero-Jimenez, M.; Campos-Sanchez, J.; Analla, M.; Muñoz-Serrano, A.; Alonso-Moraga, Á. Genotoxicity and anti-genotoxicity of some traditional medicinal herbs. *Mutat. Res.* **2005**, *585*, 147–155. [CrossRef] [PubMed]

57. Abraham, S.K.; Kesavan, P. Genotoxicity of garlic, turmeric and asafoetida in mice. *Mutat. Res.* **1984**, *136*, 85–88. [CrossRef]

58. Shukla, Y.; Taneja, P. Antimutagenic effects of garlic extract on chromosomal aberrations. *Cancer Let.* **2002**, *176*, 31–36. [CrossRef]

59. Sowjanya, B.L.; Devi, K.R.; Madhavi, D. Modulatory effects of garlic extract against the cyclophosphamide induced genotoxicity in human lymphocytes *in vitro*. *J. Environ. Boil.* **2009**, *30*, 663.

60. Chughtai, S.R.; Dhmad, M.; Khalid, N.; Mohamed, A. Genotoxicity testing of some spices in diploid yeast. *Pak. J. Bot.* **1998**, *30*, 33–38.

61. Fernández-Bedmar, Z.; Anter, J.; de La Cruz-Ares, S.; Muñoz-Serrano, A.; Alonso-Moraga, Á.; Pérez-Guisado, J. Role of citrus juices and distinctive components in the modulation of degenerative processes: Genotoxicity, antigenotoxicity, cytotoxicity, and longevity in drosophila. *J. Toxicol. Environ. Health* **2011**, *74*, 1052–1066. [CrossRef]

62. Park, J.-H.; Park, Y.K.; Park, E. Antioxidative and antigenotoxic effects of garlic (*Allium sativum* L.) prepared by different processing methods. *Plant Foods Human Nutr.* **2009**, *64*, 244. [CrossRef]

63. Benkeblia, N. Free-radical scavenging capacity and antioxidant properties of some selected onions (allium cepa l.) and garlic (*Allium sativum* L.) extracts. *Braz. Arch. Boil. Technol.* **2005**, *48*, 753–759. [CrossRef]

64. Prakash, D.; Singh, B.N.; Upadhyay, G. Antioxidant and free radical scavenging activities of phenols from onion (allium cepa). *Food Chem.* **2007**, *102*, 1389–1393. [CrossRef]

65. Kim, S.H.; Jung, E.Y.; Kang, D.H.; Chang, U.J.; Hong, Y.-H.; Suh, H.J. Physical stability, antioxidative properties, and photoprotective effects of a functionalized formulation containing black garlic extract. *J. Photochem. Photobiol. B* **2012**, *117*, 104–110. [CrossRef]

66. Bingqiao, Z.; Haige, W.; Yuanyuan, L. The antioxidation of black garlic. *Food Res. Dev.* **2008**, *29*, 56–61.

67. Huang, C.-H.; Hsu, F.-Y.; Wu, Y.-H.; Zhong, L.; Tseng, M.-Y.; Kuo, C.-J.; Hsu, A.-L.; Liang, S.-S.; Chiou, S.-H. Analysis of lifespan-promoting effect of garlic extract by an integrated metabolo-proteomics approach. *J. Nutr. Biochem.* **2015**, *26*, 808–817. [CrossRef] [PubMed]

68. Momoucm, T.; Takashina, P. Prolongation of life span and improved learning in the senescence accelerated mouse produced by aged garlic extract. *Biol. Pharm. Bull.* **1994**, *7*, l1589–l1594.

69. Prowse, G.M.; Galloway, T.S.; Foggo, A. Insecticidal activity of garlic juice in two dipteran pests. *Agric. For. Entomol.* **2006**, *8*, 1–6. [CrossRef]

70. Soh, J.W.; Hotic, S.; Arking, R. Dietary restriction in drosophila is dependent on mitochondrial efficiency and constrained by pre-existing extended longevity. *Mech. Ageing Dev.* **2007**, *128*, 581–593. [CrossRef] [PubMed]

71. Islam, M.; Kusumoto, Y.; Al-Mamun, M.A. Cytotoxicity and cancer (HeLa) cell killing efficacy of aqueous garlic (*Allium sativum*) extract. *J. Sci. Res.* **2011**, *3*, 375–382. [CrossRef]

72. Lau, B.H.; Tadi, P.P.; Tosk, J.M. Allium sativum (garlic) and cancer prevention. *Nutr. Res.* **1990**, *10*, 937–948. [CrossRef]

73. Seki, T.; Tsuji, K.; Hayato, Y.; Moritomo, T.; Ariga, T. Garlic and onion oils inhibit proliferation and induce differentiation of hl-60 cells. *Cancer Let.* **2000**, *160*, 29–35. [CrossRef]

74. Wang, X.; Jiao, F.; Wang, Q.-W.; Wang, J.; Yang, K.; Hu, R.-R.; Liu, H.-C.; Wang, H.-Y.; Wang, Y.-S. Aged black garlic extract induces inhibition of gastric cancer cell growth in vitro and in vivo. *Mol. Med. Rep.* **2012**, *5*, 66–72. [CrossRef]
75. Katsuki, T.; Hirata, K.; Ishikawa, H.; Matsuura, N.; Sumi, S.-I.; Itoh, H. Aged garlic extract has chemopreventative effects on 1,2-dimethylhydrazine-induced colon tumors in rats. *J. Nutr.* **2006**, *136*, 847S–851S. [CrossRef]
76. Wyllie, A.; Kerr, J.; Currie, A. Apoptosis: The significance of cell death. *Int. Rev. Cytol.* **1980**, *68*, 251–306. [PubMed]
77. Hong, Y.-S.; Ham, Y.-A.; Choi, J.-H.; Kim, J. Effects of allyl sulfur compounds and garlic extract on the expression of Bcl-2, Bax, and p53 in non small cell lung cancer cell lines. *Exp. Mol. Med.* **2000**, *32*, 127. [CrossRef] [PubMed]

© 2019 by the authors. Licensee MDPI, Basel, Switzerland. This article is an open access article distributed under the terms and conditions of the Creative Commons Attribution (CC BY) license (http://creativecommons.org/licenses/by/4.0/).

Article

Toxicological Studies of Czech Beers and Their Constituents

Tania Merinas-Amo [1,*], Rocío Merinas-Amo [1], Victoria García-Zorrilla [1], Alejandro Velasco-Ruiz [1], Ladislav Chladek [2], Vladimir Plachy [3], Mercedes del Río-Celestino [4], Rafael Font [4], Ladislav Kokoska [5] and Ángeles Alonso-Moraga [1]

[1] Department of Genetics, University of Córdoba, 14071 Córdoba, Spain
[2] Research and Teaching Brewery, Department of Technological Equipment of Buildings, Faculty of Engineering, Czech University of Life Sciences Prague, Kamýcká 129, 16500 Pargue, Czech Republic
[3] Department of Microbiology, Nutrition and Dietetics, Faculty of Agrobiology, Food and Natural Resources, Czech University of Life Sciences Prague, Kamýcká 129, 16500 Pargue, Czech Republic
[4] Agri-Food Laboratory, CAGPDS, Avda. Menéndez Pidal s/n, 14080, Córdoba, Spain
[5] Department of Crop Sciences and Agroforestry, Faculty of Tropical AgriSciences, Czech University of Life Sciences Prague, Kamýcká 129, 16500 Pargue, Czech Republic
* Correspondence: tania.meram@gmail.com; Tel.: +34-957-218-674

Received: 25 June 2019; Accepted: 6 August 2019; Published: 8 August 2019

Abstract: Background: Czech beers are unique because they are brewed using specific technology at a particular latitude and for being entirely produced in the area of the Czech Republic. The purpose of this work is the evaluation of toxicological effects of a variety of freeze-dried Czech beers, their raw materials (malts, hops and yeast) and processed-beer (wort, hopped wort and young beer). Methods: In vivo assays to evaluate the safety and protective effects in the *Drosophila melanogaster* eukaryotic system, and the in vitro evaluations of chemopreventive and DNA damage activity using the HL-60 tumour human cell line were carried out. Results: The safe effects for all the analysed substances and general protective effects against H_2O_2 were shown both at the individual and genomic level in the *Drosophila* animal model, with some exceptions. Moreover, all the substances were able to inhibit the tumour cell growth and to induce DNA damage in the HL-60 cells at different levels (proapoptotic, single/double strands breaks and methylation status). Conclusions: The promising effects shown by freeze-dried Czech beers due to their safety, protection against a toxin, chemopreventive potential and the induction of DNA damage in tumour cells, allow the proposition of Czech beer as a beverage with nutraceutic potential.

Keywords: Czech beer; brewing; raw material; *Drosophila melanogaster*; HL-60 cell line; toxicity; antitoxicity; longevity; cytotoxicity; DNA damage

---

1. Introduction

Beer is one of the oldest known beverages. Approximately 6000 years ago, ancient texts revealed the first beer production by the Sumerians: "the sweetest grain, if baked, left out, moistened, forgotten, and then eaten, would produce an uplifting, cheerful feeling" [1]. The different contributions like the use of hops, yeast, the different fermentation methods and the high technology development in the last century have contributed to the growth, improvement and variety of brewing as is currently known [1–3].

Beer is not only considered as a beverage, but it provides a valuable added value to the diet. It acts as a nutritional supplement with many bioactive molecules, as it contains readily available starches and sugars, various minerals and valuable B vitamins [2]. The brewing industry is a global business in

traditional European brewing regions with local varieties of beer still maintained mostly in central Europe [4].

The beer industry of the Czech Republic (Bohemia) has a long history since the first Czech brewery was founded in 993 at Břevnov Monastery in Prague [5], and it was only in October 2008, when the EU allowed the Czech Republic to use the "Ceské pivo/Czech Beer" trademark [6].

Czech beer is unique because it is brewed using specific technology and at a particular latitude. Due to the specific character of Czech beer, that is, the water, malt, hops and yeast used in the brewing, Czech brewers want recognition of the uniqueness of their beer.

Many studies highlight the value and potential uses of beer and its components. Some of the beers are those derived from raw materials (water, malts, hops and yeast) and pass unchanged through the brewing process, and some others are produced during the process [7]. Chen et al. [8] emphasised the potential uses of beer compounds in dermatology, such as in the treatment of atopic eczema, contact dermatitis, pigmentary disorders, skin infections, skin ageing, skin cancers and photoprotections. Sánchez et al. [9] highlighted the common use of hops as a tranquillizer plant. Further, it is known that both the normalized and stimulated effects of the many compounds of beer have on the sleep's rhythm because of its bitter acids, especially the alpha acid components [8]. Furthermore, others studies support the benefits of moderate alcohol consumption, being that a healthy lifestyle is associated with lower rates of cardiovascular diseases [10], or the ability against kidney stone formation [11]. Moreover, Vinson et al. [12] reported other activities of lager and dark beers, like the inhibition of atherosclerosis, lowering both the total serum cholesterol and triglycerides, to act as in vivo antioxidants by decreasing the oxidizability of low-density lipoprotein cholesterol and decreasing atherosclerosis in cholesterol-fed hamsters. These activities are mainly due to the polyphenols present in both lager and dark beer. Nevertheless, the adverse effects of alcohol on health, such as increasing risk of several mouth, oesophageal and liver cancers [13], should be noted.

Despite beers being considered beverages with important properties on health [8,12,14–17], nowadays studies about the toxicological evaluation of beers are missing. Moreover, due to the traditional history of beer in the Czech Republic, the present work attempts to add new data about the toxicological properties of a wide variety of Czech beers, as well as their unique and differential raw materials (malts, hops and yeast) and further, processed-beer at three different steps during its brewing (wort, hopped wort and young beer). For that purpose, a previous assay of the safety of freeze-dried samples followed by studies on the protective effects that the tested compounds show in the *Drosophila* animal model were carried out. Nonetheless, chemopreventive studies in the HL-60 human cell line were carried out to evaluate the possible cytotoxic potential and DNA damage that our tested compounds could induce in tumour cells.

- The fruit fly *D. melanogaster* shows a consistent similarity to humans as approximately 75% of the genes involved in diseases are similar in the fly [18,19]. This well-known eukaryote with the largest scientific history in genetics has also contributed to the understanding of developmental biology, evolutionary concepts and more recently, toxicology [20–24]. Its short life cycle (10–12 days), cost efficiency and easy maintenance are characteristics that make *Drosophila* to be considered as an ideal model organism.
- The HL-60 human model cell line originated thirty two years ago from a patient with acute myeloid leukaemia [25]. After culturing peripheral blood leukocytes from this female patient, a growth-factor-independent immortal cell line with distinct myeloid characteristics was obtained [26]. This promyelocytic leukaemia cell line has been used world-wide for many toxicity and cancer purposes.

2. Materials and Methods

2.1. Samples Preparation and Sample Compounds

For the present study, 16 samples of lager type beers commercially produced in Czech Republic were purchased at a local market (see Table 1). Moreover, 6 raw materials and 3 processed-beer samples were obtained from the Research and Teaching Brewery of the Czech University of Life Sciences, Prague. This consists of a two-roll mill, brew house (steam heated kettle 12 hl, and lauter tun 14 hl), whirlpool 12 hl, wort cooler (10 hl/h), 6 fermentation tanks (24 hl each), and 14 lager tanks (22 hl). The standard brewing conditions were as follows: 200 kg Pilsen malt, 2 mash process, 3 kg of Saazer hop variety, fermentation at 12 °C for 7 days, yeast W96, maturation at 0 °C for 3 months. The annually brewery production is 300–500 hl beer. Although Suchdol Jenik is not a commercial brewery, it is built on the tradition of beer brewing in Prague's Suchdol using the common raw materials farmed in the Czech Republic. A total of 400 mL of every sample of beer was used to be freeze-dried. The detailed characteristics of all 25 different samples analysed in this study are summarised in Table 1. The beer samples were lyophilised at −110 °C until the samples were totally dried using a Scanvac CoolSafe 110-4 freeze-dried (LaboGene ApS, Lynge, Denmark) before carrying out the assays and stored in dark and dried conditions until use. The lyophilised samples were stored in a sterilised plastic bottle with the correspondent information (name, date and quantity of beer). All lyophilised beers were dissolved and distilled.

The concentrations of the tested beers were established according to the average daily intake and body weight of *Drosophila* [27]. The concentration range was stablished to make it equivalent to the 192 mL of total beer/day in humans [28].

Table 1. The samples of Czech beers and their ingredients used in the present research.

Beer Trade Mark	Named in the Manuscript	Beer Type/Ingredient Origin
Pilsner Urquell	Pilsner Urquell	Blond
Staropramen Lezak	Staropramen Lezak	Blond
Staropramen Cerny Lezak	Staropramen Cerny	Stout
Staropramen Nealko	Staropramen Nealko	Alcohol-free
Staropramen Nefiltrovany	Staropramen Nefiltrovany	Blond
Lobkowicz Lezak Premium	Lobkowicz Lezak	Blond
Lobkowicz Cerny Premium	Lobkowicz Cerny	Stout
Lobkowicz Nealkoholicky Premium	Lobkowicz Nealko	Alcohol-free
Lobkowicz Psenicny Premium	Lobkowicz Psenicny	Wheat
Budweiser Budvar B:Original, Svetly lezak	Budweiser Lezak	Blond
Budweiser Budvar B:Free, Nealkoholicke pivo	Budweiser Nealko	Alcohol-free
Budweiser Budvar B:Cherry, Tmavy lezak s visni	Budweiser Cherry	Cherry flavoured
Herold Světly Breznicky Lezak	Herold Lezak	Blond
Herold Tmave Speciálni Pivo	Herold Cerny	Stout
Herold Psenicny Kvasnicovy Lezak	Herold Psenicny	Wheat
Herold Polotmave Specialni Pivo	Herold Polotmave	Pomegranate flavoured
Rest of mash [1]	Rest of mash	CULS [8] Prague brewery
Blond malt	Blond malt	CULS Prague brewery
Dark malt	Dark malt	CULS Prague brewery
Sladek hop [2]	Sladek hop	CULS Prague brewery
Saazer hop [3]	Saazer hop	CULS Prague brewery
Saccharomyces uvarum [4]	*S. uvarum*	CULS Prague brewery
Wort [5]	Wort	CULS Prague brewery
Hopped wort [6]	Hopped wort	CULS Prague brewery
Young beer [7]	Young beer	CULS Prague brewery

[1] Rest of malt after boiling process. [2] Hybrid aroma varieties of hop. [3] Fine aroma hop. [4] Lager yeasts sink to the bottom of the beer and ferment more slowly, preferring colder temperatures around 7–15 °C [29]. [5] Sugar solution obtained after filtration of barley malt boiling. [6] Liquid obtained after addition and boiling of hops to the wort, cooling and removal of spent hops. [7] Liquid obtained after yeast addition and fermentation during a week. [8] Czech University of Life Science.

2.2. In Vivo Animal Model and In Vitro Cellular Culture Used

2.2.1. Drosophila Melanogaster

Two *Drosophila* genetic backgrounds were used in our studies:

- *mwh/mwh*: Carries a recessive mutation *mwh* (*multiple wing hairs*) which produces several tricomas instead of one per cell when it is in homozygosis [30].
- *flr³/In (3LR) TM3, rip^p sep bx^{34e} e^s Bd^S*: Carries a *flr³* (*flare*) marker which is a recessive lethal mutation when it is in homozygosis producing deformed tricomas, though it is viable in homozygous somatic cells if and when the larvae start the development [31].

The strains were routinely maintained in the laboratory at 25 °C in a homemade meal, and serial changes were made.

The direct and reciprocal crosses between both strains of *Drosophila* were stablished and the use of the emerging trans-heterozygous adults (*mwh flr⁺/mwh⁺ flr³*) were used for the in vivo toxicity, antitoxicity, genotoxicity, antigenotoxicity and lifespan assays.

2.2.2. HL-60 Cell Cultures

The cells were grown in a conditioned RPMI-1640 medium (Sigma, Darmstadt, Germany, R5886) which was complemented with decomplemented foetal bovine serum (Linus, S01805), L-glutamine (Sigma, G7513) and an antibiotic-antimycotic solution (Sigma, A5955). Routinely, the cells were grown in a 37 °C and 5% CO_2 [26] in a Shell Lab CO_2 incubator. The cultures were seeded at 2.5×10^4 cells/mL density and renewed three times per week.

2.3. Safety In Vivo Assays of Freeze-Dried Samples

2.3.1. Toxicity

The number of emerged adults in each treatment with respect to the negative control was analysed at the following ranks of concentrations for the different lyophilised beers: 3.125–50 mg/mL. For the malts, hops and yeast, the concentrations used were those equivalent to the quantity of each compound in the full beer (0.0625–1 mg/mL for blond and dark malts; 0.00156–0.025 mg/mL for Sladek hop; 0.000625–0.01 mg/mL for Saazer hop; 0.00023–0.00375 mg/mL for *S. uvarum*; 3.125–50 mg/mL for processed-beers). The tester tubes contained *Drosophila* Instant Medium (DIM) (Formula 4–24, Carolina Biological Supply, Burlington, NC, USA) and 4 mL of the different assayed concentrations. Moreover, the negative concurrent controls were prepared with DIM and distilled water.

The toxicity significances were assessed with the non-parametric Chi-square test [32].

2.3.2. Genotoxicity

Due to the similar pattern observed for the different beers studied in the toxicity assay, a blond beer was selected to study it effects on the *Drosophila* genome. It was chosen according to today's most known and oldest Czech breweries: Pilsner Urquell (1842) [33]. Moreover, the studies with raw materials and processed-beer at different concentrations were carried out in this animal model.

For the genotoxicity assays, the method described by Graf et al. [34] were applied. As in the toxicity test, the assayed tubes contained DIM and the solutions of the substances to be tested (3.125 and 50 mg/mL of lyophilised beer; 0.0625 and 1 mg/mL for blond and dark malts; 0.00156 and 0.025 mg/mL for Sladek hop; 0.000625 and 0.01 mg/mL for Saazer hop; 0.00023 and 0.00375 mg/mL for *S. uvarum*; 3.125 and 50 mg/mL for processed-beers). A positive control, consisting of DIM, water and 0.12 M H_2O_2 as a genotoxicant, was tested to check the accuracy of the assay. After emergence, the adult flies were preserved in a solution of 70% ethanol until the next wing step.

The mutations were scored at 400× in the trans-heterozygous mounted wings. Given the substantial knowledge about the origin of the mutations, the total mutations were spliced into different categories according to their size and the type of mutation: *mwh* phenotype with small single spots (1–2 cells) that occur in the late stages of the mitotic division; *mwh* or *flr³* phenotype with large single spots (≥3 cells) that occur in the early mitotic divisions of larval development; and both *mwh* and *flr³* twin juxtaposing spots. On the one hand, small and large spots are induced by somatic point mutations, chromosome aberrations, and somatic recombinations, whereas it is assumed that twin spots appear only by somatic recombinations between the centromere and the *flr³* locus.

A total of 736 wings were mounted and analysed for the genotoxicity treatments. The number of mutations were compared to the water control and a double decision test was applied [35], and the binomial Kastenbaum and Bowman statistic test were applied [36]. The non-parametric Mann Whitney *U*-test ($\alpha = \beta = 0.05$) was used to solve the inconclusive and positive results.

2.4. Protection In Vivo Tests

2.4.1. Antitoxicity

The antitoxicity tests in the study consisted of the viability percentage of *Drosophila* treated with the same concentrations as in the above described toxicity assays but combined with 0.12 M H_2O_2 as a toxicant [37].

The statistical significances assessing the effects of the tested compounds on the survival of flies were given by the Chi-square test in Microsoft Office Excel 2007 [32].

2.4.2. Antigenotoxicity

The antigenotoxicity trials were carried out as Graf et al. [38] described, and consisted of the combined treatments with 0.12 M H_2O_2 as a genotoxin and the tested compounds (using the same conditions as in the genotoxicity assay). After the emergence of the adult flies, the same protocol of genotoxicity was followed for the mounted and scored wings. A total of 741 wings were mounted and analysed. The inhibition percentages of the mutagenic activity (IP) in the antigenotoxicity trials were obtained from the total spots per wing following the Abraham (1994) algorithm [39]:

$$\text{IP} = [(\text{genotoxin} - \text{combined treatment})/\text{genotoxin}] \times 100$$

2.4.3. Longevity

Based on the similar pattern observed for the different beers studied in the toxicity and antitoxicity assays, a blond and a stout beer was selected to study their effects on the *Drosophila* life extension. It was chosen according to today's most known and oldest Czech breweries: Pilsner Urquell (1842) [33]. Moreover, the studies with raw materials and processed-beer at different concentrations were carried out in the model organism.

All longevity assays were carried out at a constant temperature (25 °C) following the protocol described by Tasset-Cuevas et al. [40]. Briefly, the groups of 25 individuals were collected and placed into vials that contain 0.21 g of DIM and 1 mL of the different concentrations to be tested for each compound (3.125–50 mg/mL of lyophilised beer; 0.0625–1 mg/mL for blond and dark malts; 0.00156–0.025 mg/mL for Sladek hop; 0.000625–0.01 mg/mL for Saazer hop; 0.00023–0.00375 mg/mL for *S. uvarum*; 3.125–50 mg/mL for processed-beers). Four replicas were carried out during the longevity assay for each control and the concentration established, recording the number of alive animals and renewing the culture media every 3/4 days.

In addition, the quality of life of treated flies was known by studying the upper 25% of the survival curve. This part of the longevity curves is considered as the healthspan, defined as a low and more or less constant age-specific mortality rate value [41].

The statistical analysis of the survival data was carried out by applying the Kaplan-Meier methodology (SPSS Statistics 17.0 software, SPSS, Inc., Chicago, IL, USA). The Log-Rank (Mantel-Cox) post hoc method was used to solve the differences among the survival curves.

2.5. Chemopreventive In Vitro Tests

2.5.1. Cytotoxicity Assay

The HL-60 cells at 2×10^4 cells/mL density were placed in 96 well culture plates and treated for 72 h with the different lyophilised beers (1.25–250 mg/mL) and the equivalent concentrations for the raw materials (0.156–20 mg/mL for both malts; 0.00097–0.125 mg/mL for Sladek hop; 0.00039–0.05 mg/mL for Saazer hop; 0.00029–0.0375 mg/mL for *S. uvarum*) and the processed-beers (1.25–250 mg/mL). This wide range was selected to evaluate the in vitro cytotoxic doses using the inhibitory concentration of 50 (IC_{50}) when possible.

The trypan blue dye exclusion test was used to determine the cell viability. Trypan blue (Sigma-Aldrich, St. Louis, MO, USA, T8154) was added to the cells at a 1:1 volume ratio, loaded into a Neubauer chamber, counted at 100× magnification (AE30/31, Motic microscope, Cabrera de Mar Barcelona, Spain). Finally, the curves were represented graphically as a survival percentage from three independent experiments with respect to their concurrent control.

2.5.2. Tumour Cells DNA Damage Evaluation

Due to the similar pattern observed for the different beers studied in the in vivo assays, a blond beer was selected to study the DNA damage evaluation. Pilsner Urquell (1842) was chosen according to today's most known and oldest Czech breweries [33]. Moreover, the studies with raw materials and processed-beer at different concentrations were carried out in human leukaemia cells.

DNA Fragmentation

The treated HL-60 cells (1×10^6/mL) during 5 h with blond selected lyophilised beer (1.25–250 mg/mL) and its correspondent concentration for the raw materials (0.156–20 mg/mL for both malts; 0.00097–0.125 mg/mL for Sladek hop; 0.00039–0.05 mg/mL for Saazer hop; 0.00029–0.0375 mg/mL for *S. uvarum*) and the processed-beers (1.25–250 mg/mL) were centrifuged (3000 rpm for 5 min). DNA was extracted according to the protocol described by Merinas-Amo et al. [42].

For the quantification of DNA, a spectrophotometer (Nanodrop ND-1000) was used. Finally, the quantity of 1200 ng of DNA was subjected to an agarose gel (2%) electrophoresis at conditions of 85 mA for 25 min, stained with GelRed^{TM} and visualized under UV light.

Comet Assay

The cells were incubated for 5 h with different concentrations of the selected lyophilised beer (7.56, 31.25 and 62.5 mg/mL) and its equivalent concentrations for the raw materials (0.625, 2.5 and 5 mg/mL for both malts; 0.0039, 0.0156 and 0.03125 mg/mL for Sladek hop; 0.00156, 0.00625 and 0.0125 mg/mL for Saazer hop; 0.00117, 0.0047 and 0.0094 mg/mL for *S. uvarum*) and the processed-beers (7.56, 31.25 and 62.5 mg/mL). As in the DNA fragmentation assay, two representative blond (Pilsner Urquell) and stout (Staropramen Cerny) Czech beers were selected for this study.

After completing the washing steps in PBS (Phosphated-Buffered Saline), 6.25×10^5 cells/mL were mixed with low melting point agarose (Sigma, A4018) at 0.75% and immediately transferred to frosted-end slides. Next, according to the Mateo-Fernández et al. [43] protocol, the steps of lysis, alkaline electrophoresis, neutralization and drying were carried out. Finally, 50–100 single cells were visualized by treating the slides with 7 µL of propidium iodide (10 µg/mL) (Sigma, P4170). The DNA comet images were analysed (400×) in a Leica DM 2500 fluorescence microscope with a green filter and an attached camera (JAI CV-M4CL, Barcelona, Spain). The single cell parameters were calculated using the OpenComet plugging from ImageJ (NIH).

The statistical analysis of the tail moment was evaluated by an ANOVA, and a post hoc Tukey's test was applied to assess the effect that the different compounds had on the DNA integrity. Significance was considered at $p \leq 0.05$ (SPSS Statistic 17.0 software).

Methylation Status Evaluation

For the evaluation of the methylation status of the treated tumour cells, firstly the genomic DNA was isolated in the same way as described in the DNA fragmentation section. Secondly, a bisulphite-modified DNA step from the different treatments selected for the lyophilised beer (7.81 and 125 mg/mL), raw materials (0.625 and 10 mg/mL for blond and dark malts; 0.0039 and 0.0625 mg/mL for Sladek hop; 0.00156 and 0.025 mg/mL for Saazer hop; 0.00117 and 0.0188 mg/mL for *S. uvarum*) and processed-beer (7.81 and 125 mg/mL) was carried out using the EZ DNA Methylation-Gold Kit. Thirdly, a quantitative Methylation-Specific PCR (qMSP) was carried out following the methodology described by Merinas-Amo et al. [44] in a MiniOpticon Real-Time PCR System (MJ Mini Personal Thermal Cycler, Bio-Rad Laboratories Inc., Hercules, CA, USA) and analysed by the Bio-Rad CFX Manager 3.1 Software. The reaction mixture consisted of: 2 µL of deionised water, 5 µM of each forward and reverse primer, 2 µl of iTaq™ Universal SYBR® Green Supermix (Bio-Rad, containing antibody-mediated hot-start iTaq DNA polymerase, dNTPs, $MgCl_2$, SYBR® Green I dye, enhancers, stabilizers and a blend of passive reference dyes including ROX and fluorescein) and 25 ng of bisulphite converted genomic DNA up to a 10 µL final volume and submitted to specific qMSP conditions (95 °C for 3 min, 45 cycles at 95 °C for 10 s, 60 °C for 15 s, 72 °C for 15 s, a step at 95 °C for 30 s following by a 65 °C step during 30 s and finally a boost step from 65 °C to 95 °C for 95 s increasing 0.5 °C each 0.05 s).

A wide range of human genomic DNA could be evaluated across repetitive sequences. Alu and LINE sequences belong to the interspersed repeat sequences throughout the genome, while the satellites are located in the centromere areas [45–48]. The Alu M1, LINE-1 and Sat-α sequences were obtained from Isogen Life Science (see Table 2 for further information [49]).

Table 2. Primers information.

Reaction ID	GenBank Number	Amplicon Start	Amplicon End	Forward Primer Sequence 5′ to 3′ (N)	Reverse Primer Sequence 5′ to 3′ (N)	GC-Content (%)	
						Forward	Reverse
Alu C4	Consensus Sequence	1	98	GGTTAGGTATAGT GGTTTATATTTGTAA TTTTAGTA (36)	ATTAACTAAAC TAATCTTAAACT CCTAACCTCA (33)	25	27.3
Alu M1	Y07755	5059	5164	ATTATGTTAGTT AGGATGGTTT CGATTTT (29)	CAATCGACC GAACGCGA (17)	27.6	58.8
LINE-1	X52235	251	331	GGACGTATTTGG AAAATCGGG (21)	AATCTCGCGAT ACGCCGTT (19)	47.6	52.6
Sat-α	M38468	139	260	TGATGGAGTATT TTTAAAATATAC GTTTTGTAGT (34)	AATTCTAAAAA TATTCCTCTTCA ATTACGTAAA (33)	23.5	21.2

Source: Weisenberger, Campan, Long, Kim, Woods, Fiala, Ehrlich and Laird [49].

The relative expressions obtained from three replicas of each sample were normalised with the Alu C4 housekeeping sequence using the Nikolaidis et al. [50] and Liloglou et al. [51] comparative C_T method. To evaluate the statistical differences between the control and each treated group, the one-way ANOVA and post hoc Tukey's tests were used.

3. Results and Discussion

3.1. Safety of Freeze-Dried Samples

The toxicity of Czech lyophilised beers, raw materials and processed-beer has been carried out in the *Drosophila* in vivo model. The relative percentage of emerging flies after the treatments with different concentrations of the substances is shown in Figure 1. None of the substances reached the LD_{50} thus all of them can be considered non-toxic in the *Drosophila* animal model. Nevertheless, the viability rates were different.

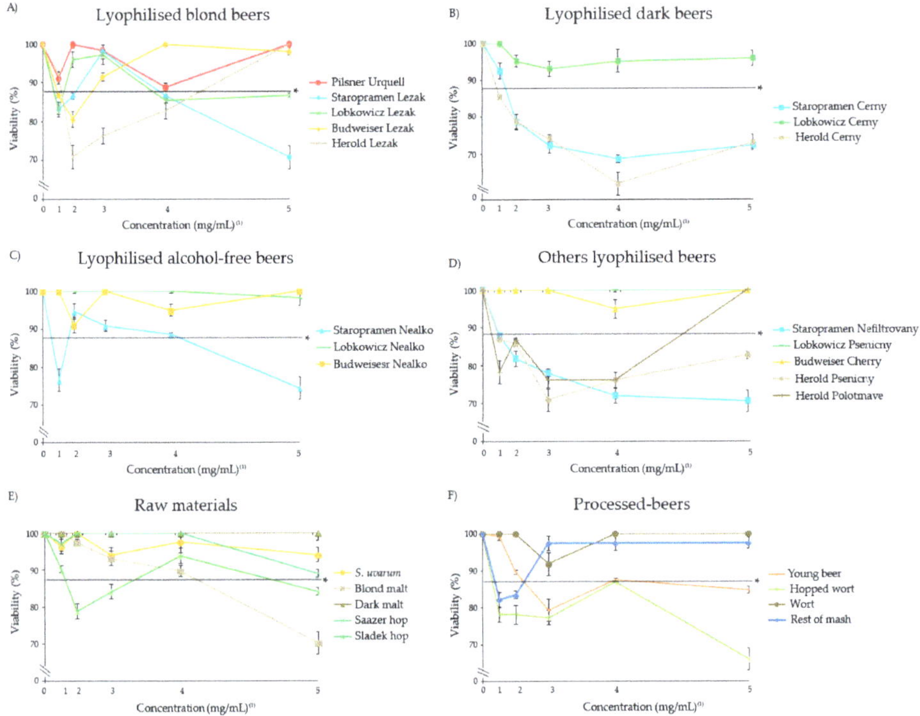

Figure 1. The toxicity levels of lyophilised Czech beers (**A–D**), raw materials (**E**) and processed-beers (**F**) in *D. melanogaster*. The data represent the percentage of surviving adults with respect to 300 control untreated 72-h-old larvae from three independent experiments ± SE treated with five concentrations of lyophilised Czech beers, raw materials and processed-beers. *: Chi-square value ($p < 0.05$). The values 1, 2, 3, 4 and 5 correspond to the concentrations of 3.125, 6.25, 12.5, 25 and 50 mg/mL for beers and their respective concentrations for each single component.

The highest reduction of *Drosophila* viability caused by the tested beer samples and their constituents reached a 63.4% of survival, compared with the control. Taking into account that a toxic effect is considered when survival reaches a reduction of under 50% of survival, all our samples exhibited a non-toxic/safe effect on the animal model. On the other hand, if the survival of flies is analysed with a $p < 0.05$, the results showed different patterns between the control and the treatments: (i) The samples with a safety potential significantly similar to the control in all the concentrations (Pilsner Urquell; Lobkowicz Cerny; Lobkowicz Nealko; Lobkowicz Psenicny; Budweiser Nealko; Budweiser Cherry; Dark malt; Saazer hop; *S. uvarum* and Wort); (ii) the samples with a safety potential at the lowest/medium concentrations (Staropramen Cerny; Herold Lezak; Blond malt; Sladek hop and Young beer); (iii) the samples with safety properties at the highest/medium concentrations (Rest of mash; Herold Lezak; Herold Polotmave; Budweiser Lezak; Sladek hop); (iv) the samples with safety properties at medium concentrations (Staropramen Lezak; Staropramen Nealko; Lobkowicz Lezak) and (v) the samples with no safety properties at any concentrations with respect to the control (Staropramen Nefiltrovany; Herold Cerny; Herold Psenicny; Hopped wort).

To the best of the authors' knowledge, this is the first study on the toxicity of Czech beers using *D. melanogaster*. However, some studies carried out with prokaryotes using the Ames test which tested other types of beers suggested the possibility that the antimutagenic activity showed by these drinks is a result of an accumulated effect of their individual constituents [52]. Moreover, phenolics and some B vitamins have been argued to be health promoter compounds of beer in reference to degenerative diseases [53]. Our results agree with alleged safety properties of freeze-dried beer and its components adding more information about the promising properties that specifically Czech beers show as nutraceutic substances in a model organism.

3.2. Genotoxicity of Freeze-Dried Samples

Table 3 shows the results of genotoxicity in the somatic mutation and recombination test (SMART) of *D. melanogaster*. The negative control showed a rate of 0.158 mutations/wing [54] which falls into the historical range for the wing spot test. All results showed non-genotoxic effects at the assayed concentration with a non-significant frequency of mutations per wing ranging between 0.125 and 0.368 with respect to the negative control. Moreover, a positive control to check the assay accuracy was carried out with a combined treatment prepared with DIM, water and 0.12 M H_2O_2 as an oxidative genotoxicant. As indicated in the antigenotoxicity section, the results fall into the established range as genotoxic substances [54].

Table 3. The genotoxicity of a lyophilised Czech beer, raw materials and processed-beers in the *Drosophila* wing spot test.

Compound	N° of Wings	Clones per Wing (n° Spots) [1]				
		Small Single Clones (1–2 Cells) m = 2	Large Simple Clones (>2 Cells) m = 5	Twin Clones m = 5	Total Clones m = 2	U-Test [2]
H_2O	38	0.105 (4)	0.053 (2)	0	0.158 (6)	
H_2O_2	40	0.2 (8)	0.2 (8)	0	0.400 (16) +	
Pilsner Urquell (mg/mL)						
3.125	40	0.175 (7)	0.100 (4)	0	0.275 (11) i	Δ
50	40	0.225 (9)	0.075 (3)	0	0.300 (12) i	Δ
Blond malt (mg/mL)						
0.0625	40	0.225 (9)	0.025 (1)	0	0.250 (10) i	Δ
1	40	0.250 (10)	0.075 (3)	0	0.325 (13) i	Δ
Dark malt (mg/mL)						
0.0625	40	0.200 (8)	0.025 (1)	0	0.225 (9) i	Δ
1	40	0.100 (4)	0.025 (1)	0	0.125 (5) i	Δ
Sladek hop (mg/mL)						
0.00156	40	0.125 (5)	0.05 (2)	0	0.175 (7) i	Δ
0.025	38	0.316 (12)	0.052 (2)	0	0.368 (14) i	Δ
Saazer hop (mg/mL)						
0.000625	38	0.211 (8)	0.052 (2)	0	0.263 (10) i	Δ
0.01	38	0.131 (5)	0.079 (3)	0	0.210 (8) i	Δ
S. uvarum **(mg/mL)**						
0.00023	40	0.200 (8)	0.075 (3)	0	0.275 (11) i	Δ
0.00375	34	0.294 (10)	0.059 (2)	0	0.353 (12) i	Δ
Wort (mg/mL)						
3.125	40	0.300 (12)	0	0.025 (1)	0.325 (13) i	Δ
50	38	0.132 (5)	0.026 (1)	0	0.158 (6) i	Δ
Hopped wort (mg/mL)						
3.125	40	0.200 (8)	0.025 (1)	0	0.225 (9) i	Δ
50	38	0.131 (5)	0.158 (6)	0	0.289 (11) i	Δ
Young beer (mg/mL)						
3.125	40	0.255 (9)	0	0	0.255 (9) i	Δ
50	34	0.353 (12)	0	0	0.353 (12) i	Δ

[1] Statistical significances: + (positive); i (inconclusive) and + (positive) versus negative control. m multiplication factor. Kastenbaum-Bowman Test, error levels $\alpha = \beta = 0.05$. [2] Inconclusive results were resolved by Mann Whitney U-test. Delta marker (Δ) means no significant differences between treatment and concurrent control [35].

No information has been found about genotoxicity related to raw materials and processed beers. Hops seems to be one the most genoprotective compounds among the assayed ones, and may be conferring beer its genoprotective properties. This fact is in agreement with other authors, who pointed that Xanthohumol, a main flavonoid found in hops, has a strong antimutagenic activity [36,55]. Furthermore, Arimoto-Kobayashi et al. [56] suggested that beer components act in a protective capacity against the genotoxic effects of heterocyclic amines in vivo. A previous study with a lyophilised blond lager beer on *Drosophila* suggested that the values of a total mutation frequency were similar or lower than the negative control with no significant differences between them [36].

3.3. Antitoxicity

The antitoxicity assays revealed the ability of the Czech lyophilised beers, raw materials and processed-beers to protect the individuals against an oxidative stress at the tested concentrations. Figure 2 shows that the positive control H_2O_2 is toxic at 0.12 M, with respect to the negative water control, with an average survival percentage of 39.2%.

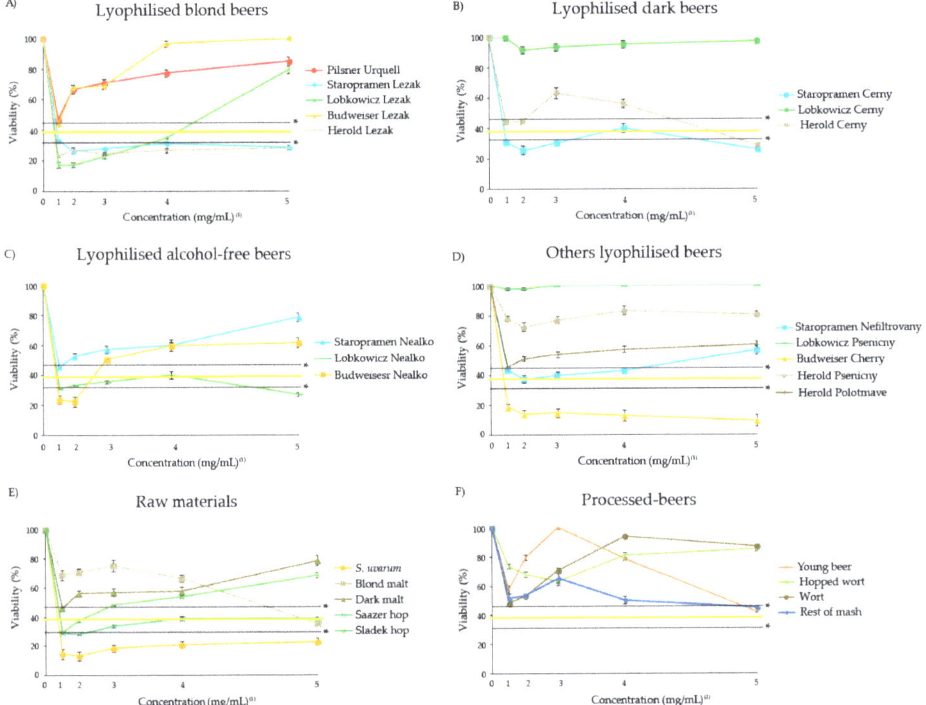

Figure 2. The antitoxicity levels of different lyophilised Czech beers (**A–D**), raw materials (**E**) and processed-beers (**F**) in *D. melanogaster*. The data represent the percentage of surviving adults with respect to 300 control untreated 72-hour-old larvae from three independent experiments ± SE treated with different concentrations of lyophilised Czech beers, raw materials and processed-beers combined with 0.12 M H_2O_2. *: Chi-square value ($p < 0.05$). The value 1, 2, 3, 4 and 5 correspond to concentrations of 3.125, 6.25, 12.5, 25 and 50 mg/mL for beers and their respective concentrations for each single component. The positive control of H_2O_2 0.12 M is represented as yellow line showing a toxic effect with an average survival rate of 39.2% with respect to the negative control.

All compounds showed a protective effect with respect to their concurrent control except *S. uvarum*, Budweiser Cherry and Staropramen Nefiltrovany at all the tested concentrations. Moreover, no antitoxic activity against H_2O_2 was shown at lower concentrations of Sladek hop, Staropramen Nealko and Staropramen Cerny, and at higher concentrations of Herold Cerny, Lobkowicz Nealko and Staropramen Cerny. On the other side, a significant protective effect was shown at combined treatments with Pilsner Urquell, Lobkowicz Cerny, Lobkowicz Psenicny, Budweiser Lezak, Herold Lezak, Herold Psenicny, Herold Polotmave, rest of mash, blond malt, dark malt, wort, hopped wort and young beer; at lower concentrations of Staropramen Lezak; at medium/higher concentrations of Herold Cerny; and at higher concentrations of Staropramen Nealko, Lobkowicz Lezak, Sladek hop, Saazer hop, Budweiser Nealko.

Due to the ability of beers to increase antioxidant capacity in humans, many studies focused on the evaluation of their possible benefits, since oxidative stress and consequently the presence of reactive oxygen species are closely associated with diverse types of diseases [57]. To the authors' knowledge, this is the first study on the antitoxicity of Czech beers and its constituents using *D. melanogaster*.

It is known that the antioxidant capacity of beer is mainly derived from polyphenols, Maillard compounds and vitamin C from malt and hops [17,58–62]. Nevertheless, it is also known that the Maillard reaction can cause both deterioration of food quality [63–65], by the formation of anti-nutritional and toxic products (MRPs) with mutagenic [66], carcinogenic [67] and cytotoxic [68] effects. Our results agree with this fact as both malts and hops showed protective effects against H_2O_2 with respect to their concurrent controls, except at the lowest concentration of Sladek hop tested.

The antioxidant power of beer depends on a wide variety of parameters involved in brewing, like the malt variety and the malting process, the mashing temperature and pH, sparing, boiling, the hops variety added during wort boiling and yeast fermentation [17]. According to Rivero et al. [69], dark beer has a higher content of antioxidants than lager or alcohol-free beer, which are responsible for its colour. Moreover, because of the brewing process, the antioxidant capacity of alcohol-free beer is considerably lower than lager [17]. No relationship has been found between the antioxidant activity of phenols and the degree of browning [70]. Our results did not establish any pattern about the characteristics of beers and their antioxidant potential: A stout, an unfiltered, an alcohol-free and a special cherry beer, together with *Saccharomyces* were the only four substances among a total of 26 samples studied which did not show protective effects against the hydrogen peroxide oxidative model toxin. This study hypothesised that antioxidant activities shown by beers are caused by the synergic activities of the different phenols that compose them. Previous results supported our result that no correlation was established by the different styles of beer studied in the in vivo assays. However, better and safe properties for lyophilized beer are shown by the blond ones, followed by alcohol-free and dark beers [71,72].

The antioxidants present in hops are also important for the brewing industry. A study confirms the different composition of both Saazer and Sladek hop resins [73]. Saazer fine aroma hops has a similar composition of hop resins owing to genetic affinity. The content of α-bitter acids ranged between 3–4.5% *w/w* and was lower than the content of β-bitter acids (4.5–6.5% *w/w*). Sladek hybrid aroma varieties hop resins have a composition of 4–8% *w/w* of α-bitter acids and 3.5–8% *w/w* of β-bitter acids. Moreover, different relative percentages of antioxidant activity has been observed in these hops which correlated with the content of polyphenol substances [74]. The highest antioxidant activity within 70 to 80% rel. was exhibited by the variety Saazer, followed at a relatively considerable distance by the Sladek variety with an antioxidant activity within 40 to 60% rel. Our results agree with this characterisation of hops, showing a stronger antitoxic potential for Saazer hop at the lowest concentrations tested.

3.4. Antigenotoxicity

The results of antigenotoxicity obtained in the SMART test are shown in Table 1. Oxygen peroxide (0.12 M) behaves as a potent mutagen able to induce somatic mutations in *D. melanogaster* [75] at a rate of 0.388 spots/wing.

Table 4. Antigenotoxicity of a lyophilised Czech beer, raw materials and processed-beers in the *Drosophila* wing spot test.

Compound	N° of Wings	Small Single Clones (1–2 Cells) $m = 2$	Large Simple Clones (More Than 2 Cells) $m = 5$	Twin Clones $m = 5$	Total Clones $m = 2$	Mann Whitney U-Test [2]	Inhibition Percentage (%) [3]
H$_2$O	40	0.175 (7)	0	0	0.175 (7)		
H$_2$O$_2$	40	0.2 (8)	0.2 (8)	0	0.400 (16) +		
Pilsner Urquell (mg/mL)							
3.125	38	0.263 (10)	0.026 (1)	0	0.289 (11) β		27.75
50	40	0.200 (8)	0.025 (1)	0	0.225 (9) β		43.75
Blond malt (mg/mL)							
0.0625	40	0.225 (9)	0.050 (2)	0	0.275 (11) β		31.25
1	38	0.315 (12)	0.106 (4)	0	0.421 (16) λ	Δ	−5.25
Dark malt (mg/mL)							
0.0625	40	0.175 (7)	0.05 (2)	0	0.225 (9) β		43.75
1	38	0.131 (5)	0.052 (2)	0	0.183 (7) β		54.25
Sladek hop (mg/mL)							
0.00156	38	0.131 (5)	0.079 (3)	0	0.210 (8) β		47.50
0.025	40	0.325 (13)	0.075 (3)	0	0.400 (16) λ	Δ	0.00
Saazer hop (mg/mL)							
0.000625	38	0.079 (3)	0.052 (2)	0	0.131 (5) β		67.25
0.01	40	0.050 (2)	0.050 (2)	0	0.100 (4) β		75.00
***S. uvarum* (mg/mL)**							
0.00023	18	0.278 (5)	0.222 (4)	0	0.500 (9) λ	Δ	−25.00
0.00375	27	0.222 (6)	0.125 (3)	0	0.333 (9) β		16.75
Wort (mg/mL)							
3.125	40	0.125 (5)	0.025 (1)	0	0.150 (6) β		62.50
50	40	0.250 (10)	0.025 (1)	0	0.275 (11) β		31.25
Hopped wort (mg/mL)							
3.125	40	0.325 (13)	0.075 (3)	0	0.400 (16) λ	Δ	00.00
50	40	0.100 (4)	0.100 (4)	0	0.200 (8) β		50.00
Young beer (mg/mL)							
3.125	40	0.425 (17)	0.075 (3)	0	0.500 (20) λ	Δ	−25.00
50	26	0.500 (13)	0.192 (5)	0	0.692 (18) λ	Δ	−73.00

[1] Statistical significance: + (positive) versus negative control; β (significantly different) and λ (inconclusive) versus positive control. m: multiplication factor. Kastenbaum-Bowman Test, error levels α = β = 0.05 [35]. [2] Inconclusive results were resolved by Mann Whitney U-test. Delta marker (Δ) means no differences between the treatments and the concurrent control. [3] The inhibition percentage was calculated according to Abraham [39].

The results showed that all compounds were able to inhibit the genotoxic activity of H$_2$O$_2$, with the exception of the highest concentration of Blond malt and Sladek hop, the lowest concentration of *S. uvarum* and Hopped wort, and both tested concentrations of Young beer that showed no-significant differences with respect to the positive control. An inhibition percentage that ranged between 16.75 and 75 was induced by the compounds corresponding to the lowest concentration of *S. uvarum* and the highest concentrations of Saazer hop, respectively. In addition, the most antigenotoxic potential against hydrogen peroxide was shown by Saazer hop followed by Dark malt, Wort, Pilsner Urquell, Hopped wort, Sladek hop, Blond malt, *S. uvarm* and finally, by Young beer as shown in Table 4.

To the authors' knowledge, this is the first time that the antigenotoxic effects of raw materials and processed-beers in the in vivo tests have been studied. From the results obtained, beers and their constituents are not only not genotoxicants, but also protect from the damage to DNA caused by oxidative genotoxicants. A previous study with a lyophilised blond lager beer and some of its compounds indicated an anigenotoxic effect supporting our results [42].

Moreover, it is known that individual nutrients or other chemicals derived from food have effects on chronic diseases which may not show the same effects as the whole nutrient [76]. This may the reason why no-relationship could be stablished between the compounds in this study.

3.5. Longevity

The longevity parameters assayed in the *D. melanogaster* model for the four compounds tested are shown in Figure 3 and Table 5. The survival curves showed a significant expansion of life at both extreme concentrations of blond malt, at two-lowest concentrations of Sladek hop and the highest concentration of *S. uvarum* tested. A significant reduction in lifespan was only induced by the lowest and fourth-to high concentrations of Pilsner Urquell, at second and third-to low concentrations of dark malt and young beer, at medium concentrations of Saazer hop and Hopped wort, at lowest concentrations of *S. uvarum* and at the lowest and two-to high concentrations of wort. The rest of the concentrations assayed did not show significant differences with respect to their concurrent controls.

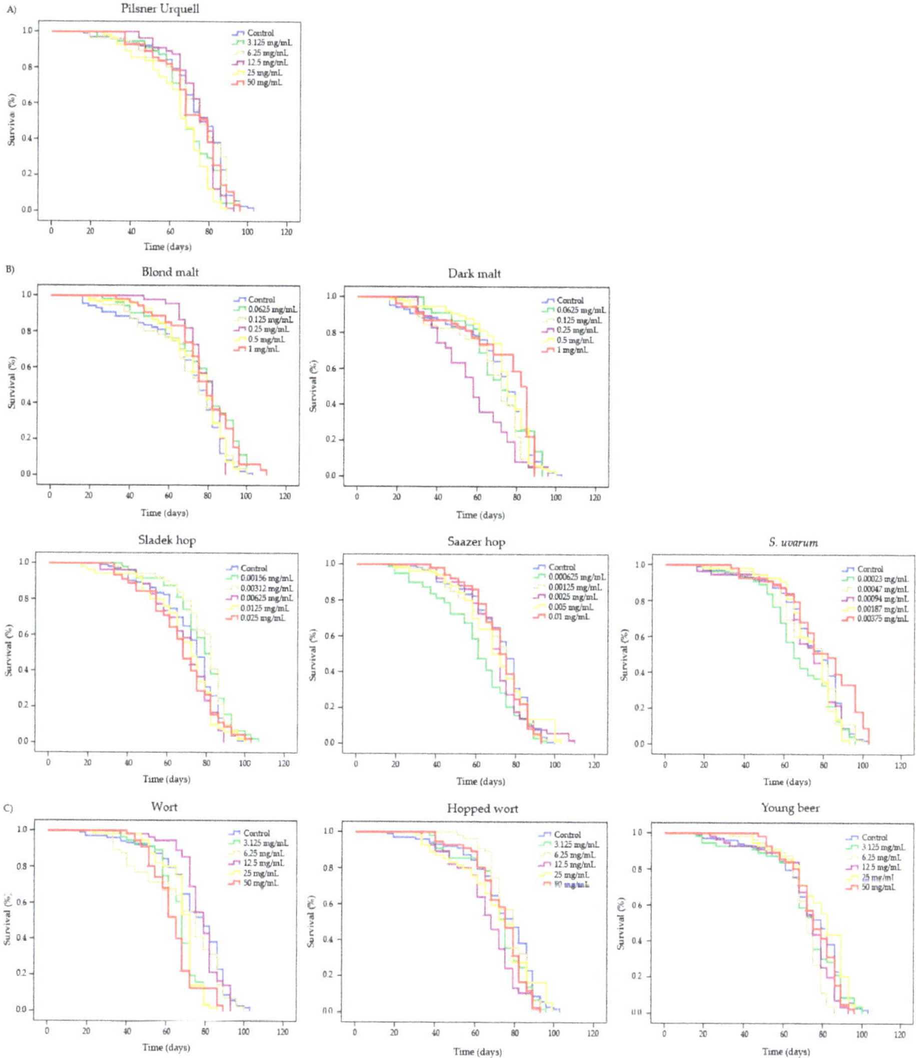

Figure 3. The survival parameters of *D. melanogaster* fed with different concentrations of a lyophilised Czech beer (**A**), raw materials (**B**) and processed-beers (**C**). The curves were obtained by the Kaplan-Meier method and the significances of the curves were determined by the Log-Rank method (Mantel-cox).

Table 5. The mean and significances of lifespan and healthspan.

Compound	Concentration	Mean Lifespan [1] (Days)		Mean Healthspan [1] (Days)	
Pilsner Urquell	Control	74.319		47.048	
	3.125 mg/mL	69.096	*	48.615	ns
	6.25 mg/mL	73.179	ns	45.385	ns
	12.5 mg/mL	74.571	ns	57.923	*
	25 mg/mL	64.857	*	44.765	ns
	50 mg/mL	72.413	ns	49.462	ns
Blond malt	Control	69.888		35.959	
	0.0625 mg/mL	76.349	*	47.769	*
	0.125 mg/mL	70.412	ns	39.714	ns
	0.25 mg/mL	78.448	ns	65.617	*
	0.5 mg/mL	73.514	ns	48.067	*
	1 mg/mL	77.946	*	53.583	*
Dark malt	Control	69.888		35.959	
	0.0625 mg/mL	70.557	ns	50.129	*
	0.125 mg/mL	66.314	*	34.135	ns
	0.25 mg/mL	58.281	*	35.692	ns
	0.5 mg/mL	73.077	ns	44.001	ns
	1 mg/mL	72.433	ns	38.176	ns
Sladek hop	Control	72.107		44.211	
	0.00156 mg/mL	77.439	*	50.792	ns
	0.00312 mg/mL	78.083	*	54.987	*
	0.00625 mg/mL	68.774	ns	45.750	ns
	0.0125 mg/mL	68.280	ns	42.154	ns
	0.025 mg/mL	67.556	ns	36.667	*
Saazer hop	Control	72.107		44.211	
	0.000625 mg/mL	60.714	*	29.929	*
	0.00125 mg/mL	69.993	ns	43.947	ns
	0.0025 mg/mL	70.314	ns	48.909	ns
	0.005 mg/mL	70.143	ns	44.754	ns
	0.01 mg/mL	72.098	ns	53.091	*
S. uvarum	Control	74.319		47.048	
	0.00023 mg/mL	67.590	*	42.545	*
	0.00047 mg/mL	73.738	ns	53.167	ns
	0.00094 mg/mL	71.876	ns	51.533	*
	0.00187 mg/mL	73.223	ns	55.292	*
	0.00375 mg/mL	79.060	*	54.393	*
Wort	Control	74.319		47.048	
	3.125 mg/mL	66.179	*	51.929	ns
	6.25 mg/mL	68.399	ns	38.143	*
	12.5 mg/mL	77.628	ns	64.238	*
	25 mg/mL	67.464	*	52.571	ns
	50 mg/mL	63.885	*	48.250	ns
Hopped wort	Control	74.319		47.048	
	3.125 mg/mL	72.232	ns	50.077	ns
	6.25 mg/mL	76.922	ns	63.375	*
	12.5 mg/mL	66.256	*	41.167	ns
	25 mg/mL	71.095	ns	40.541	*
	50 mg/mL	73.004	ns	53.833	ns
Young beer	Control	74.319		47.048	
	3.125 mg/mL	71.696	ns	47.357	ns
	6.25 mg/mL	70.747	*	57.357	*
	12.5 mg/mL	72.117	*	52.224	*
	25 mg/mL	78.426	ns	57.786	*
	50 mg/mL	75.314	ns	58.314	*

Means calculated by the Kaplan-Meier method and significances of the curves determined by the Log-Rank method (Mantel-cox). [1] ns: non-significant ($p > 0.05$), *: significant ($p < 0.05$) with respect to their concurrent controls.

Many factors affect longevity, among them are gender, genetic variation, environmental factors like diet, access to health care and traditions [77]. Fleming et al. [78] pointed out that Drosophila was a choice animal model to investigate longevity-promoting activities of a variety substances, like nutraceuticals, as flies seem to show many of the cellular senescence characteristics observed in mammals. Schriner et al. [79] showed that barley malts had the ability to increase the lifespan of Drosophila individuals, our results being in agreement for the lowest and highest concentrations tested showing an expansion of life of 7 and 8 days respectively. Unfortunately, no previous studies about lifespan have been found for the rest of the studied samples.

Regarding the healthspan results, all freeze-dried substances were able to improve the quality of life of flies at some concentrations tested, with respect to their concurrent control, except the highest concentration of Sladek hop, the lowest concentration of Saazer hop and *S. uvarum*, the second-to low concentration of wort and the second-to high concentration of Hopped wort that reduced the healthspan of *Drosophila* with respect to their concurrent controls. To the authors' knowledge, this is the first time that healthspan has been studied for the tested compounds in the in vivo model organism, but there are previous studies with blond beer that showed beneficial effects in healthspan [42].

3.6. Cytotoxicity

All substances assayed showed cytotoxic activity against the HL-60 cells except rest of mash which did not reach the inhibitory concentration of 50 (IC_{50}) at any assayed concentration (Figure 4).

According to the IC_{50}, lyophilised Czech beers are classified as follows: (i) Pilsner Urquell and Budweiser Lezak with a half maximal inhibitory concentration lower than 1.25 mg/mL; (ii) Lobkowicz Cerny followed by Budweiser Nealko, then Lobkowicz Lezak and finally Budweiser Cherry, with a IC_{50} between 1.25 and 3.5 mg/mL; (iii) Lobkowicz Nealko and Herold Cerny followed by Staropramen Cerny and Herold Polotmave, then Lobkowicz Nefiltrovany, followed by Herold Lezak, then Staropramen Nefiltrovany and finally Staropramen Nealko and Herold Psenicny with a IC_{50} ranging between 3.5 and 7.81 mg/mL. Despite this ranking, all beers were able to completely inhibit the tumour cells growth in a concentration that ranged between 7.81 and 15.625 mg/mL. The most cytotoxic raw material was Sladek hop followed by *S. uvarum*, Saazer hop, blond malt and dark malt reaching an IC_{50} at concentrations lower than the corresponding 15.625 mg/mL in beer.

Scientific studies indicate the cancer-prevention activity of beer. However, a variety of levels in the inhibition of tumour cell growth has been observed for different brands of beer, probably caused by the wide range in the components and in the antioxidant activities [53,80,81]. Nozawa et al. [82] showed that a moderate consumption of beer reduced the formation of colorectal tumours and preneoplastic lesions in azoxymethane to a significant level when experimental in vivo carcinogenesis was induced in male rats. Furthermore, the studies with a type of lyophilised blond lager beer showed a chemopreventive dose dependent response, reaching an IC_{50} at 125 mg/mL in the HL-60 cell line and an IC_{50} at 25 mg/mL in the immortal NIH3T3 cells [42].

Our results agree with the cytotoxic effects of beers, showing that Czech beers exhibit a stronger chemopreventive potential compared to other brands of beers as they reached an IC_{50} at lower concentrations than other beers. Moreover, although hops has been poorly studied by itself, some of its compounds like flavonols [83] and Xanthohumol [84] have shown anticarcinogenic effects.

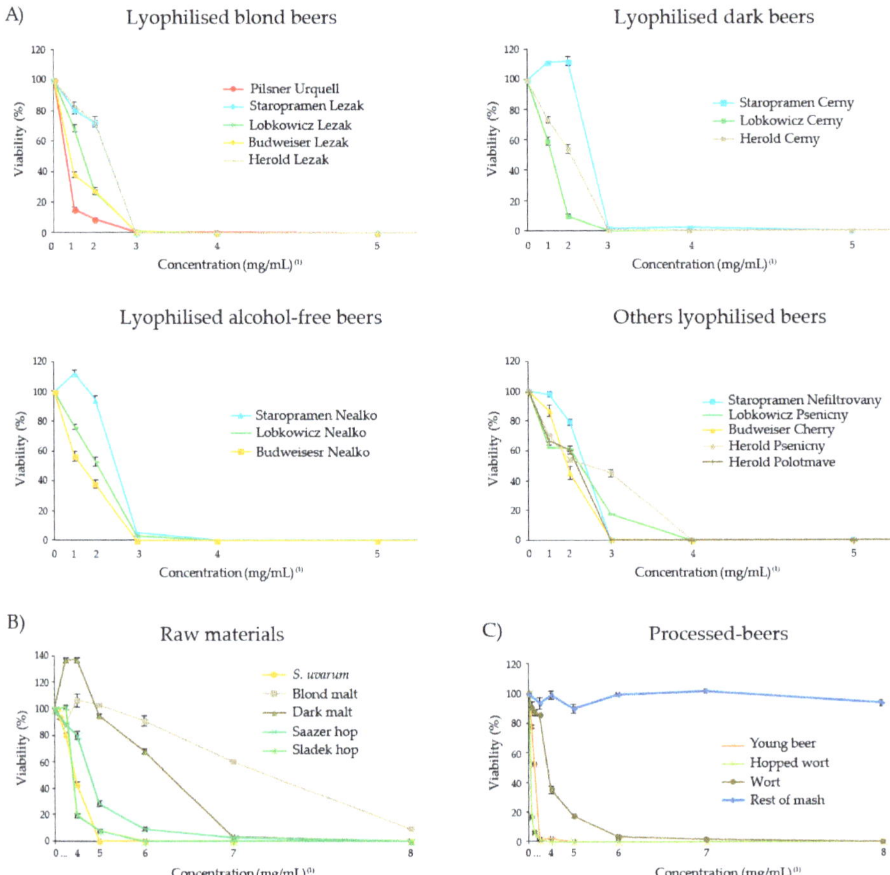

Figure 4. The effect of lyophilised Czech beers (**A**), raw materials (**B**) and processed-beers (**C**) on cell viability. The viability in the leukaemia cells (HL-60) treated with different concentrations of Czech lyophilised beers (**A**), raw materials (**B**) and processed-beers (**C**) for 72 h. Each point represents the growing percentage with respect to its control. The values indicate the mean from three independent experiments ± SE. [1] The value 1, 2, 3, 4, 5, 6, 7 and 8 correspond to the concentrations of 1.95, 3.9, 7.81, 15.625, 31.25, 62.5, 125 and 250 mg/mL for beers and their respective concentrations for each single component.

3.7. DNA Fragmentation

According to Wyllie, et al. [85], DNA internucleosomal breakage can be detected by a DNA band pattern related to the activation of the apoptotic way in cancer cells, being a hallmark of the genomic stability. The genomic integrity results of the HL-60 cells treated with lyophilised blond and stout Czech beers, raw materials and processed-beers are shown in Figure 5.

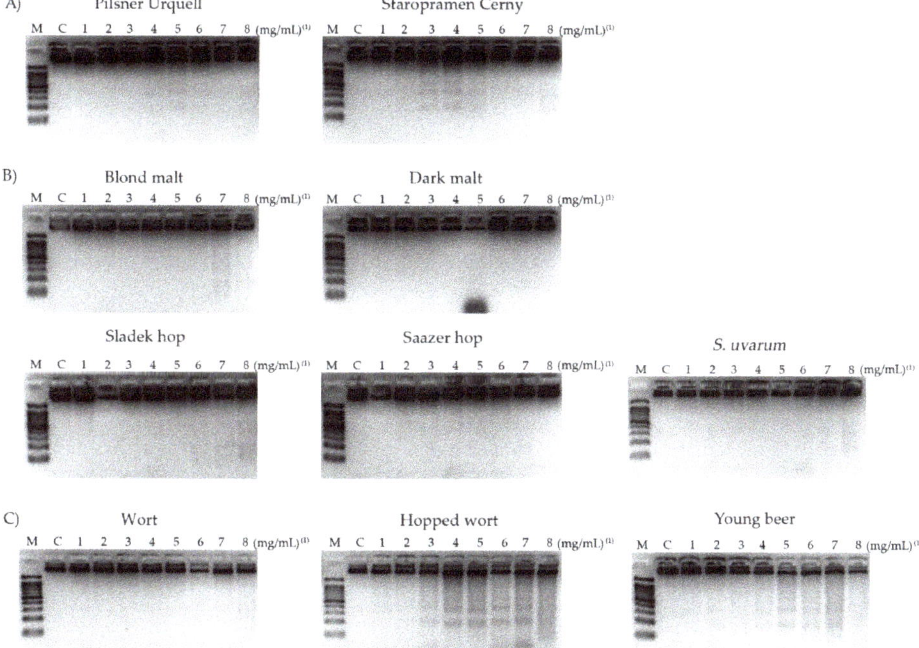

Figure 5. DNA damage induced by internucleosomal DNA fragmentation in promyelocytic HL-60 cells treated with different concentrations of a lyophilised Czech beer (**A**), raw materials (**B**) and processed-beers (**C**) for 5 h. M: DNA size marker. C: Control treatment. [1] The value 1, 2, 3, 4, 5, 6, 7 and 8 correspond to concentrations of 1.95, 3.9, 7.81, 15.625, 31.25, 62.5, 125 and 250 mg/mL for beers and their respective concentrations for each single component.

Pilsner Urquell showed a slight pattern of fragmentation when the cells were treated at concentrations of 15.625, 31.25 and 62.5 mg/mL (Figure 5A). A similar study with another brand of a lyophilised blond lager beer showed that beer induced DNA fragmentation damage at 15.625, 31.25 and 62.5 mg/mL concentrations [42]. The results obtained are similar to the results for Czech beers.

With respect to raw materials, all of them showed DNA damage at the concentration tested (Figure 5B), except blond malt and *S. uvarum* that induced a very light pattern of fragmentation at concentrations of 10 and 0.375 mg/mL (lane 7 and lane 8 in the agarose gel, respectively). Finally, processed-beer showed DNA fragmentation in a dose-dependent manner for hopped wort and young beer (Figure 5C).

Our study with the raw materials and processed-beers allow the suggestion of possible correlations between the modifications of beer and its compound properties during the brewing development. The most abundant phenols during the hopped wort process are xanthohumol and humulone, hop-phenols characterised by an induction of DNA damage as other studies have indicated [42,71]. During the beer maturing process, the characteristics of *Saccharomyces* and other phenols like ferulic acid, isoxanthohumol, vanillic acid, tyrosol, epicatechin gallate, among others, are involved in the final outcome of beers, reducing the proapoptotic DNA fragmentation activity as shown by matured beers [86]. In addition, some information about the prevention of the DNA damage capacity of hops could be found, but it was mainly due to one of its flavonoids [87,88], supporting in general our results of there being no induction of DNA-damage. Unfortunately, no prior information about malts on internucleosomal DNA fragmentation activity could be found.

3.8. Comet Assay

The averages and significances of the tail moment (TM) of HL-60 cells treated with blond and stout Czech lyophilised beers, raw materials and processed-beers are shown in Figure 6. The comet assay allows the ready identification of the apoptotic/necrotic cells based on changes in the nuclei morphology attributable to DNA damage at a unicellular level [89].

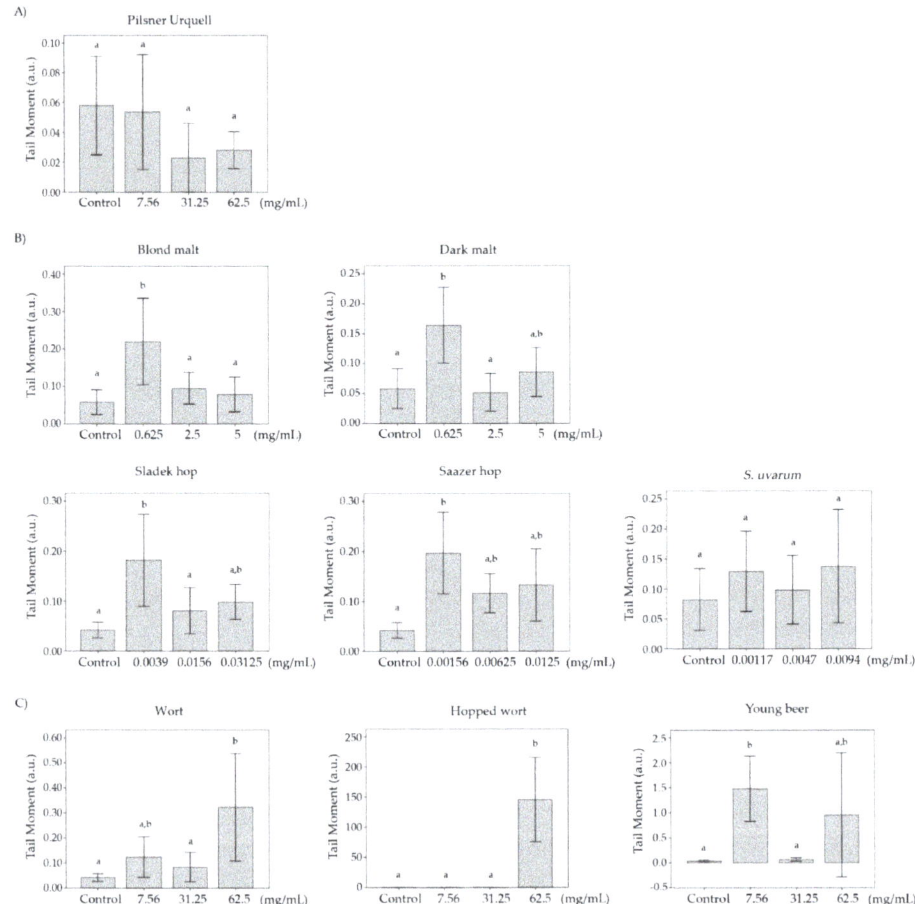

Figure 6. DNA strand breaks induction in promyelocytic HL-60 cells treated with different concentrations of a lyophilised Czech beer (**A**), raw materials (**B**) and processed-beers (**C**) for 5 h. Alkaline comet assay (pH > 13) of HL-60 cells treated. DNA migration is reported as the mean TM. The values are the mean ± SE. The different letters in the treatments mean the differences with respect to the negative control after a one-way ANOVA and post hoc Tukey's test.

According to Fabiani et al. [90], the frequency distribution of DNA damage can be divided into five classes depending on the TM values as follows: Class 0, 1, 2, 3 and 4 with TM values <1; 1–5, 5–10, 10–20 and >20 corresponding to no damage, slightly damaged, medium damage, highly damaged and completely damaged DNA, respectively. Taking into account this classification, all concurrent control values fell into class 0 (TM < 1) showing no DNA damage in the untreated HL-60 cells. Pilsner Urquell, the raw materials and wort did not induce DNA damage at single and/or double strand breaks showing TM values lower than 1. On the other hand, young beer showed a slight damage at the lowest concentration assayed (7.56 mg/mL) with a TM value included in class 1. Moreover, Staropramen Cerny and hopped wort completely induced DNA damage at the medium and highest concentration respectively (31.25 and 62.5 mg/mL), with a TM value higher than 100 compared with their concurrent controls.

The results of DNA fragmentation fit with those obtained in the comet assay since the TM values higher than 30 mean that apoptosis mechanisms have been induced [91]. The highest concentration (62.5 mg/mL) of hopped wort tested showed a TM > 30 confirming the death of leukaemia cells by apoptosis as was observed in lane 5 and 6 of the fragmentation (Figure 5). The rest of compounds and concentrations tested showed a light or null DNA damage at double/single strands levels inducing cell death by a necrosis mechanism.

To the authors' knowledge, no studies about the effects of beer and its components at an individual cellular level have been carried out. A study with a different brand of blond lager beer by Merinas-Amo et al. [42] showed significant damage with a TM higher than 100 when the HL-60 cells were treated with different concentrations of beer. However, this study did not obtain the same results for the lager beer. A similar clastogenic pattern was observed in the stout beer and hopped wort at the highest concentrations tested. Moreover, the different composition of beers allows the observation of a strong cytotoxic potential of full Czech beers compared with the studied blond lager beer, but not causing DNA damage at the cells during the 5 h of treatment and inducing total cell death at 72 h. It could be mainly because of the raw materials used (water, malt, hop and yeast), as those produced in the Czech Republic area are considered as some of the highest-quality materials in the world [6].

3.9. Methylation Status

The results about the relative normalized expression of the Alu M1, LINE-1 and Sat-α repetitive sequences in the HL-60 cells treated with the compounds showed a general hypermethylation level has been observed, with some exceptions (Figure 7): Hopped wort showed a significant hypomethylating activity in the Alu M1 sequence at both studied concentrations, the highest concentration of Sladek hop and the highest concentration of Young beer treatments. Moreover, Dark malt exhibited no significant variation in the methylation level of the repetitive sequences studied. Contrarily, a significant hypermethylation was induced by all the Wort tested concentrations in all the genomic sequences studied: By Pilsner Urquell, Saazer hop and *S. uvarum* at the highest concentration of for all the sequences tested; by Sladek hop at lowest concentration for all the sequences studied; by Blond malt at lowest concentration assayed of Sat-α; and at the lowest concentration of Young beer for LINE and Sat-α repetitive sequences.

The control mechanisms of epigenetic DNA modifications are complex [92]. Endoparasitic repetitive sequences refrain from jumping around due to the repression of DNA methylation [93]. Most of the DNA methylation in humans occurs in the cytosine of CpG dinucleotides (CpG islands). In a normal cell, they are unmethylated, and the gene is expressed if the required transcription factors are present [93]. Due to the effects that methylation causes in the repetitive sequences of the genome, Pilsner Urquell, Saazer hop, *S. uvarum* and Wort at their highest concentration tested for all repetitive sequences studied, as well as lowest the concentration of Slaldek hop, Blond malt and Young beer, prevents DNA epigenetic damage, as methylation of the repetitive sequences is understood as a genomic protective mechanism [49,94]. These effects are not shown in the HL-60 cells when treated with Dark malt and at some concentrations and repetitive sequences of Sladek hop, Hopped wort and Young beer. Bearing in mind the idea of tumoral cell demethylation, it is suggested that beer and its compound act as preventive agents cutting off the repetitive sequences of tumour cells [94].

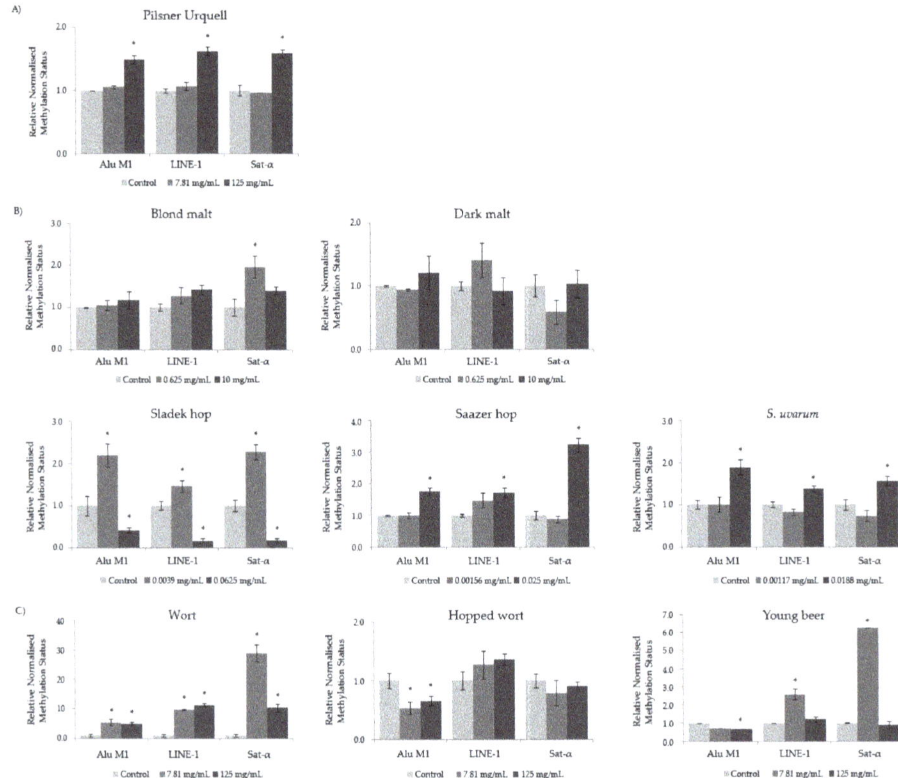

Figure 7. The relative normalised expression data of each repetitive element in treated HL-60 cells with different concentrations of a lyophilised Czech beer (**A**), raw materials (**B**) and processed-beers (**C**). The relative normalised expression data of each repetitive element (Alu M1, LINE-1 and Sat-α). The values represent the mean ± SE from three independent experiments. The untreated cells grown in RPMI were used as a control.

Unfortunately, there are no previous studies about the methylation status of lyophilised beers or raw materials. Merinas-Amo et al. [42] demonstrated that a lyophilised blond Lager beer induced hypermethylation in the DNA of tumour cells in a wide range of repetitive sequences. Fang et al. [95] suggested that the effect of a single polyphenol is not significant in normal dietary consumption, whereas the combination of polyphenols with dietary histone deacetylase inhibitors and the additive effect of different dietary chemicals may produce some effects. Furthermore, an excessive consumption of polyphenols in dietary supplements may affect the DNA methylation status [95]. The single or combined phenols contained in food should be taken into account in epigenetic-focussed therapies.

4. Conclusions

The new data corpus of our study contributes and supports the benefits showed by this lyophilized beverage, due to its safety, protection against an oxidative toxin, chemopreventive potential and the induction of DNA damage in tumour cells. A wide range of freeze-dried concentrations of beers and beer-processing substances have been tested and different biological activities have been observed in a multi-level set of in vivo and in vitro assays. Although several classes of volatile compounds like alcohol, ester, aldehydes, centone, etc. are lost in the lyophilisation [96], polyphenols and other bioactive components remain in the final sample [42]. All samples suffered the same lyophilisation process which makes it possible to properly compare the results among them in the different tests assayed.

Our work results indicate that raw materials and brewing conditions are key points for the final biological activities of beers. Despite the promising properties showed for all Czech beers, Pilsner Urquell, Budweiser Lezak, Lobkowicz Cerny and Budweiser Nealko were the beverages with the best results obtained in the in vivo and in vitro assays. The possible cause could be found in the absence of a Maillard reaction during the colouring of malt, which could induce a deterioration and an enhancement of food quality by the formation of antinutritional and toxic Maillard reaction products or beneficial compounds with antioxidant capacity.

Furthermore, the toxicological activities attributed to beer consumption cannot be related to a particular constituent because of their wide complexity and variability in their polyphenolic pattern, and also because the brewing is influenced by the raw materials, wort composition, yeast strains and fermentation conditions, among other factors.

On the whole, the positive properties in the in vivo (no toxic, no genotoxic, antitoxic, antigenotoxic, longevity and healthspan) and in the in vitro (cytotoxic and induction of DNA damage at different levels) tests are shown by the freeze-dried samples at different concentrations.

This study suggests that Czech beers as a drink have no severe potential adverse effects. However, further investigations are needed to clarify the effects of beer to other diets, as well as its important role in the prevention of chronic diseases, which mainly are related to the intake of antioxidants. Moreover, and despite the promising results obtained for the different freeze-dried beers and its materials, their consumption must be moderate due to the known negative effects induced by alcohol.

Author Contributions: Conceptualization, Á.A.-M.; Formal analysis, T.M.-A.; Investigation, R.M.-A.; Methodology, V.G.-Z., and A.V.-R.; Resources: L.C., V.P., L.K., funding acquisition, M.d.R.-C. and R.F.; Supervision, Á.A.-M. and L.K.; Writing - original draft, T.M.-A.

Funding: This research received no external funding.

Conflicts of Interest: The authors declare no conflicts of interest.

References

1. Damerow, P. Sumerian beer: The origins of brewing technology in ancient Mesopotamia. *Cuneif. Digit. Libr. J.* **2012**, *2*, 1–20.
2. Hornsey, I.S. *A History of Beer and Brewing*; Royal Society of Chemistry: London, UK, 2003; Volume 34.
3. Wolf, A.; Bray, G.A.; Popkin, B.M. A short history of beverages and how our body treats them. *Obes. Rev.* **2008**, *9*, 151–164. [CrossRef]
4. Marová, I.; Parilova, K.; Friedl, Z.; Obruca, S.; Duronova, K. Analysis of Phenolic Compounds in Lager Beers of Different Origin: A Contribution to Potential Determination of the Authenticity of Czech Beer. *Chromatographia* **2011**, *73*, 83–95. [CrossRef]
5. Wanninayake, W.; Chovancova, M. Consumer Ethnocentrism and Attitudes towards Foreign Beer Brands: With Evidence from Zlin Region in the Czech Republic. *J. Compet.* **2012**, *4*, 3–19. [CrossRef]
6. Commiccion, A Publication of an Application Pursuant to Article 6 (2) of Council Regulation (EC) No 510/2006 on the Protection of Geographical Indications and Designations of Origin for Agricultural Products and Foodstuffs. Available online: http://eur-lex.europa.eu/LexUriServ/LexUriServ.do?uri=OJ:C:2008:016:0014:0022:CS (accessed on 31 March 2017).
7. Buiatti, S. *Chapter 20-Beer Composition: An Overview*; Academic Press: Cambridge, MA, USA, 2009; pp. 212–225.
8. Chen, W.; Becker, T.; Qian, F.; Ring, J. Beer and beer compounds: Physiological effects on skin health. *J. Eur. Acad. Dermatol. Venereol.* **2013**, *28*, 142–150. [CrossRef]
9. Sanchez, C.L.; Franco, L.; Bravo, R.; Rubio, C.; Rodríguez, A.; Barriga, C.; Cubero, J. Cerveza y salud, beneficios en el sueño. *Rev. Española Nutr. Comunitaria* **2010**, *16*, 160–163. [CrossRef]
10. Rimm, E.; Giovannucci, E.; Willett, W.; Colditz, G.; Ascherio, A.; Rosner, B.; Stampfer, M.; Colditz, G. Prospective study of alcohol consumption and risk of coronary disease in men. *Lancet* **1991**, *338*, 464–468. [CrossRef]

11. Curhan, G.C.; Willett, W.C.; Rimm, E.B.; Spiegelman, D.; Stampfer, M.J. Prospective Study of Beverage Use and the Risk of Kidney Stones. *Am. J. Epidemiol.* **1996**, *143*, 240–247. [CrossRef]

12. Vinson, J.A.; Mandarano, M.; Hirst, M.; Trevithick, J.R.; Bose, P. Phenol Antioxidant Quantity and Quality in Foods: Beers and the Effect of Two Types of Beer on an Animal Model of Atherosclerosis. *J. Agric. Food Chem.* **2003**, *51*, 5528–5533. [CrossRef]

13. World Cancer Research Fund and American Institute for Cancer Research. *Food, Nutrition and the Prevention of Cancer: A Global Perspective*; American Institute for Cancer Research: Washington, DC, USA, 1997.

14. Preedy, V.R. *Beer in Health and Disease Prevention*; Academic Press: Cambridge, MA, USA, 2011.

15. Granato, D.; Branco, G.F.; Faria, J.D.A.F.; Cruz, A.G. Characterization of Brazilian lager and brown ale beers based on color, phenolic compounds, and antioxidant activity using chemometrics. *J. Sci. Food Agric.* **2011**, *91*, 563–571. [CrossRef]

16. Reddy, L.; Odhav, B.; Bhoola, K. Natural products for cancer prevention: A global perspective. *Pharmacol. Ther.* **2003**, *99*, 1–13. [CrossRef]

17. Saura-Calixto, F.; Serrano, J.; Pérez-Jiménez, J. Chapter 42-What contribution is beer to the intake of antioxidants in the diet? In *Beer in Health and Disease Prevention*; Academic Press: Cambridge, MA, USA, 2008.

18. Bier, E. Drosophila, the golden bug, emerges as a tool for human genetics. *Nat. Rev. Genet.* **2005**, *6*, 9–23. [CrossRef]

19. Lloyd, T.E.; Taylor, J.P. Flightless Flies: Drosophila models of neuromuscular disease. *Ann. N. Y. Acad. Sci.* **2010**, *1184*, 1–20. [CrossRef]

20. Bhargav, D.; Singh, M.P.; Murthy, R.C.; Mathur, N.; Misra, D.; Saxena, D.K.; Chowdhuri, D.K. Toxic potential of municipal solid waste leachates in transgenic *Drosophila melanogaster* (hsp70-lacZ): hsp70 as a marker of cellular damage. *Ecotoxicol. Environ. Saf.* **2008**, *69*, 233–245. [CrossRef]

21. Coulom, H.; Birman, S. Chronic exposure to rotenone models sporadic Parkinson's disease in *Drosophila melanogaster*. *J. Neurosci.* **2004**, *24*, 10993–10998. [CrossRef]

22. Dean, B.J. Recent findings on the genetic toxicology of benzene, toluene, xylenes and phenols. *Mutat. Res. Genet. Toxicol.* **1985**, *154*, 153–181. [CrossRef]

23. Hosamani, R. Acute Exposure of *Drosophila melanogaster* to Paraquat Causes Oxidative Stress and Mitochondrial Dysfunction. *Arch. Insect Biochem. Physiol.* **2013**, *83*, 25–40. [CrossRef]

24. Siddique, Y.H.; Fatima, A.; Jyoti, S.; Naz, F.; Khan, W.; Singh, B.R.; Naqvi, A.H. Evaluation of the toxic potential of graphene copper nanocomposite (GCNC) in the third instar larvae of transgenic *Drosophila melanogaster* (hsp70-lacZ) Bg 9. *PLoS ONE* **2013**, *8*, e80944. [CrossRef]

25. Collins, S.J. The HL-60 promyelocytic leukemia cell line: Proliferation, differentiation, and cellular oncogene expression. *Blood* **1987**, *70*, 1233–1244.

26. Gallagher, R.; Collins, S.; Trujillo, J.; McCredie, K.; Ahearn, M.; Tsai, S.; Metzgar, R.; Aulakh, G.; Ting, R.; Ruscetti, F.; et al. Characterization of the continuous, differentiating myeloid cell line (HL-60) from a patient with acute promyelocytic leukemia. *Blood* **1979**, *54*, 713–733.

27. Ja, W.W.; Carvalho, G.B.; Mak, E.M.; De La Rosa, N.N.; Fang, A.Y.; Liong, J.C.; Brummel, T.; Benzer, S. Prandiology of Drosophila and the CAFE assay. *Proc. Natl. Acad. Sci. USA* **2007**, *104*, 8253–8256. [CrossRef]

28. Kirin. Globar Beer Consumption by Country in 2011. Available online: http://www.kirinholdings.co.jp/english/news/2012/1226_01.html (accessed on 31 March 2017).

29. Goldammer, T. Hop varieties. In *The Brewers' Handbook—The Complete Book to Brewing Beer*; KVP Publishers: Huntsville, AL, USA, 2000; Volume 53.

30. Yan, J.; Huen, D.; Morely, T.; Johnson, G.; Gubb, D.; Roote, J.; Adler, P.N. The multiple-wing-hairs Gene Encodes a Novel GBD–FH3 Domain-Containing Protein That Functions Both Prior to and After Wing Hair Initiation. *Genetics* **2008**, *180*, 219–228. [CrossRef]

31. Ren, N.; Charlton, J.; Adler, P.N. The flare Gene, Which Encodes the AIP1 Protein of Drosophila, Functions to Regulate F-Actin Disassembly in Pupal Epidermal Cells. *Genetics* **2007**, *176*, 2223–2234. [CrossRef]

32. Merinas-Amo, T.; Tasset-Cuevas, I.; Díaz-Carretero, A.M.; Alonso-Moraga, Á.; Calahorro, F. Role of Choline in the Modulation of Degenerative Processes: In Vivo and In Vitro Studies. *J. Med. Food* **2017**, *20*, 223–234. [CrossRef]

33. Kozák, V. *Analysis of Reasons for Beer Consuption Drop in the Czech Republic*; Technical University of Liberec: Liberec, Czechia, 2013.

34. Graf, U.; Würgler, F.E.; Katz, A.J.; Frei, H.; Juon, H.; Hall, C.B.; Kale, P.G. Somatic mutation and recombination test in *Drosophila melanogaster*. *Environ. Mutagen.* **1984**, *6*, 153–188. [CrossRef]

35. Frei, H.; Würgler, F. Statistical methods to decide whether mutagenicity test data from Drosophila assays indicate a positive, negative, or inconclusive result. *Mutat. Res. Mutagen. Relat. Subj.* **1988**, *203*, 297–308. [CrossRef]

36. Frei, H.; Würgler, F.E. Optimal experimental design and sample size for the statistical evaluation of data from somatic mutation and recombination tests (SMART) in Drosophila. *Mutat. Res. Mutagen. Relat. Subj.* **1995**, *334*, 247–258. [CrossRef]

37. Anter, J.; Romero-Jiménez, M.; Fernández-Bedmar, Z.; Villatoro-Pulido, M.; Analla, M.; Alonso-Moraga, Á.; Muñoz-Serrano, A. Antigenotoxicity, Cytotoxicity, and Apoptosis Induction by Apigenin, Bisabolol, and Protocatechuic Acid. *J. Med. Food* **2011**, *14*, 276–283. [CrossRef]

38. Graf, U.; Abraham, S.K.; Guzmán-Rincón, J.; Würgler, F.E. Antigenotoxicity studies in *Drosophila melanogaster*. *Mutat. Res. Mol. Mech. Mutagen.* **1998**, *402*, 203–209. [CrossRef]

39. Abraham, S.K. Antigenotoxicity of coffee in the Drosophila assay for somatic mutation and recombination. *Mutagenesis* **1994**, *9*, 383–386. [CrossRef]

40. Tasset-Cuevas, I.; Fernández-Bedmar, Z.; Lozano-Baena, M.D.; Campos-Sánchez, J.; De Haro-Bailón, A.; Muñoz-Serrano, A.; Alonso-Moraga, Á. Protective Effect of Borage Seed Oil and Gamma Linolenic Acid on DNA: In Vivo and In Vitro Studies. *PLoS ONE* **2013**, *8*, e56986. [CrossRef] [PubMed]

41. Soh, J.W.; Hotic, S.; Arking, R. Dietary restriction in Drosophila is dependent on mitochondrial efficiency and constrained by pre-existing extended longevity. *Mech. Ageing Dev.* **2007**, *128*, 581–593. [CrossRef] [PubMed]

42. Merinas-Amo, T.; Tasset-Cuevas, I.; Díaz-Carretero, A.M.; Alonso-Moraga, Á.; Calahorro, F. In vivo and in vitro studies of the role of lyophilised blond Lager beer and some bioactive components in the modulation of degenerative processes. *J. Funct. Foods* **2016**, *27*, 274–294. [CrossRef]

43. Mateo-Fernández, M.; Merinas-Amo, T.; Moreno-Millán, M.; Alonso-Moraga, Á.; Demyda-Peyrás, S. In vivo and in vitro genotoxic and epigenetic effects of two types of cola beverages and caffeine: A multi-assay approach. *BioMed Res. Int.* **2016**, *2016*, 15. [CrossRef] [PubMed]

44. Merinas-Amo, T.; Merinas-Amo, R.; Alonso-Moraga, A. A clinical pilot assay of beer consumption: Modulation in the methylation status patterns of repetitive sequences. *Sylwan* **2017**, *161*, 134–156.

45. Deininger, P.L.; Moran, J.V.; A Batzer, M.; Kazazian, H.H. Mobile elements and mammalian genome evolution. *Curr. Opin. Genet. Dev.* **2003**, *13*, 651–658. [CrossRef]

46. Ehrlich, M. DNA Hypomethylation, Cancer, the Immunodeficiency, Centromeric Region Instability, Facial Anomalies Syndrome and Chromosomal Rearrangements. *J. Nutr.* **2002**, *132*, 2424–2429. [CrossRef]

47. Lee, C.; Wevrick, R.; Fisher, R.; Ferguson-Smith, M.; Lin, C. Human centromeric DNAs. *Hum. Genet.* **1997**, *100*, 291–304. [CrossRef]

48. Weiner, A.M. SINEs and LINEs: The art of biting the hand that feeds you. *Curr. Opin. Cell Boil.* **2002**, *14*, 343–350. [CrossRef]

49. Weisenberger, D.J.; Campan, M.; Long, T.I.; Kim, M.; Woods, C.; Fiala, E.; Ehrlich, M.; Laird, P.W. Analysis of repetitive element DNA methylation by MethyLight. *Nucleic Acids Res.* **2005**, *33*, 6823–6836. [CrossRef]

50. Nikolaidis, G.; Raji, O.Y.; Markopoulou, S.; Gosney, J.R.; Bryan, J.; Warburton, C.; Walshaw, M.; Sheard, J.; Field, J.K.; Liloglou, T. DNA methylation biomarkers offer improved diagnostic efficiency in lung cancer. *Cancer Res.* **2012**, *72*, 5692–5701. [CrossRef] [PubMed]

51. Liloglou, T.; Bediaga, N.G.; Brown, B.R.; Field, J.K.; Davies, M.P.; Davies, M.P.A. Epigenetic biomarkers in lung cancer. *Cancer Lett.* **2014**, *342*, 200–212. [CrossRef] [PubMed]

52. Arimoto-Kobayashi, S.; Sugiyama, C.; Harada, N.; Takeuchi, M.; Takemura, M.; Hayatsu, H. Inhibitory Effects of Beer and Other Alcoholic Beverages on Mutagenesis and DNA Adduct Formation Induced by Several Carcinogens. *J. Agric. Food Chem.* **1999**, *47*, 221–230. [CrossRef] [PubMed]

53. Gerhäuser, C. Beer constituents as potential cancer chemopreventive agents. *Eur. J. Cancer* **2005**, *41*, 1941–1954. [CrossRef] [PubMed]

54. Anter, J.; Campos-Sánchez, J.; El Hamss, R.; Rojas-Molina, M.; Muñoz-Serrano, A.; Analla, M.; Alonso-Moraga, Á. Modulation of genotoxicity by extra-virgin olive oil and some of its distinctive components assessed by use of the Drosophila wing-spot test. *Mutat. Res. Toxicol. Environ. Mutagen.* **2010**, *703*, 137–142. [CrossRef] [PubMed]

55. Wang, X.; Yang, L.; Yang, X.; Tian, Y. In vitro and in vivo antioxidant and antimutagenic activities of polyphenols extracted from hops (*Humulus lupulus* L.). *J. Sci. Food Agric.* **2014**, *94*, 1693–1700. [CrossRef] [PubMed]

56. Arimoto-Kobayashi, S.; Ishida, R.; Nakai, Y.; Idei, C.; Takata, J.; Takahashi, E.; Okamoto, K.; Negishi, T.; Konuma, T. Inhibitory Effects of Beer on Mutation in the Ames Test and DNA Adduct Formation in Mouse Organs Induced by 2-Amino-1-methyl-6-phenylimidazo[4,5-b]pyridine (PhIP). *Biol. Pharm. Bull.* **2006**, *29*, 67–70. [CrossRef] [PubMed]

57. Halliwell, B.; Gutteridge, J.M.C. *Free Radicals in Biology and Medicine*; Oxford University Press (OUP): Oxford, UK, 2015.

58. Callemien, D.; Collin, S. Structure, organoleptic properties, quantification methods, and stability of phenolic compounds in beer a review. *Food Rev. Int.* **2010**, *26*, 1–84. [CrossRef]

59. Callemien, D.; Jerkovic, V.; Rozenberg, R.; Collin, S. Hop as an Interesting Source of Resveratrol for Brewers: Optimization of the Extraction and Quantitative Study by Liquid Chromatography/Atmospheric Pressure Chemical Ionization Tandem Mass Spectrometry. *J. Agric. Food Chem.* **2005**, *53*, 424–429. [CrossRef]

60. Goupy, P.; Hugues, M.; Boivin, P.; Amiot, M.J. Antioxidant composition and activity of barley (*Hordeum vulgare*) and malt extracts and of isolated phenolic compounds. *J. Sci. Food Agric.* **1999**, *79*, 1625–1634. [CrossRef]

61. Vinson, J.A.; Jang, J.; Yang, J.; Dabbagh, Y.; Liang, X.; Serry, M.; Proch, J.; Cai, S. Vitamins and Especially Flavonoids in Common Beverages Are Powerful in Vitro Antioxidants Which Enrich Lower Density Lipoproteins and Increase Their Oxidative Resistance after ex Vivo Spiking in Human Plasma. *J. Agric. Food Chem.* **1999**, *47*, 2502–2504. [CrossRef]

62. Woffenden, H.M.; Ames, J.M.; Chandra, S. Relationships between Antioxidant Activity, Color, and Flavor Compounds of Crystal Malt Extracts. *J. Agric. Food Chem.* **2001**, *49*, 5524–5530. [CrossRef]

63. Friedman, M. Food Browning and Its Prevention: An Overview†. *J. Agric. Food Chem.* **1996**, *44*, 631–653. [CrossRef]

64. Henle, T. Protein-bound advanced glycation endproducts (AGEs) as bioactive amino acid derivatives in foods. *Amino Acids* **2005**, *29*, 313–322. [CrossRef]

65. Silván, J.M.; Van De Lagemaat, J.; Olano, A.; Del Castillo, M.D. Analysis and biological properties of amino acid derivates formed by Maillard reaction in foods. *J. Pharm. Biomed. Anal.* **2006**, *41*, 1543–1551. [CrossRef]

66. Yen, G.C.; Chau, C.F.; Lii, J.D. Isolation and characterization of the most antimutagenic Maillard reaction products derived from xylose and lysine. *J. Agric. Food Chem.* **1993**, *41*, 771–776. [CrossRef]

67. Yen, G.-C.; Tsai, L.-C. Antimutagenicity of a partially fractionated Maillard reaction product. *Food Chem.* **1993**, *47*, 11–15. [CrossRef]

68. Vagnarelli, P.; De Sario, A.; Cuzzoni, M.T.; Mazza, P.; De Carli, L. Cytotoxicity and clastogenic activity of ribose-lysine browning model system. *J. Agric. Food Chem.* **1991**, *39*, 2237–2239. [CrossRef]

69. Rivero, D.; Perez-Magariño, S.; Gonzalez-SanJose, M.L.; Valls-Belles, V.; Codoñer, P.; Muñiz, P. Inhibition of Induced DNA Oxidative Damage by Beers: Correlation with the Content of Polyphenols and Melanoidins. *J. Agric. Food Chem.* **2005**, *53*, 3637–3642. [CrossRef]

70. Morales, F.J.; Babbel, M.-B. Antiradical Efficiency of Maillard Reaction Mixtures in a Hydrophilic Media. *J. Agric. Food Chem.* **2002**, *50*, 2788–2792. [CrossRef]

71. Merinas-Amo, T.; Martinez-Jurado, M.; Merinas-Amo, R.; Mateo-Fenandez, M.; Alonso-Moraga, Á. In vivo and in vitro toxicological and biological activities of alpha-acid humulone, isoxanthohumol and stout beer. *Toxicol. Lett.* **2014**, *229*, S179–S180. [CrossRef]

72. Merinas-Amo, T.; Martínez-Jurado, M.; Merinas-Amo, R.; García-Zorrilla, V.; Fernández-Bedmar, Z.; Alonso-Moraga, A.; Mateo-Fernández, M. Comparison of in vivo and in vitro activities of alcohol and non-alcohol beer. *Toxicol. Lett.* **2015**, *238*, 65–66. [CrossRef]

73. Krofta, K. Comparison of quality parameters of Czech and foreign hop varieties. *Plant Soil Environ.* **2003**, *49*, 261–268. [CrossRef]

74. Krofta, K.; Mikyška, A.; Hašková, D. Antioxidant characteristics of hops and hop products. *J. Inst. Brew.* **2008**, *114*, 160–166. [CrossRef]

75. Romero-Jiménez, M.; Campos-Sánchez, J.; Analla, M.; Muñoz-Serrano, A.; Alonso-Moraga, Á. Genotoxicity and anti-genotoxicity of some traditional medicinal herbs. *Mutat. Res. Toxicol. Environ. Mutagen.* **2005**, *585*, 147–155. [CrossRef]

76. Jacobs, D.R.; Tapsell, L.C. Food synergy: The key to a healthy diet. *Proc. Nutr. Soc.* **2013**, *72*, 200–206. [CrossRef]
77. Larsson, S.; Kaluza, J.; Wolk, A. Combined impact of healthy lifestyle factors on lifespan: Two prospective cohorts. *J. Intern. Med.* **2017**, *282*, 209–219. [CrossRef]
78. Fleming, J.; Reveillaud, I.; Niedzwiecki, A. Role of oxidative stress in Drosophila aging. *Mutat. Res.* **1992**, *275*, 267–279. [CrossRef]
79. Schriner, S.E.; Coskun, V.; Hogan, S.P.; Nguyen, C.T.; Lopez, T.E.; Jafari, M. Extension of Drosophila Lifespan by Rhodiola rosea Depends on Dietary Carbohydrate and Caloric Content in a Simplified Diet. *J. Med. Food* **2016**, *19*, 318–323. [CrossRef]
80. Hevia, D.; Mayo, J.C.; Quiros, I.; Sainz, R. Beer constituents inhibit prostate cancer cells proliferation. *Eur. J. Cancer Suppl.* **2008**, *6*, 142. [CrossRef]
81. Stevens, J.F.; E Page, J. Xanthohumol and related prenylflavonoids from hops and beer: To your good health! *Phytochemistry* **2004**, *65*, 1317–1330. [CrossRef]
82. Nozawa, H.; Yoshida, A.; Tajima, O.; Katayama, M.; Sonobe, H.; Wakabayashi, K.; Kondo, K. Intake of beer inhibits azoxymethane-induced colonic carcinogenesis in male Fischer 344 rats. *Int. J. Cancer* **2004**, *108*, 404–411. [CrossRef]
83. Maietti, A.; Brighenti, V.; Bonetti, G.; Tedeschi, P.; Prencipe, F.P.; Benvenuti, S.; Brandolini, V.; Pellati, F. Metabolite profiling of flavonols and in vitro antioxidant activity of young shoots of wild *Humulus lupulus* L. (hop). *J. Pharm. Biomed. Anal.* **2017**, *142*, 28–34. [CrossRef]
84. Luzak, B.; Kassassir, H.; Rój, E.; Stanczyk, L.; Watala, C.; Golanski, J. Xanthohumol from hop cones (*Humulus lupulus* L.) prevents ADP-induced platelet reactivity. *Arch. Physiol. Biochem.* **2017**, *123*, 54–60. [CrossRef]
85. Wyllie, A.; Kerr, J.; Currie, A. Cell Death: The Significance of Apoptosis. *Adv. Clin. Chem.* **1980**, *68*, 251–306.
86. Merinas-Amo, T.; Merinas-Amo, R.; Valenzuela-Gómez, F.; Bailón, A.D.H.; Lozano-Baena, M.; Mateo-Fernández, M.; Fernández-Bedmar, Z.; Alonso-Moraga, Á. Toxicological in vivo and in vitro evaluation of three diet abundant phenols: Tyrosol, ferulic and vanillic acids. *Toxicol. Lett.* **2016**, *258*, 177–178. [CrossRef]
87. Weiskirchen, R.; Mahli, A.; Weiskirchen, S.; Hellerbrand, C. The hop constituent xanthohumol exhibits hepatoprotective effects and inhibits the activation of hepatic stellate cells at different levels. *Front. Physiol.* **2015**, *6*, 140. [CrossRef]
88. Pichler, C.; Ferk, F.; Al-Serori, H.; Huber, W.; Jäger, W.; Waldherr, M.; Mišík, M.; Kundi, M.; Nersesyan, A.; Herbacek, I. Xanthohumol prevents DNA damage by dietary carcinogens: Results of a human intervention trial. *Cancer Prev. Res.* **2017**, *10*, 153–160. [CrossRef]
89. Poe, B.S.; O'Neill, K.L. Caffeine modulates heat shock induced apoptosis in the human promyelocytic leukemia cell line HL-60. *Cancer Lett.* **1997**, *121*, 1–6. [CrossRef]
90. Fabiani, R.; Rosignoli, P.; De Bartolomeo, A.; Fuccelli, R.; Morozzi, G. Genotoxicity of alkene epoxides in human peripheral blood mononuclear cells and HL60 leukaemia cells evaluated with the comet assay. *Mutat. Res. Toxicol. Environ. Mutagen.* **2012**, *747*, 1–6. [CrossRef]
91. Fairbairn, D.W.; Olive, P.L.; O'Neill, K.L. The comet assay: A comprehensive review. *Mutat. Res. Genet. Toxicol.* **1995**, *339*, 37–59. [CrossRef]
92. Herman, J.G.; Baylin, S.B. Gene Silencing in Cancer in Association with Promoter Hypermethylation. *N. Engl. J. Med.* **2003**, *349*, 2042–2054. [CrossRef]
93. Esteller, M. DNA methylation and cancer therapy: New developments and expectations. *Curr. Opin. Oncol.* **2005**, *17*, 55–60. [CrossRef]
94. Román-Gómez, J.; Jiménez-Velasco, A.; Agirre, X.; Castillejo, J.A.; Navarro, G.; José-Enériz, E.S.; Garate, L.; Cordeu, L.; Cervantes, F.; Prosper, F.; et al. Repetitive DNA hypomethylation in the advanced phase of chronic myeloid leukemia. *Leuk. Res.* **2008**, *32*, 487–490. [CrossRef]
95. Yang, C.S.; Fang, M.; Chen, D. Dietary Polyphenols May Affect DNA Methylation. *J. Nutr.* **2007**, *137*, 223S–228S.
96. Lindsay, R.C.; Withycombe, D.A.; Micketts, R.J. Comparison of Gas Chromatographic Methods for Analysis of Beer Flavors. *Proc. Annu. Meet. Am. Soc. Brew. Chem.* **1972**, *30*, 4–7. [CrossRef]

© 2019 by the authors. Licensee MDPI, Basel, Switzerland. This article is an open access article distributed under the terms and conditions of the Creative Commons Attribution (CC BY) license (http://creativecommons.org/licenses/by/4.0/).

Article

Biological Effects of Food Coloring in In Vivo and In Vitro Model Systems

Rocío Merinas-Amo, María Martínez-Jurado, Silvia Jurado-Güeto, Ángeles Alonso-Moraga and Tania Merinas-Amo *

Department of Genetics, University of Córdoba, 14071 Córdoba, Spain; rocio.merinas@gmail.com (R.M.-A.); martinezjurado.maria@gmail.com (M.M.-J.); silviajuradogueto@gmail.com (S.J.-G.); ge1almoa@uco.es (Á.A.-M.)
* Correspondence: tania.meram@gmail.com

Received: 18 April 2019; Accepted: 22 May 2019; Published: 24 May 2019

Abstract: (1) Background: The suitability of certain food colorings is nowadays in discussion because of the effects of these compounds on human health. For this reason, in the present work, the biological effects of six worldwide used food colorings (Riboflavin, Tartrazine, Carminic Acid, Erythrosine, Indigotine, and Brilliant Blue FCF) were analyzed using two model systems. (2) Methods: In vivo toxicity, antitoxicity, and longevity assays using the model organism *Drosophila melanogaster* and in vitro cytotoxicity, DNA fragmentation, and methylation status assays using HL-60 tumor human cell line were carried out. (3) Results: Our in vivo results showed safe effects in *Drosophila* for all the food coloring treatments, non-significant protective potential against an oxidative toxin, and different effects on the lifespan of flies. The in vitro results in HL-60 cells, showed that the tested food colorings increased tumor cell growth but did not induce any DNA damage or modifications in the DNA methylation status at their acceptable daily intake (ADI) concentrations. (4) Conclusions: From the in vivo and in vitro studies, these results would support the idea that a high chronic intake of food colorings throughout the entire life is not advisable.

Keywords: additives; food coloring; *Drosophila melanogaster*; leukemia cells; toxicity; antitoxicity; longevity; cytotoxicity; DNA damage; methylation status

---

1. Introduction

A food coloring is a dye, pigment, or substance that, when added to food, drugs, or cosmetics, is able to provide color. The Food and Drugs Administration (FDA) is responsible for regulating dyes to assure their safety. Dyes are classified on the basis of their necessity of certification. According to the FDA, dyes are used to confer color to food that has lost it and to improve the color or provide it to uncolored food to make it attractive [1].

A food additive is defined as "any substance not normally consumed as food by itself and not normally used as a typical ingredient of food, whether or not it has nutritive value, the intentional addition of which to food for a technological (including organoleptic) purpose in the manufacture, processing, preparation, treatment, packing, packaging, transport, or holding of such food results, or may be reasonably expected to result (directly or indirectly), in it or its by-products becoming a component or otherwise affecting the characteristics of such food" [2].

Additives are found in many types of food that we often consume not knowing that they are present, so it is very important to study the biological consequences of using food coloring. Moreover, because of the well-known relationship between diet and health and the increasing awareness of people about their quality of life, a great deal of studies have been performed to determine which dyes may be harmful for health, promoting, for instance, childhood hyperactivity, urticaria, asthma [3], and rhinitis [4]. Information about the most consumed food coloring is reported below:

- Riboflavin (E-101) is part of the vitamin B group. It is a yellow-orange solid substance with poor solubility in water. This food coloring is present in a wide range of foods, with liver, milk, meat, and fish being the most important sources [5]. Riboflavin can be obtained by controlled fermentation using a genetically modified strain of *Bacillus subtilis* or the fungus *Ashbya gossypii* [6]. Riboflavin was evaluated by the Joint FAO/WHO Expert Committee on Food Additives (JECFA) in 1969, which established an acceptable daily intake (ADI) of 0.5 mg/kg·body weight (bw)/day on the basis of limited data [5]. No adverse toxic, genotoxic, cytotoxic, or allergic effects have been related to Riboflavin in different organisms [7,8].

- Tartrazine (E-102) is a synthetic lemon-yellow azo dye primarily used as a food coloring. Its presence is allowed in various foodstuffs and beverages [9]. Both the JECFA and the EU Scientific Committee for Food (SCF) established an ADI of 7.5 mg/kg·bw/day in 1996 [10]. Controversial studies about the effects of Tartrazine on health have been reported. The most adverse effects have been related to DNA damage [11], hyperactivity [12], changes in the central nervous system [13], and allergic reactions [14–18].

- Carminic Acid (E-120) is a natural red colorant which comes from *Datylopius coccus*, an insect which lives on *Opuntia coccinellifer*. In order to obtain this dye, it is necessary to dry and spray the body of pregnant females of these insects [19]. The JECFA and SCF committees established an ADI of 5 mg/kg·bw/day for Carminic Acid [20]. This dye is called by the FDA "cochineal extract" or "carmine" and is classified as exempt from certification. According to the FDA, it is used in food, drugs, and cosmetics [1]. Despite the absence of genotoxic or cytotoxic effects described for Carminic Acid, it has been related to anaphylactic reactions, asthma, urticaria, and angioedema [19,21–23]. Furthermore, impairment in renal function has been demonstrated in male albino rats [24].

- Erythrosine (E-127) is a cherry-pink synthetic food colorant with a polyiodinated xanthene structure [25]. It is widely used to color children's sweets [26], as well as to determine the presence of dental plate in Odontology [27]. The ADI of Erythrosine was established by the JECFA and SCF in 0.1 mg/kg·bw/day [28]. Regarding the FDA, it allows the use of Erythrosine both for food and drugs [1]. Some studies suggested a relationship between Erythrosine consumption and altered cognition and behavior in children, which could be due to the inhibition of dopamine receptors [29]. Moreover, different studies suggested the induction of chromosome aberrations and an increase in the incidence of thyroid tumors by Erythrosine consumption [11,30,31].

- Indigotine (E-132) is one of the earliest known natural dyes. Originally, it was obtained from the leaves of the plants *Indigofera tinctoria*, *Indigofera suifruticosa*, and *Isatis tinctoria*, where it occurs as indican, a glycoside of indoxyl [32]. In 1975, the JECFA and SCF established an ADI of 5 mg/kg·bw/day for this blue additive [33]. Only a subacute toxicity study performed with adult male Swiss albino mice showed severe adverse effects of Indigotine on the testis [34].

- Brilliant Blue FCF (E-133) is a triarylmethane synthetic food coloring authorized as a food additive. In 2017, the JECFA revised the ADI to 6 mg/kg·bw/day for this blue additive [35]. Brilliant Blue FCF has recently been evaluated and approved as a cosmetic colorant by the Scientific Committee on Cosmetic Products (SCCP) [35]. Current databases show no adverse effects of Brilliant Blue FCF in any organism assayed for any biological test carried out [11,36–39].

Considering the available information about the toxicological effects of food coloring on health, our main goal was to evaluate the biological and nutritional effects that the mentioned additives have on time-related degenerative processes, as well as to add new scientific data. For that purpose, an integrative study of the biological activity at the individual, cellular, and molecular levels based on in vivo and in vitro assays was carried out using two model systems. The *Drosophila* animal model is known to have more than 75% of human disease homologous genes [40] related to different human degenerative illnesses, such as Parkinson's and Alzheimer's diseases, and allergic diseases, among others. For this reason, it is a reliable system to test toxicity, antitoxicity, longevity, and many other processes [41]. Moreover, using an in vitro model of human leukemia cells (HL-60), we studied the effect of this compound on cell growth inhibition, DNA damage (internucleosomal fragmentation as double-strand breaks leading to DNA laddering associated with the activation of the apoptotic pathway in cells), and the modulation of the methylation status. The purpose of the present study was to extend knowledge and provide new scientific data in this area for future clinical studies.

2. Materials and Methods

2.1. Samples

Six different types of food coloring were selected for this study according to their high consume and abundance in the diet. A range of six concentrations were tested for each food coloring in order to better understand their biological activity at different endpoints in in vivo and in vitro assays.

The concentrations of the food colorings were established taking into account the average daily food intake of *Drosophila melanogaster* (1 mg/day) and the average body weight of *D. melanogaster* individuals (1 mg) [42]. The concentration range for all tested substances was calculated in order to make it comparable with their ADI in humans, as it summarized in Table 1.

Table 1. Food coloring information.

Food Coloring		ADI (Mg/Kg)	Tested Concentrations in *Drosophila* (mg/mL) *					
			1	2	3	4	5	6
E-101	Riboflavin	0.5	0.0000025	0.000025	0.00025	0.0025	0.025	0.25
E-102	Tartrazine	7.5	0.0000375	0.000375	0.00375	0.0375	0.375	3.75
E-120	Carminic Acid	5	0.000025	0.00025	0.0025	0.025	0.25	2.5
E-127	Erythrosine	0.1	0.0000005	0.000005	0.00005	0.0005	0.005	0.05
E-132	Indigotine	5	0.000025	0.00025	0.0025	0.025	0.25	2.5
E-133	Brilliant Blue FCF	6	0.0000005	0.000005	0.00005	0.0005	0.005	0.05

* numbers 1 to 6 represent the value, in mg/mL, of the different dilutions assayed in the in vivo and in vitro assays for each food coloring; the concentration corresponding to number 3 is the equivalent quantity of ADI in humans.

2.2. In Vivo Assays

The value of using *Drosophila* to investigate fundamental biological processes is increasingly evident. This organism is revealing itself as an appropriate system as it is a complex multicellular organism in which many aspects of gene expression are parallel to those of humans. *Drosophila* substitute mammals in experiments with the distinct goal of uncovering insights directly relevant to human beings, because it is a model for many human diseases, including cancer and ageing [43–45].

In the present study, two *Drosophila* strains were used, each with a hair marker in the third chromosome: (i) *mwh/mwh*, carrying the recessive mutation *mwh* (multiple wing hairs) that in homozygosis produces multiple tricomas per cell instead of one per cell [46], and (ii) flr^3/In *(3LR) TM3, rip^{p}sep* bx^{34e} e^s Bd^S, where the flr^3 (flare) marker is a homozygous recessive lethal mutation that produces deformed tricomas but is viable in homozygous somatic cells once larvae start development. All in vivo treatments were carried using the offspring of the reciprocal crosses of the two strains, to finally use the emerging trans-heterozygous individuals ($mwh \cdot flr^+/mwh^+ \cdot flr^3$) for the different toxicity, antitoxicity, and longevity assays [47].

2.2.1. Toxicity and Antitoxicity Assays

The survival percentages of treated *Drosophila* were determined in toxicity assays ((number of individuals born in each treatment group/number of individuals born in the negative control group) × 100). The antitoxicity tests consisted of combining treatments with food colorings at the same concentrations as in the toxicity assays with H_2O_2 at 0.12 M (Sigma; H1009) [48]. The negative controls were prepared with *Drosophila* Instant Medium (Formula 4-24, Carolina Biological Supply, Burlington, NC) and distilled water, and the positive controls with medium and H_2O_2.

Three independent experiments were carried out for each assay. Chi-square test in Microsoft Office Excel 2007 was used to determine if the tested compounds significantly affected fly survival, with respect to the control. In the toxicity assay, statistical chi-square values ($p < 0.05$) for the different concentrations tested were obtained by comparing the effects of different concentrations with those of the negative control, whereas statistical chi-square values of antitoxicity assays were obtained by comparing the effects of the different concentrations with those of the positive control [49].

A wide range of researches are found on the effects of hydrogen peroxide: it can interact directly with DNA or modulate transcription and suppress genomic repair pathways; induce microsatellite instability in germ cells of *D. melanogaster* [50]; produce genetic damage due to the generated electrophilic compounds [51]. Also, it is well established that hydrogen peroxide is an endogenous mutagen responsible for some of the highest cancer risks associated with persistent inflammation [52,53]. Oxy-radicals derived from hydrogen peroxide can act on the genome either directly, causing chromosome damage that induces oncogenic mutations [54,55], or indirectly, by modulating gene transcription [56,57] or by suppressing genome repair pathways [58,59]. Moreover, a study of genotoxicity induced by hydrogen peroxide using the in vivo *Drosophila* assay [60] indicated that the oxidative agent is able to induce somatic mutations and mitotic recombination (concentrations ranged from 0.12 M to 0.48 M). The relative contribution of the recombinational events to the total clone induction was estimated by comparing the frequency of *mwh* spots on the marker wings with the frequency of *mwh* spots in the balancer wings, concluding that an average of 60% of clones showed a genetic recombinational origin.

2.2.2. Lifespan Assays

All experiments were carried out at 25 °C according to the procedure described in Tasset-Cuevas, et al. [61]. Sets of 25 individuals of the same gender were selected and placed into sterile vials containing 0.21 g of *Drosophila* Instant Medium and 1 mL of different concentrations of solutions of the food coloring to be tested. Two replicates were followed during the complete life extension for each control and concentration established. Alive animals were counted, and the respective nourishment renewed twice a week.

In order to know the quality of life of the treated *Drosophila* in the longevity trials, the upper 25% of lifespan survival curves was studied. This part of the lifespan is considered as the healthspan of a curve, characterized by low and more or less constant age-specific mortality rate values [62].

The statistical treatment of the survival data for each control and concentration was carried out with the SPSS Statistics 17.0 software (SPSS, Inc., Chicago, IL, USA), applying the Kaplan–Meier method to obtain the survival curves. The significance of the curves was determined using the Log-Rank method (Mantel-Cox).

2.3. In Vitro Assays

The in vitro model of human leukemia cells (HL-60) was used to study the effect of food coloring on growth inhibition of the tumor cells, DNA damage (internucleosomal fragmentation as double-strand breaks leading to DNA laddering associated with the activation of the apoptotic pathway), and modulation of DNA methylation status.

The promyelocytic human leukemia HL-60 cell line was grown in RPMI-1640 medium (Sigma, R5886) supplemented with heat-inactivated fetal bovine serum (Linus, S01805), L-glutamine at 200 mM (Sigma, G7513), and an antibiotic–antimycotic solution (Sigma, A5955). The cells were incubated at 37 °C in a humidified atmosphere with 5% CO_2 [63]. The cultures were plated at 2.5×10^4 cells/mL density in 10 mL culture bottles and passed every two days.

2.3.1. Cytotoxicity Assays

HL-60 cells were placed in 96-well culture plates (2×10^4 cells/mL), cultured for 72 h, and supplemented with the food colorings at different concentrations. This allowed the assessment of a wide range of concentrations in the in vitro cytotoxicity assays, with the aim to predict acute in vivo lethality. Although a continuous evaluation of the cytotoxic effects was studied, only the results at 72 h allowed us to acquire more knowledge about the in vitro lethality of the tested food colorings at the different concentrations assayed, because the IC_{50} was reached for most of them at that time-point.

Cell viability was determined by the trypan blue dye exclusion test. Trypan blue (Sigma-Aldrich, St. Louis, MO, USA, T8154) was added to the cell cultures at a 1:1 volume ratio, and 20 µL of cell suspension was loaded into a Neubauer chamber. The cells were counted with an inverted microscope at 100× magnification (AE30/31, Motic). Curves were plotted as the average survival percentage of three independent experiments with respect to the control growing for 72 h.

2.3.2. Determination of DNA fragmentation

HL-60 cells (1×10^6/mL) were treated with different concentrations of food coloring for 5 h.

The treated cells were collected and centrifuged at 3000 rpm for 5 min, and DNA was extracted according to the procedure described in Merinas-Amo, et al. [64]. Briefly, the cell pellet was resuspended in lysis buffer and incubated in an SDS 10% and proteinase K solution. DNA precipitation with NaCl and isopropanol was followed by washing with 70% ethanol DNA and incubation with RNAse overnight. For the negative control, RPMI was used as the cell medium; as a routine positive control, a concentration of 62.5 mg/mL of a lyophilized blond beer (LBB) was used [61].

DNA was quantified with a spectrophotometer (Nanodrop ND-1000), and 1200 ng of DNA was subjected to 2% agarose gel electrophoresis at 85 mA for 25 min, stained with GelRed, and visualized under UV light.

2.3.3. Methylation Status

Genomic DNA was isolated in the same way as described in the DNA fragmentation section. Bisulphite-modified DNA from food coloring treatments, using the EZ DNA Methylation-Gold Kit, was used as a template for fluorescence-based real-time quantitative Methylation-Specific PCR (qMSP). qMSP was carried out according to the protocol described by Merinas-Amo, et al. [65] in 48-well plates in a MiniOpticon Real-Time PCR System (MJ Mini Personal Thermal Cycler, Bio-Rad Laboratories Inc., Hercules, CA, USA) and was analyzed by the Bio-Rad CFX Manager 3.1 Software. Briefly, the final reaction mixture with a total volume of 10 µL consisted of: 2 µL of deionized water, 5 µM of each forward and reverse primer, 2 µL of iTaq™ Universal SYBR® Green Supermix (Bio-Rad, containing antibody-mediated hot-start iTaq DNA polymerase, dNTPs, $MgCl_2$, SYBR® Green I dye, enhancers, stabilizers, and a blend of passive reference dyes including ROX and fluorescein), and 25 ng of bisulphite-converted genomic DNA. qMSP conditions were as follows: one step at 95 °C for 3 min, 45 cycles at 95 °C for 10 s, 60 °C for 15 s, 72 °C for 15 s, another step at 95 °C for 30 s, followed by a 65 °C step during 30 s and finally a boost step from 65 °C to 95 °C for 95 s, increasing the temperature of 0.5 °C each 0.05 s.

Repetitive elements were selected in order to analyze a wide range of human genomic DNA. While Alu and LINE sequences are interspersed throughout the genome, satellites are confined to the centromere areas [66–69]. All sequences were obtained from Isogen Life Science. Alu M1, LINE-1, and Sat-α sequences were used (see Table 2 for detailed information) [70].

The relative yield results were normalized with respect to the housekeeping sequence Alu C4 using the Nikolaidis, et al. [71] and Liloglou, et al. [72] comparative C_T method:

- C_T of the target gene was normalized with respect to the referent gene (ΔC_T).
- ΔC_T of each experimental sample or reference ($\Delta C_{T,r}$) were compared with ΔC_T of the calibrator sample ($\Delta C_{T,cb}$), $\Delta\Delta C_T$.
- The relative value of each sample was defined using the formula:

$$2^{-(\Delta C_{T,r} - \Delta C_{T,cb})} = 2^{-\Delta\Delta C_T}$$

Each sample was analyzed in triplicate. One-way ANOVA and post hoc Tukey's tests were used to evaluate the differences among the tested compounds, repetitive elements, and concentrations.

Table 2. Primers information.

Reaction ID	GenBank Number	Amplicon Start	Amplicon End	Forward Primer Sequence 5′ to 3′ (N)	Reverse Primer Sequence 5′ to 3′ (N)	GC Content (%)	
						Forward	Reverse
Alu C4	Consensus Sequence	1	98	GGTTAGGTATAGTGGTTTATATTTGTAATTTTAGTA (36)	ATTAACTAAACTAATCTTAAACTCCTAACCTCA (33)	25	27.3
Alu M1	Y07755	5059	5164	ATTATGTTAGTTAGGATGGTTTCGATTTT (29)	CAATCGACCGAACGCGA (17)	27.6	58.8
LINE-1	X52235	251	331	GGACGTATTTGGAAAATCGGG (21)	AATCTCGCGATACGCCGTT (19)	47.6	52.6
Sat-α	M38468	139	260	TGATGGAGTATTTTTAAAATATACGTTTTGTAGT (34)	AATTCTAAAAATATTCCTCTTCAATTACGTAAA (33)	23.5	21.2

Source: Weisenberger, Campan, Long, Kim, Woods, Fiala, Ehrlich and Laird [70].

3. Results

3.1. In Vivo

3.1.1. Toxicity and Antitoxicity

Figure 1 shows the relative percentage of emerging adults after toxicity treatments with different concentrations of food colorings. Our results showed that Riboflavin and Indigotine were non-toxic at any assayed concentration. Tartrazine showed a significant dose-independent survival percentage at the assayed concentrations, being toxic at the fourth highest concentrations with respect to the control. Moreover, a significant survival rate compared with the control was shown for individuals treated with the red additives, except for the concentration numbered as 3, with a decreasing rate of *Drosophila* survival lower than 80%. Brilliant Blue FCF also showed a significant diminution of the survival of *Drosophila* at the two highest and the two lowest concentrations tested with respect to the control.

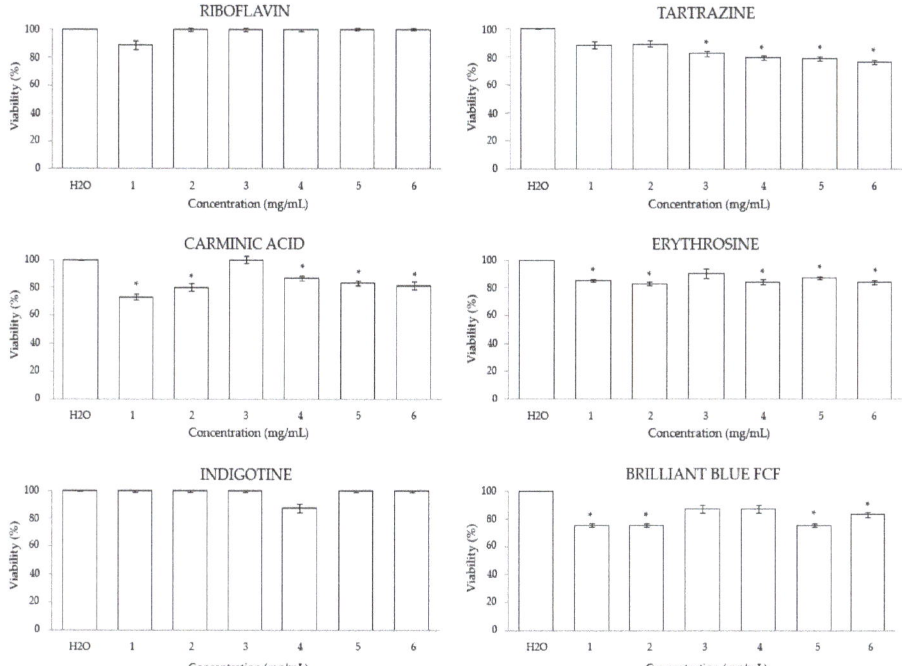

Figure 1. Toxicity levels of food coloring in *Drosophila melanogaster*. Data are expressed as percentage of surviving adults with respect to 300 untreated 72 h-old larvae from three independent experiments treated with different concentrations of food colorings. Values represent the mean ± SE from three independent experiments. * Indicates significant differences with respect to the control. 1– 6 numbers indicate the different dilutions tested (see Table 1).

On the whole, any food coloring at any assayed concentration reached the lethal dose 50 (LD_{50}), which is considered toxic. This fact confirms in the *Drosophila* in vivo eukaryotic model that the ADI (concentration numbered as 3) established by the JECFA for each food coloring is a safe dose [5,10,20,28,33,35].

The antitoxicity results showed in Figure 2 revealed the ability of the blue additives to protect individuals against stress, although only at three-highest concentrations assayed. Furthermore, Tartrazine and Carminic Acid showed no significant effects in the combined treatments at any concentrations tested with respect to the positive control. On the other hand, extremes concentrations of Riboflavin and Erythrosine and the lowest concentration of Brilliant Blue FCF showed a toxic synergic effect when combined with the oxidative toxicant hydrogen peroxide in *Drosophila*.

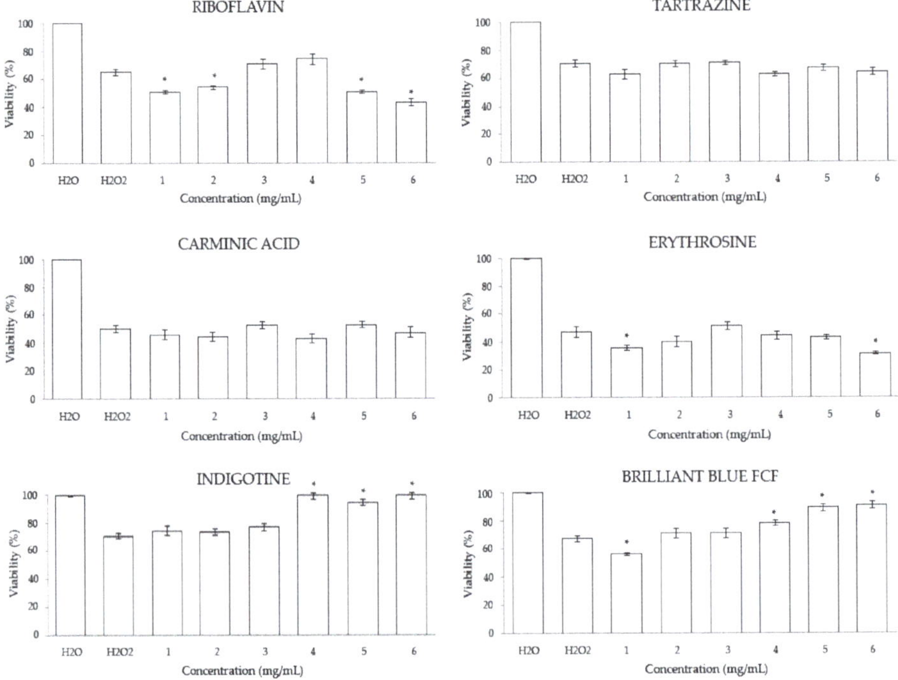

Figure 2. Antitoxicity levels of food coloring in *D. melanogaster*. Data are expressed as percentage of surviving adults with respect to 300 untreated 72 h-old larvae from three independent experiments treated with different concentrations of food colorings combined with 0.12 M H_2O_2. Values represent the mean ± SE from three independent experiments. * Indicates significant differences with respect to the positive control. 1–6 numbers indicate the different dilutions tested (see Table 1).

The absence of a relationship between the toxicity and the antitoxicity results in Tartrazine, Carminic Acid, Erythrosine, and Brilliant Blue FCF could be due to the fact that each substance might exhibit antioxidant or prooxidant activities in a competitive manner against the effect of hydrogen peroxide when combined with it [73].

3.1.2. Lifespan

The entire lifespan curves obtained by the Kaplan–Meier method for each substance and concentration are shown in Figure 3. Tartrazine and Brilliant Blue FCF induced a lifespan extension in *D. melanogaster* at the three highest concentrations tested and at the concentrations numbered 2 to 4, corresponding to 5–10 and 5–8 days, respectively, with respect to their control (Table 3). On the other hand, all concentrations of Carminic Acid and Erythrosine, except the lowest one, showed a significant decrease of longevity corresponding to 9–14 and 12–13 days, respectively, compared with their control, except for the lowest concentration (Table 3). With respect to Riboflavin and Indigotine, no significant effect on *Drosophila* longevity was observed at any assayed concentration.

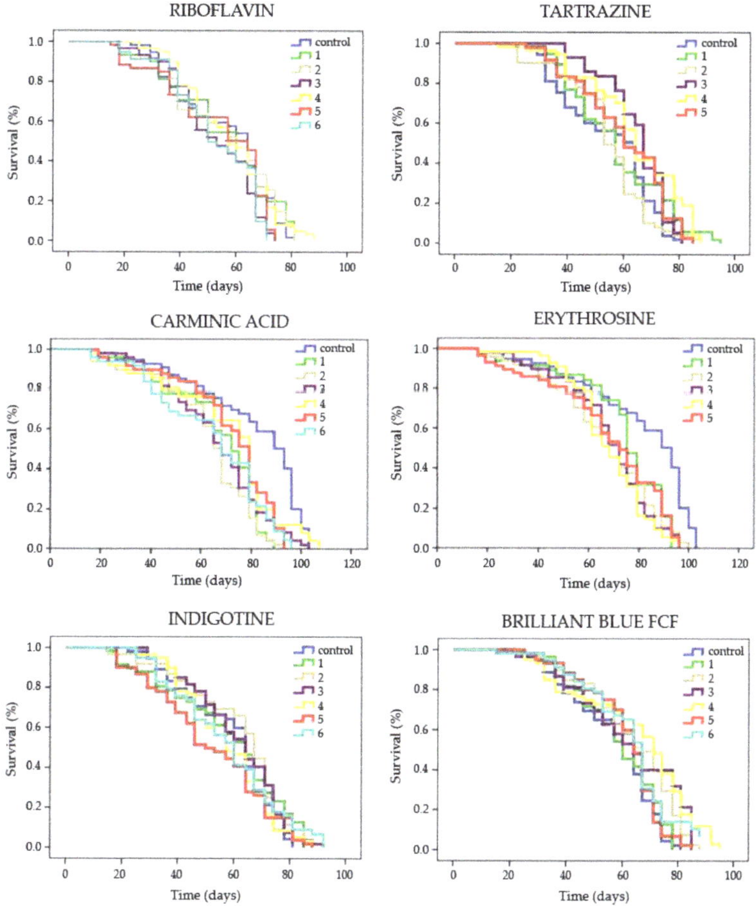

Figure 3. Complete survival curves of *D. melanogaster* fed with different concentrations of food colorings. The numbers 1–6 indicate the different dilutions tested (see Table 1). Curves were obtained by the Kaplan–Meier method, and significance was determined by the Log-Rank method (Mantel-cox).

The healthspan results (upper 25% portion of the lifespan curves) are shown in Table 3. Tartrazine induced a significant increase of healthspan in *D. melanogaster* when compared with the negative control, with the exception of the concentration numbered as 2, whose effect was similar to that of the control. The value of mean survival time of this additive ranged between 4 and 12 days. On the other hand, the highest concentration of Erythrosine assayed and the concentrations numbered 1 and 5 of Indigotine showed a significant reduction in the quality of life of *D. melanogaster* after 16 and 6 days, respectively. The remaining food colorings showed no significant differences in the mean value of healthspan of the treated flies with respect to their concurrent controls.

A non-visible dose–effect relationship has been shown by the different food colorings studied, suggesting a threshold level, rather than a degree of variation, in the significant cohorts. The data provided to the research community by the present study could be related to the controversial results observed in the database about food coloring [5,10,20,28,33,35].

Table 3. Mean and significances of lifespan and healthspan curves.

Food Coloring	Concentration	Mean Lifespan [1] (days)		Mean Healthspan [1] (days)	
Riboflavin	Control	55.985		31.399	
	1	55.019	ns	27.607	ns
	2	55.864	ns	29.110	ns
	3	52.534	ns	29.966	ns
	4	57.067	ns	32.714	ns
	5	53.341	ns	25.500	ns
	6	52.660	ns	27.222	ns
Tartrazine	Control	54.375		31.399	
	1	57.664	ns	36.154	*
	2	54.037	ns	32.681	ns
	3	64.618	*	43.571	*
	4	63.860	*	35.760	*
	5	59.989	*	37.252	*
Carminic Acid	Control	62.345		38.509	
	1	44.958	*	35.630	ns
	2	43.215	*	29.000	ns
	3	45.515	*	37.067	ns
	4	46.998	*	39.320	ns
	5	48.211	*	40.000	ns
	6	44.236	*	30.530	ns
Erythrosine	Control	62.345		38.015	
	1	55.487	ns	27.614	ns
	2	49.589	*	34.276	ns
	3	49.847	*	35.051	ns
	4	50.011	*	43.333	ns
	5	50.214	*	22.501	*
Indigotine	Control	58.433		32.988	
	1	57.791	ns	27.019	*
	2	61.547	ns	29.182	ns
	3	61.181	ns	33.857	ns
	4	57.067	ns	32.714	ns
	5	52.024	ns	27.189	*
	6	57.570	ns	31.068	ns
Brilliant Blue FCF	Control	57.526		32.988	
	1	58.686	ns	34.333	ns
	2	63.513	*	31.286	ns
	3	62.664	*	33.000	ns
	4	65.095	*	32.877	ns
	5	61.074	ns	30.800	ns
	6	62.466	ns	31.984	ns

Means were calculated by the Kaplan–Meier method, and significance of the curves was determined by the Log-Rank method (Mantel-Cox). [1] ns: non-significant, * significant ($p < 0.05$); numbers 1–6 indicate the different dilutions tested (see Table 1).

3.2. In Vitro

3.2.1. Cytotoxicity

In general, the red and yellow additives showed a dose-dependent response, with an increase of the cytotoxicity level according to the increased concentration of the food coloring. All food colorings reached an inhibitory concentration 50 (IC_{50}) between the concentration numbered as 5 and 6 in HL-60 cells, except for Riboflavin that was the only dye able to induce total death of the tumor cells at the concentration numbered as 5.

In relation to the blue additives, Indigotine showed a slight (51%) growth inhibition at the highest concentration tested. No inhibition was observed for Brilliant Blue FCF at any concentration tested, with respect to the control but, contrarily, a tendency to promote cell growth was observed. The concentration numbered as 4 and 5 were the closest to the established ADI for Brilliant Blue FCF; we found that, although their viability-promoting effect was higher than the corresponding effect of the control, it was nonetheless lower than the effect of the other concentrations tested for this food coloring suggesting, their low chemopreventive potential in tumor cells (Figure 4).

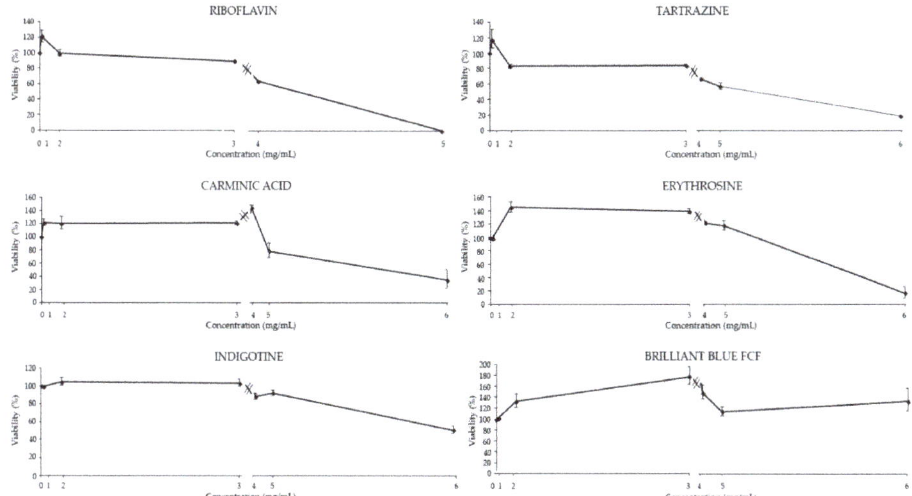

Figure 4. Effect of food coloring on HL-60 cells viability. Viability of the promyelocytic human leukemia cells (HL-60) treated with different concentrations of food colorings for 72 h. Each point represents the growing percentage with respect to its control. Values represent the mean ± SE from three independent experiment. Numbers 1–6 indicate the different dilutions tested (see Table 1).

3.2.2. DNA Fragmentation

Figure 5 shows electrophoresis experiments of the genomic DNA of HL-60 cells treated with different concentrations of food colorings. The results showed that the proapoptotic hallmark DNA internucleosomal fragmentation was only observed at the highest concentration of Riboflavin assayed. The rest of the food colorings assayed did not induce internucleosomal fragmentation.

3.2.3. Methylation Status

The relative normalized expression of three repetitive sequences (Alu M1, LINE-1, and Sat-α) studied in HL-60 cells treated with different concentrations of food colorings and RPMI as a control is shown in Figure 6. The food colorings did not modulate the methylation status at the assayed concentrations. After one-way ANOVA and post-hoc Tukey's test, the statistical results showed a methylation level in the treated samples similar to that of the normalized control.

Despite of the non-significant results in the methylation status for any food coloring tested, Riboflavin exhibited a tendency to hypomethylate the genomic randomly distributed sequences of HL-60 cells (Alu M1 and LINE-1). Taking into account that methylation of repetitive sequences is considered a genomic protective mechanism [70,74], this yellow additive could have inhibitory effects on tumor cells and could be an interesting chemopreventive compound.

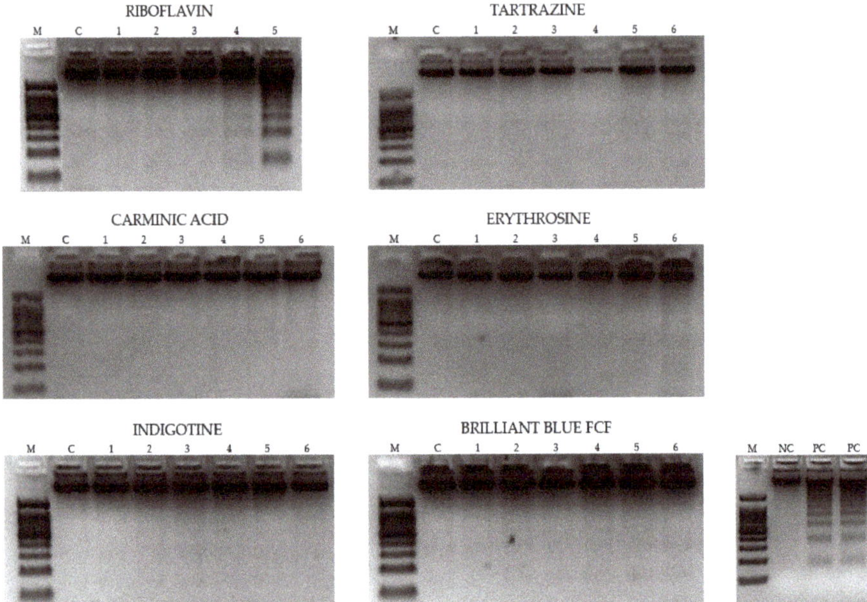

Figure 5. Internucleosomal DNA fragmentation in HL-60 cells. DNA-induced damage in promyelocytic HL-60 cells treated for 5 h with different concentrations of food colorings. M indicates the DNA size marker; NC indicates negative control treatment (RPMI); PC indicates positive control treatment (LBB). Numbers 1–6 indicate the different dilutions tested (see Table 1).

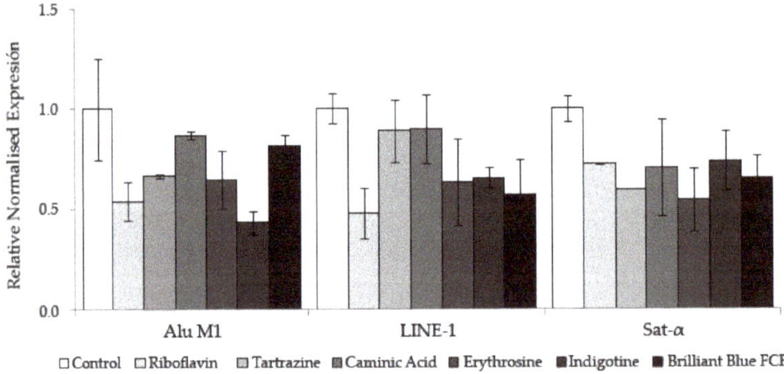

Figure 6. Methylation status of food colorings in HL-60 cells. Relative normalized expression data of each repetitive element (Alu M1, LINE-1, and Sat-α). Values represent the mean ± SE from three independent experiments. Untreated cells grown in RPMI were used as a control.

4. Discussion

4.1. In Vivo

According to the toxicity assay, none of the food colorings at any of the assayed concentrations reached the lethal dose 50 (LD_{50}) in *D. melanogaster*, which is considered as the toxic level for any substance. Our results are in agreement with the wide variety of researches showing the absence of toxic effects for Riboflavin [75], Tartrazine [11], Carminic Acid [24], Indigotine [76–78], and Brilliant Blue FCF [36,79,80] in mice, rats, rabbits, and dogs models. On the other hand, statistically significant severe adverse effects on the testis were described in a subacute toxicity study (45 day) performed on adult male Swiss albino mice treated with oral doses of Indigo of 0, 17, and 39 mg/kg·bw/day [34]. No data were found about Erythrosine toxicity.

Regarding the protective effects of food colorings, no previous data about antioxidative effects were found. Taking into account the concentration corresponding to the equivalent ADI for humans (concentration numbered as 3), no significant results were obtained for any food coloring tested. This fact is in agreement with the results of Scotter and Castle [81], who suggested that, in general, the majority of color additives are unstable in combination with oxidizing and reducing agents in food. Moreover, since color depends on the existence of a conjugated unsaturated system within a dye molecule, any substance which modifies this system (e.g., oxidizing or reducing agents, sugars, acids, and salts) will affect the color [81].

To our knowledge, no previous studies assessing the effects on lifespan and healthspan have been published. Our results indicated that the highest concentrations of Tartrazine and medium quantities of Brilliant Blue FCF induced a significant life extension with respect to the controls, whereas both red food colorings showed significantly negative effects on the longevity of *Drosophila*. Furthermore, quality of life was only improved by Tartrazine and, even, it worsened at some concentrations of Erythrosine and Indigotine.

On the whole, a non-visible dose–effect relationship for the food colorings could be appreciated in the different assays. This could be explained by the possible differential responses of the organism against each substance and by the biological level at which it was acting.

4.2. In Vitro

A dose-dependent cytotoxic effect was observed for the food colorings assayed in HL-60 cells, except for both types of blue dyes which did not reach the inhibitory concentration 50 (IC_{50}) or even increased tumor cells' growth. Our results fit with those that demonstrated that Tartrazine, Carminic Acid, and Erythrosine did not have any potential to induce tumor cells' growth. The available carcinogenicity studies have demonstrated that Tartrazine does not induce benign or malignant neoplasia [82,83]. Moreover, in a combined chronic toxicity/carcinogenicity study involving in utero exposure of Wistar rats to Carminic Acid, the general pattern of tumor incidence in the treated animals did not significantly differ statistically from those of the controls [84]. Besides, studies about Erythrosine treatments in mice [78], rats [85], and gerbils [86] showed no significant adverse effects of this food coloring. On the other hand, our Indigotine and Brilliant Blue FCF results are not in agreement with those of different researches that indicated no carcinogenetic and tumor potential of this food coloring: exposure of mice to Indigo did not demonstrate carcinogenic or toxic effects [77]; subcutaneous injections of 10 doses of Brilliant Blue FCF, 4 mg each, followed by 50 doses of 6 mg showed no tumor production after 78 weeks in mouse [35]. These controversial results may be due to differences in the organisms used, the cell line studied, or the range of concentrations tested. No data about Riboflavin cytotoxicity were found.

Effects on the DNA damage at the internucleosomal level in HL-60 cells did not appear in our study, with the exception of the highest concentration of Riboflavin. The Indigotine results are supported by in vitro studies using MCF-7 breast cancer cells [87] and the human colonic adenocarcinoma cell line (CaCo2 cells) [88], which demonstrated a lack of statistical significance in DNA damaging. Similar results were obtained in ddY male mice treated with Brilliant Blue FCF, showing not statistically significant increases in DNA damage in glandular stomach, colon, liver, kidney, urinary bladder, lung, brain, and bone marrow [11]. In contrast, our results for Tartrazine and Erythrosine are not in agreement with those of Sasaki, Kawaguchi, Kamaya, Ohshita, Kabasawa, Iwama, Taniguchi, and Tsuda [11] and Tsuda, et al. [89], who demonstrated the effect of Tartrazine on nuclear DNA electrophoretic migration in the mouse and the induction of DNA damage in the stomach at doses of 10 and 2000 mg/kg·bw without a dose–effect relationship, and a dose-related induction of DNA damage by Erythrosine in the glandular stomach, colon, and urinary bladder after oral administration of 100 mg/kg·bw and

2000 mg/kg·bw and in the lung following administration of 2000 mg/kg·bw in mice. To our knowledge, no previous results about the effects of Riboflavin and Carminic Acid on DNA damage were published.

Finally, no significant modification of the DNA methylation status was found compared with the control. This means that modifications of the DNA epigenome are not induced, which is in agreement with studies showing no chromosome aberrations upon Riboflavin [5], Tartrazine [90], and Erythrosine [91] treatments in Chinese hamster ovary cells, mice bone morrow cells, and Syrian Hamster Embryo, respectively.

To sum up, no beneficial effects of food coloring were shown in the different in vitro tests with tumor cells. These controversial data with respect to the current well-known data supporting the safe consumption of additives may be due to the variety of conditions used: cell line, in vitro conditions, range of concentrations, or even the tests conditions.

5. Conclusions

Additives are found in many types of food, and we often consume them unknowingly; therefore, it is very important to study the biological consequences of using food coloring. Nowadays, people are becoming more aware of the possible danger of these additives that have no nutritional value.

Two model systems (in vivo and in vitro) were used to carry out the different screening tests. *D. melanogaster* is a well-known insect with a large scientific history in biological sciences that has highly contributed to understanding developmental biology, evolutionary concepts, and, recently, toxicology [92–96]. The unique characteristics that *Drosophila* possesses, such as a rapid and short life cycle (10–12 days at 25 °C), reliability, cost-efficiency, easy maintenance and manipulation, and consistent genetic similarity to humans, make this eukaryote an ideal model organism [40,97]. On the other hand, the human model HL-60 cell line was originated from a female patient with acute myeloid leukemia [98]. The promyelocytic human leukemia cell line HL-60 is used worldwide for many toxicity and cancer scientific purposes [63].

In conclusion, and taking into account the concentration indicated as the equivalent ADI for humans, the in vivo toxicity assays showed safe effects for all food colorings, as shown by the fact that the LD_{50} was not achieved by any of the additives. Nevertheless, no significant differences were shown for any compound in the combined antitoxicity assays with respect to the controls, since they did not protect against oxidative damage by hydrogen peroxide. However, the longevity assays showed a differential behavior of the six food colorings, being Tartrazine and Brilliant Blue FCF the only colorants that significantly improved the longevity of *Drosophila*, whereas the red additives reduced significantly the lifespan of *Drosophila*. On the other hand, the in vitro results demonstrated that, despite the dose-dependent cytotoxic effects shown by the yellow and red additives, none of them reached the IC_{50} at their ADI concentration. Moreover, red and blue food colorings induced an increasing of tumor cell growth. Besides, no DNA damage was observed by the internucleosomal fragmentation apoptotic assay, and no methylation status modification was found for any food coloring. To our knowledge, this is the first time that an integrative study with a wide range of in vivo and in vitro screening tests has been carried out in the model systems *D. melanogaster* and HL-60 tumor cells with food colorings. Several checkpoints to evaluate the biological activity of such important food additives have been established at the molecular (DNA internucleosomal proapoptotic clastogenicity and epigenetic status), unicellular (cytotoxicity), and individual (toxicity, antitoxicity, and longevity) levels. Although more scientific researches are needed to understand the effects that these highly consumed additives could have on our health, these results represent the first step and may encourage additional studies.

On the whole and despite the safe use suggested by the different assays carried out with food colorings, the overall results would support the idea that a high chronic intake of these additives throughout the entire life is not advisable, and more research on the biological effects that different concentrations of food colorings could have in model systems is warranted.

Author Contributions: Conceptualization, Á.A.-M.; Formal analysis, T.M.-A.; Investigation, R.M.-A., M.M.-J., S.J.-G., and T.M.-A.; Methodology, R.M.-A. and T.M.-A.; Project administration, Á.A.-M. and T.M.-A.; Supervision, Á.A.-M. and T.M.-A.; Validation, Á.A.-M.; Writing – original draft, R.M.-A., Á.A.-M., and T.M.-A.

Funding: This research received no external funding.

Conflicts of Interest: The authors declare no conflict of interest.

References

1. Food and Drug Administration (FDA). Available online: https://www.fda.gov/1 (accessed on 1 May 2019).
2. Swaroop, V.; Roy, D.D.; Vijayakumar, T. Genotoxicity of synthetic food colorants. *J. Food Sci. Eng.* **2011**, *1*, 128.
3. Juhlin, L.; Michaëlsson, G.; Zetterström, O. Urticaria and asthma induced by food-and-drug additives in patients with aspirin hypersensitivity. *J. Allergy Clin. Immunol.* **1972**, *50*, 92–98. [CrossRef]
4. Vedanthan, P.K.; Menon, M.M.; Bell, T.D.; Bergin, D. Aspirin and tartrazine oral challenge: Incidence of adverse response in chronic childhood asthma. *J. Allergy Clin. Immunol.* **1977**, *60*, 8–13. [CrossRef]
5. EFSA. Scientific Opinion on the Re-Evaluation of Riboflavin (e 101 (i)) and Riboflavin 5′ Phosphate Sodium (e 101 (ii)) as Food Additives. *EFSA J.* **2013**, *11*, 3357.
6. Elmadfa, I. *European Nutrition and Health Report 2009*; Karger Medical and Scientific Publishers: Vienna, Austria, 2009; Volume 62.
7. Kale, H.; Harikumar, P.; Kulkarni, S.; Nair, P.; Netrawali, M. Assessment of the genotoxic potential of riboflavin and lumiflavin: B. Effect of light. *Mutat. Res. Genet. Toxicol.* **1992**, *298*, 17–23. [CrossRef]
8. Unna, K.; Greslin, J.G. Studies on the toxicity and pharmacology of riboflavin. *J. Pharmacol. Exp. Ther.* **1942**, *76*, 75–80.
9. Komisyonu, A. European parliament and council directive 94/36/ec of 30 june 1994 on colours for use in foodstuffs. *Off. J. Eur. Union L* **1994**, *237*, 13–29.
10. EFSA. Scientific opinion on the re-evaluation tartrazine (e 102). *EFSA J.* **2009**, *7*, 1331. [CrossRef]
11. Sasaki, Y.F.; Kawaguchi, S.; Kamaya, A.; Ohshita, M.; Kabasawa, K.; Iwama, K.; Taniguchi, K.; Tsuda, S. The comet assay with 8 mouse organs: Results with 39 currently used food additives. *Mutat. Res. Genet. Toxicol. Environ. Mutagenesis* **2002**, *519*, 103–119. [CrossRef]
12. McCann, D.; Barrett, A.; Cooper, A.; Crumpler, D.; Dalen, L.; Grimshaw, K.; Kitchin, E.; Lok, K.; Porteous, L.; Prince, E. Food additives and hyperactive behaviour in 3-year-old and 8/9-year-old children in the community: A randomised, double-blinded, placebo-controlled trial. *Lancet* **2007**, *370*, 1560–1567. [CrossRef]
13. Novembre, E.; Dini, L.; Bernardini, R.; Resti, M.; Vierucci, A. Unusual reactions to food additives. *Pediatr. Med. Chir. Med. Surg. Pediatrics* **1992**, *14*, 39–42.
14. Elhkim, M.O.; Héraud, F.; Bemrah, N.; Gauchard, F.; Lorino, T.; Lambré, C.; Frémy, J.M.; Poul, J.M. New considerations regarding the risk assessment on tartrazine: An update toxicological assessment, intolerance reactions and maximum theoretical daily intake in france. *Regul. Toxicol. Pharmacol.* **2007**, *47*, 308–316. [CrossRef]
15. Bhatia, M.S. Allergy to tartrazine in psychotropic drugs. *J. Clin. Psychiatry* **2000**, *61*, 473–476. [CrossRef] [PubMed]
16. Nettis, E.; Colanardi, M.; Ferrannini, A.; Tursi, A. Suspected tartrazine-induced acute urticaria/angioedema is only rarely reproducible by oral rechallenge. *Clin. Exp. Allergy* **2003**, *33*, 1725–1729. [CrossRef] [PubMed]
17. Worm, M.; Vieth, W.; Ehlers, I.; Sterry, W.; Zuberbier, T. Increased leukotriene production by food additives in patients with atopic dermatitis and proven food intolerance. *Clin. Exp. Allergy* **2001**, *31*, 265–273. [CrossRef] [PubMed]
18. Inomata, N.; Osuna, H.; Fujita, H.; Ogawa, T.; Ikezawa, Z. Multiple chemical sensitivities following intolerance to azo dye in sweets in a 5-year-old girl. *Allergol. Int.* **2006**, *55*, 203–205. [CrossRef]
19. Wüthrich, B.; Kägi, M.; Stücker, W. Anaphylactic reactions to ingested carmine (e120). *Allergy* **1997**, *52*, 1133–1137. [CrossRef] [PubMed]
20. EFSA. Scientific opinion on the re-evaluation of cochineal, carminic acid, carmines (e 120) as a food additive. *EFSA J.* **2015**, *13*, 4288.
21. Sarıkaya, R.; Selvi, M.; Erkoç, F. Evaluation of potential genotoxicity of five food dyes using the somatic mutation and recombination test. *Chemosphere* **2012**, *88*, 974–979. [CrossRef]

22. DiCello, M.C.; Myc, A.; Baker, J.R., Jr.; Baldwin, J.L. *Anaphylaxis after Ingestion of Carmine Colored Foods: Two Case Reports and a Review of the Literature, Allergy and Asthma Proceedings*; OceanSide Publications: East Providence, RI, USA, 1999.

23. Ferrer, Đ.; Marco, F.M.; Andreu, C.; Sempere, J.M. Occupational asthma to carmine in a butcher. *Int. Arch. Allergy Immunol.* **2005**, *138*, 243–250. [CrossRef]

24. Helal, E.G.; Zaahkouk, S.A.; Mekkawy, H.A. Effect of some food colorants (synthetic and natural products) of young albino rats. I–liver and kidney functions. *Egypt. J. Hosp. Med.* **2000**, *1*, 103–113.

25. Chequer, F.M.D.; de Paula Venâncio, V.; Bianchi, M.d.L.P.; Antunes, L.M.G. Genotoxic and mutagenic effects of erythrosine b, a xanthene food dye, on hepg2 cells. *Food Chem. Toxicol.* **2012**, *50*, 3447–3451. [CrossRef] [PubMed]

26. Tuormaa, T.E. The adverse effects of food additives on health: A review of the literature with a special emphasis on childhood hyperactivity. *J. Orthomol. Med.* **1994**, *9*, 225.

27. Mpountoukas, P.; Pantazaki, A.; Kostareli, E.; Christodoulou, P.; Kareli, D.; Poliliou, S.; Mourelatos, C.; Lambropoulou, V.; Lialiaris, T. Cytogenetic evaluation and DNA interaction studies of the food colorants amaranth, erythrosine and tartrazine. *Food Chem. Toxicol.* **2010**, *48*, 2934–2944. [CrossRef] [PubMed]

28. EFSA. Scientific opinion on the re-evaluation of erythrosine (e 127) as a food additive. *EFSA J.* **2011**, *9*, 1854. [CrossRef]

29. Ganesan, L.; Margolles-Clark, E.; Song, Y.; Buchwald, P. The food colorant erythrosine is a promiscuous protein–protein interaction inhibitor. *Biochem. Pharmacol.* **2011**, *81*, 810–818. [CrossRef] [PubMed]

30. Hagiwara, M.; Watanabe, E.; Barrett, J.C.; Tsutsui, T. Assessment of genotoxicity of 14 chemical agents used in dental practice: Ability to induce chromosome aberrations in syrian hamster embryo cells. *Mutat. Res. Genet. Toxicol. Environ. Mutagenesis* **2006**, *603*, 111–120. [CrossRef]

31. Mekkawy, H.A.; Massoud, A.; El-Zawahry, A. Mutagenic effects of the food color erythrosine in rats. *Probl. Forensic Sci.* **2000**, *43*, 184–191.

32. Steingruber, E. Indigo and indigo colorants. In *Ullmann's Encyclopedia of Industrial Chemistry*; Wiley-VCH Verlag GmbH & Co. KGaA: Weinheim, Germany, 2000.

33. EFSA. Scientific opinion on the re-evaluation of indigo carmine (e 132) as a food additive. *EFSA J.* **2014**, *12*, 3768. [CrossRef]

34. Dixit, A.; Goyal, R. Evaluation of reproductive toxicity caused by indigo carmine on male swiss albino mice. *Pharmacology* **2013**, *1*, 218–224.

35. EFSA. Scientific opinion on the re-evaluation of brilliant blue fcf (e 133) as a food additive. *EFSA J.* **2010**, *8*, 1853. [CrossRef]

36. Aboel-Zahab, H.; El-Khyat, Z.; Sidhom, G.; Awadallah, R.; Abdel-Al, W.; Mahdy, K. Physiological effects of some synthetic food colouring additives on rats. *Boll. Chim. Farm.* **1997**, *136*, 615–627. [PubMed]

37. Lau, K.; McLean, W.G.; Williams, D.P.; Howard, C.V. Synergistic interactions between commonly used food additives in a developmental neurotoxicity test. *Toxicol. Sci.* **2005**, *90*, 178–187. [CrossRef] [PubMed]

38. Borzelleca, J.; Depukat, K.; Hallagan, J. Lifetime toxicity/carcinogenicity studies of fd & c blue no. 1 (brilliant blue fcf) in rats and mice. *Food Chem. Toxicol.* **1990**, *28*, 221–234.

39. Lucarelli, M.R.; Shirk, M.B.; Julian, M.W.; Crouser, E.D. Toxicity of food drug and cosmetic blue no. 1 dye in critically ill patients. *Chest* **2004**, *125*, 793–795. [CrossRef]

40. Bier, E. *Drosophila*, the golden bug, emerges as a tool for human genetics. *Nat. Rev. Genet.* **2005**, *6*, 9. [CrossRef]

41. Graf, U.; Würgler, F.; Katz, A.; Frei, H.; Juon, H.; Hall, C.; Kale, P. Somatic mutation and recombination test in *Drosophila melanogaster*. *Environ. Mutagenesis* **1984**, *6*, 153–188. [CrossRef]

42. Ja, W.W.; Carvalho, G.B.; Mak, E.M.; de la Rosa, N.N.; Fang, A.Y.; Liong, J.C.; Brummel, T.; Benzer, S. Prandiology of *Drosophila* and the cafe assay. *Proc. Natl. Acad. Sci. USA* **2007**, *104*, 8253–8256. [CrossRef]

43. Gonzalez, C. *Drosophila melanogaster*: A model and a tool to investigate malignancy and identify new therapeutics. *Nat. Rev. Cancer* **2013**, *13*, 172–183. [CrossRef]

44. Lints, F.A.; Soliman, M.H. *Drosophila as a Model Organism for Ageing Studies*; Springer: Boston, MA, USA, 1988.

45. Rudrapatna, V.A.; Cagan, R.L.; Das, T.K. *Drosophila* cancer models. *Dev. Dyn.* **2012**, *241*, 107–118. [CrossRef]

46. Yan, J.; Huen, D.; Morely, T.; Johnson, G.; Gubb, D.; Roote, J.; Adler, P.N. The multiple-wing-hairs gene encodes a novel gbd-fh3 domain-containing protein that functions both prior to and after wing hair initiation. *Genetics* **2008**, *180*, 219–228. [CrossRef] [PubMed]

47. Ren, N.; Charlton, J.; Adler, P.N. The flare gene, which encodes the aip1 protein of *Drosophila*, functions to regulate f-actin disassembly in pupal epidermal cells. *Genetics* **2007**, *176*, 2223–2234. [CrossRef] [PubMed]

48. Anter, J.; Campos-Sanchez, J.; Hamss, R.E.; Rojas-Molina, M.; Munoz-Serrano, A.; Analla, M.; Alonso-Moraga, A. Modulation of genotoxicity by extra-virgin olive oil and some of its distinctive components assessed by use of the *Drosophila* wing-spot test. *Mutat. Res.* **2010**, *703*, 137–142. [CrossRef] [PubMed]

49. Merinas-Amo, T.; Tasset-Cuevas, I.; Díaz-Carretero, A.M.; Alonso-Moraga, A.; Calahorro, F. Role of choline in the modulation of degenerative processes: In vivo and in vitro studies. *J. Med. Food* **2017**, *20*, 223–234. [CrossRef]

50. López, A.; Xamena, N.; Marcos, R.; Velázquez, A. Germ cells microsatellite instability: The effect of different mutagens in a mismatch repair mutant of *Drosophila* (spel1). *Mutat. Res. Genet. Toxicol. Environ. Mutagenesis* **2002**, *514*, 87–94. [CrossRef]

51. Allen, R.; Tresini, M. Oxidative stress and gene regulation. *Free Radic. Biol. Med.* **2000**, *28*, 463–499. [CrossRef]

52. Veal, E.A.; Day, A.M.; Morgan, B.A. Hydrogen peroxide sensing and signaling. *Mol. Cell* **2007**, *26*, 1–14. [CrossRef] [PubMed]

53. Fitzpatrick, F.A. Inflammation, carcinogenesis and cancer. *Int. Immunopharmacol.* **2001**, *1*, 1651–1667. [CrossRef]

54. Burcham, P.C. *Genotoxic Lipid Peroxidation Products: Their DNA Damaging Properties and Role in Formation of Endogenous DNA Adducts*; Oxford University Press: Oxford, UK, 1998.

55. Feig, D.I.; Reid, T.M.; Loeb, L.A. Reactive oxygen species in tumorigenesis. *Cancer Res.* **1994**, *54*, 1890–1894.

56. Vivancos, A.P.; Castillo, E.A.; Biteau, B.; Nicot, C.; Ayte, J.; Toledano, M.B.; Hidalgo, E. A cysteine-sulfinic acid in peroxiredoxin regulates h2o2-sensing by the antioxidant pap1 pathway. *Proc. Natl. Acad. Sci. USA* **2005**, *102*, 8875–8880. [CrossRef]

57. Cerda, S.; Weitzman, S. Influence of oxygen radical injury on DNA methylation. *Mutat. Res. Rev. Mutat. Res.* **1997**, *386*, 141–152. [CrossRef]

58. Ghosh, R.; Mitchell, D.L. Effect of oxidative DNA damage in promoter elements on transcription factor binding. *Nucleic Acids Res.* **1999**, *27*, 3213–3218. [CrossRef] [PubMed]

59. Hu, J.J.; Dubin, N.; Kurland, D.; Ma, B.L.; Roush, G.C. The effects of hydrogen peroxide on DNA repair activities. *Mutat. Res.* **1995**, *336*, 193–201. [CrossRef]

60. Romero-Jiménez, M.; Campos-Sánchez, J.; Analla, M.; Muñoz-Serrano, A.; Alonso-Moraga, Á. Genotoxicity and anti-genotoxicity of some traditional medicinal herbs. *Mutat. Res. Genet. Toxicol. Environ. Mutagenesis* **2005**, *585*, 147–155. [CrossRef] [PubMed]

61. Tasset-Cuevas, I.; Fernandez-Bedmar, Z.; Lozano-Baena, M.D.; Campos-Sanchez, J.; de Haro-Bailon, A.; Munoz-Serrano, A.; Alonso-Moraga, A. Protective effect of borage seed oil and gamma linolenic acid on DNA: In vivo and in vitro studies. *PLoS ONE* **2013**, *8*, e56986. [CrossRef] [PubMed]

62. Soh, J.W.; Hotic, S.; Arking, R. Dietary restriction in *Drosophila* is dependent on mitochondrial efficiency and constrained by pre-existing extended longevity. *Mech. Ageing Dev.* **2007**, *128*, 581–593. [CrossRef] [PubMed]

63. Gallagher, R.; Collins, S.; Trujillo, J.; McCredie, K.; Ahearn, M.; Tsai, S.; Metzgar, R.; Aulakh, G.; Ting, R.; Ruscetti, F.; et al. Characterization of the continuous, differentiating myeloid cell line (hl-60) from a patient with acute promyelocytic leukemia. *Blood* **1979**, *54*, 713–733.

64. Merinas-Amo, T.; Tasset-Cuevas, I.; Díaz-Carretero, A.M.; Alonso-Moraga, Á.; Calahorro, F. In vivo and in vitro studies of the role of lyophilised blond lager beer and some bioactive components in the modulation of degenerative processes. *J. Funct. Foods* **2016**, *27*, 274–294. [CrossRef]

65. Merinas-Amo, T.; Merinas-Amo, R.; Alonso-Moraga, A. A clinical pilot assay of beer consumption: Modulation in the methylation status patterns of repetitive sequences. *Sylwan* **2017**, *161*, 134–156.

66. Deininger, P.L.; Moran, J.V.; Batzer, M.A.; Kazazian, H.H. Mobile elements and mammalian genome evolution. *Curr. Opin. Genet. Dev.* **2003**, *13*, 651–658. [CrossRef]

67. Ehrlich, M. DNA hypomethylation, cancer, the immunodeficiency, centromeric region instability, facial anomalies syndrome and chromosomal rearrangements. *J. Nutr.* **2002**, *132*, 2424S–2429S. [CrossRef] [PubMed]

68. Lee, C.; Wevrick, R.; Fisher, R.; Ferguson-Smith, M.; Lin, C. Human centromeric dnas. *Hum. Genet.* **1997**, *100*, 291–304. [CrossRef] [PubMed]

69. Weiner, A.M. Sines and lines: The art of biting the hand that feeds you. *Curr. Opin. Cell Biol.* **2002**, *14*, 343–350. [CrossRef]

70. Weisenberger, D.J.; Campan, M.; Long, T.I.; Kim, M.; Woods, C.; Fiala, E.; Ehrlich, M.; Laird, P.W. Analysis of repetitive element DNA methylation by methylight. *Nucleic Acids Res.* **2005**, *33*, 6823–6836. [CrossRef]

71. Nikolaidis, G.; Raji, O.Y.; Markopoulou, S.; Gosney, J.R.; Bryan, J.; Warburton, C.; Walshaw, M.; Sheard, J.; Field, J.K.; Liloglou, T. DNA methylation biomarkers offer improved diagnostic efficiency in lung cancer. *Cancer Res.* **2012**, *72*, 5692–5701. [CrossRef]

72. Liloglou, T.; Bediaga, N.G.; Brown, B.R.; Field, J.K.; Davies, M.P. Epigenetic biomarkers in lung cancer. *Cancer Lett.* **2014**, *342*, 200–212. [CrossRef] [PubMed]

73. Anter, J.; Romero-Jimenez, M.; Fernandez-Bedmar, Z.; Villatoro-Pulido, M.; Analla, M.; Alonso-Moraga, A.; Munoz-Serrano, A. Antigenotoxicity, cytotoxicity, and apoptosis induction by apigenin, bisabolol, and protocatechuic acid. *J. Med. Food* **2011**, *14*, 276–283. [CrossRef]

74. Roman-Gomez, J.; Jimenez-Velasco, A.; Agirre, X.; Castillejo, J.A.; Navarro, G.; San Jose-Eneriz, E.; Garate, L.; Cordeu, L.; Cervantes, F.; Prosper, F. Repetitive DNA hypomethylation in the advanced phase of chronic myeloid leukemia. *Leuk. Res.* **2008**, *32*, 487–490. [CrossRef] [PubMed]

75. Fukuwatari, T.; Kuzuya, M.; Satoh, S.; Shibata, K. Effects of excess vitamin b1 or vitamin b2 on growth and urinary excretion of water-soluble vitamins in weaning rats. *Shokuhin Eiseigaku Zasshi J. Food Hyg. Soc. Jpn.* **2009**, *50*, 70–74. [CrossRef]

76. Oettel, H.; Frohberg, H.; Nothdurft, H.; Wilhelm, G. Die prüfung einiger synthetischer farbstoffe auf ihre eignung zur lebensmittelfärbung. *Arch. Toxicol.* **1965**, *21*, 9–29.

77. Borzelleca, J.; Hogan, G. Chronic toxicity/carcinogenicity study of fd & c blue no. 2 in mice. *Food Chem. Toxicol.* **1985**, *23*, 719–722. [PubMed]

78. Borzelleca, J.; Hallagan, J. Lifetime toxicity/carcinogenicity study of fd & c red no. 3 (erythrosine) in mice. *Food Chem. Toxicol.* **1987**, *25*, 735–737. [PubMed]

79. Tsujita, J. Comparison of protective activity of dietafy fiber against the toxicities of various food colors in rats. *Nutr. Rep. Int.* **1979**, *20*, 635–642.

80. Hansen, W.; Fitzhugh, O.; Nelson, A.; Davis, K. Chronic toxicity of two food colors, brilliant blue fcf and indigotine. *Toxicol. Appl. Pharmacol.* **1966**, *8*, 29–36. [CrossRef]

81. Scotter, M.; Castle, L. Chemical interactions between additives in foodstuffs: A review. *Food Addit. Contam.* **2004**, *21*, 93–124. [CrossRef] [PubMed]

82. Maekawa, A.; Matsuoka, C.; Onodera, H.; Tanigawa, H.; Furuta, K.; Kanno, J.; Jang, J.; Hayashi, Y.; Ogiu, T. Lack of carcinogenicity of tartrazine (fd & c yellow no. 5) in the f344 rat. *Food Chem. Toxicol.* **1987**, *25*, 891–896.

83. Moutinho, I.; Bertges, L.; Assis, R. Prolonged use of the food dye tartrazine (fd&c yellow n° 5) and its effects on the gastric mucosa of wistar rats. *Braz. J. Biol.* **2007**, *67*, 141–145.

84. Ford, G.; Gopal, T.; Grant, D.; Gaunt, I.; Evans, J.; Butler, W. Chronic toxicity/carcinogenicity study of carmine of cochineal in the rat. *Food Chem. Toxicol.* **1987**, *25*, 897–902. [CrossRef]

85. Hiasa, Y.; Ohshima, M.; Kitahori, Y.; Konishi, N.; Shimoyama, T.; Sakaguchi, Y.; Hashimoto, H.; Minami, S.; Kato, Y. The promoting effects of food dyes, erythrosine (red 3) and rose bengal b (red 105), on thyroid tumors in partially thyroidectomized n-bis (2-hydroxypropyl)-nitrosamine-treated rats. *Jpn. J. Cancer Res.* **1988**, *79*, 314–319. [CrossRef]

86. Collins, T.; Long, E. Effects of chronic oral administration of erythrosine in the mongolian gerbil. *Food Cosmet. Toxicol.* **1976**, *14*, 233–248. [CrossRef]

87. Masannat, Y.A.; Hanby, A.; Horgan, K.; Hardie, L.J. DNA damaging effects of the dyes used in sentinel node biopsy: Possible implications for clinical practice. *J. Surg. Res.* **2009**, *154*, 234–238. [CrossRef]

88. Davies, J.; Burke, D.; Olliver, J.; Hardie, L.; Wild, C.; Routledge, M. *The Induction of DNA Damage by Methylene Blue but Not by Indigo Carmine in Human Colonocytes In Vitro and In Vivo, Mutagenesis*; Oxford Univ Press Great Clarendon St: Oxford, UK, 2006.

89. Tsuda, S.; Murakami, M.; Matsusaka, N.; Kano, K.; Taniguchi, K.; Sasaki, Y.F. DNA damage induced by red food dyes orally administered to pregnant and male mice. *Toxicol. Sci.* **2001**, *61*, 92–99. [CrossRef]

90. Durnev, A.; Oreshchenko, A.; Kulakova, A. Analysis of cytogenetic activity of food dyes. *Vopr. Meditsinskoi khimii* **1995**, *41*, 50–53.

91. Miyachi, T.; Tsutsui, T. Ability of 13 chemical agents used in dental practice to induce sister-chromatid exchanges in syrian hamster embryo cells. *Odontology* **2005**, *93*, 24–29. [CrossRef] [PubMed]

92. Bhargav, D.; Singh, M.P.; Murthy, R.C.; Mathur, N.; Misra, D.; Saxena, D.K.; Chowdhuri, D.K. Toxic potential of municipal solid waste leachates in transgenic *Drosophila melanogaster* (hsp70-lacz): Hsp70 as a marker of cellular damage. *Ecotoxicol. Environ. Saf.* **2008**, *69*, 233–245. [CrossRef] [PubMed]

93. Coulom, H.; Birman, S. Chronic exposure to rotenone models sporadic parkinson's disease in *Drosophila melanogaster*. *J. Neurosci.* **2004**, *24*, 10993–10998. [CrossRef]

94. Dean, B.J. Recent findings on the genetic toxicology of benzene, toluene, xylenes and phenols. *Mutat. Res. Genet. Toxicol.* **1985**, *154*, 153–181. [CrossRef]

95. Hosamani, R. Acute exposure of *Drosophila melanogaster* to paraquat causes oxidative stress and mitochondrial dysfunction. *Arch. Insect Biochem. Physiol.* **2013**, *83*, 25–40. [CrossRef]

96. Siddique, Y.H.; Fatima, A.; Jyoti, S.; Naz, F.; Khan, W.; Singh, B.R.; Naqvi, A.H. Evaluation of the toxic potential of graphene copper nanocomposite (gcnc) in the third instar larvae of transgenic *rDosophila melanogaster* (hsp70-lacz) bg 9. *PLoS ONE* **2013**, *8*, e80944. [CrossRef]

97. Lloyd, T.E.; Taylor, J.P. Flightless flies: *Drosophila* models of neuromuscular disease. *Ann. N. Y. Acad. Sci.* **2010**, *1184*. [CrossRef]

98. Collins, S.J. The hl-60 promyelocytic leukemia cell line: Proliferation, differentiation, and cellular oncogene expression. *Blood* **1987**, *70*, 1233–1244. [PubMed]

© 2019 by the authors. Licensee MDPI, Basel, Switzerland. This article is an open access article distributed under the terms and conditions of the Creative Commons Attribution (CC BY) license (http://creativecommons.org/licenses/by/4.0/).

Article

Food Safety and Nutraceutical Potential of Caramel Colour Class IV Using In Vivo and In Vitro Assays

Marcos Mateo-Fernández [1,*], Pilar Alves-Martínez [1], Mercedes Del Río-Celestino [2], Rafael Font [2], Tania Merinas-Amo [1] and Ángeles Alonso-Moraga [1]

[1]   Department of Genetic, Rabanales Campus, University of Córdoba (UCO), 14071 Córdoba, Spain
[2]   Agri-Food Laboratory, Council of Agriculture, Fisheries and Rural Development of Andalusia (CAPDER), 14004 Córdoba, Spain
*   Correspondence: mercedes.rio.celestino@juntadeandalucia.es

Received: 28 June 2019; Accepted: 23 August 2019; Published: 5 September 2019

Abstract: Nutraceutical activity of food is analysed to promote the healthy characteristics of diet where additives are highly used. Caramel is one of the most worldwide consumed additives and it is produced by heating natural carbohydrates. The aim of this study was to evaluate the food safety and the possible nutraceutical potential of caramel colour class IV (CAR). For this purpose, in vivo toxicity/antitoxicity, genotoxicity/antigenotoxicity and longevity assays were performed using the *Drosophila melanogaster* model. In addition, cytotoxicity, internucleosomal DNA fragmentation, single cell gel electrophoresis and methylation status assays were conducted in the in vitro HL-60 human leukaemia cell line. Our results reported that CAR was neither toxic nor genotoxic and showed antigenotoxic effects in *Drosophila*. Furthermore, CAR induced cytotoxicity and hipomethylated sat-α repetitive element using HL-60 cell line. In conclusion, the food safety of CAR was demonstrated, since Lethal Dose 50 (LD_{50}) was not reached in toxicity assay and any of the tested concentrations induced mutation rates higher than that of the concurrent control in *D. melanogaster*. On the other hand, CAR protected DNA from oxidative stress provided by hydrogen peroxide in *Drosophila*. Moreover, CAR showed chemopreventive activity and modified the methylation status of HL-60 cell line. Nevertheless, much more information about the mechanisms of gene therapies related to epigenetic modulation by food is necessary.

Keywords: caramel colour E150d-Class IV (CAR); nutraceutical potential; food safety

---

1. Introduction

From the beginnings, humanity has been searching for different methods in order to improve our feeding. Adding molecules to increase the flavour or to get a better preservation of food is one of these methods. Its consequence is the appearance of a problem: the quality of food which has been altered with additives. According to the "Codex Alimentary", an additive is "any substance not normally consumed as a food by itself and not normally used as a typical ingredient of the food, whether or not it has nutritive value, the intentional addition of which to food for a technological (including organoleptic) purpose in the manufacture, processing, preparation, treatment, packing, transport or holding of such food results, or may be reasonably expected to result (directly or indirectly), in it or its by-products becoming a component of or otherwise affecting the characteristics of such foods. The term does not include contaminants or substances added to food for maintaining or improving nutritional qualities".

Natural food additives, such as salt, vinegar, wine and spices have been extensively used in order to preserve foods and improve their organoleptic properties. However, synthetic additives are the most used nowadays. The food additives consumption is regulated for maintaining the quality and health security of food. New additives must undergo toxicity and carcinogenesis studies before entering in the market. These in vivo assays are performed in order to determine the carcinogenic potential of these compounds. This is the reason why some new additives are safer than other compounds used for years. With this in mind, coal tar has been used since 1956, although carcinogenic risks for the consumers were noticed only when researchers began studying it in in vivo assays [1], and more recently, studies have showed that some molecules incorporated in food could cause cancer [2]. For this reason, a food safety evaluation is needed [3].

It is known that dietary compounds are related to the induction or prevention of several diseases. As a proof of that, patients from developing and underdeveloped countries suffer from different kinds of cancer that could be related to the increase of the food additives consumption [1]. Therefore, the relationship between diet and health is very close [4], and food genotoxicologic assays have been widely used through time in order to evaluate the healthy properties of diet before being considered as nutraceutic substances [5,6]. Nowadays, there is a great deal of scientific evidence based on nutraceuticals, supporting the idea that a deliberate consume of certain food can be a health promoter.

Caramel is one of the most worldwide consumed additives and is produced by heating carbohydrates from vegetable sources (glucose, sucrose, invert sugar, etc.) in the presence of caramelisation promoters (ammonia or ammonium in class III and IV, respectively). The result is a complex mixture which is responsible for the aromatic and colourant characters of caramels [7,8], and it has been used in food and beverages to provide some properties such as colour, taste, smell and texture, and for its ability for stabilisation of colloidal systems, as well as for its emulsifying properties, facilitating the dispersion of water-insoluble materials, retarding flavour changes and preserving the shelf-life of beverages exposed to light [9]. Besides these, caramel has recently also been highlighted as beneficial in nonenzymatic browning inhibition [10].

Drosophila is a reliable model to evaluate the toxicity, genotoxicity and other degenerative processes of food or chemical structures [11]. The results obtained in this eukaryote organism are considered translational and highly specific, as more than 80% of genes related to human disease are homologous in *Drosophila* [12]. The SMART (Somatic Mutation and Recombination Test) is a useful tool for in vivo genetic studies with *D. melanogaster*. The SMART is based on the genetic alterations produced in the cells of imaginal discs of larvae. These alterations can phenotypically be expressed in adult tissues after clonal expansion and metamorphosis. The SMART has shown to be able to detect genotoxic activity of various compounds with different chemical structures, either as direct mutagens or promutagens, and with different genotoxic action modes, such as alkylating, intercalating agents, adducts formers [11] as well as complex mixtures [13–15]. The detection of genotoxic and antigenotoxic agents is important since they can be considered as carcinogenic or anticarcinogenic substances, respectively.

In addition, *Drosophila melanogaster* is an excellent model for the study of aging, because adults show many similarities with the cellular senescence observed in mammals [16]. This is the reason why this particular model is frequently used to understand the relationship between nutrient metabolism and aging mechanisms [17], and further substantial contributions in this sense are expected [18].

In parallel, in vitro cytotoxicity assays are used to assess the chemopreventive potential of compounds [19]. Taking into account that cancer therapies are highly toxic and mainly unspecific, an alternative strategy could be the use of agents able to induce cell differentiation in cancerous cells [20]. HL-60 human leukaemia cell line is widely used to detect the capability of inducing cell differentiation and proapoptotic mechanisms of the compounds to be tested. These compounds could be considered as chemopreventive agents [21,22].

DNA methylation is an epigenetic mark that shows the transcriptional gene silencing and that plays a vital role during development and in the genome defence against transposable elements [23]. The methylation status of these transposable sequences is relevant for understanding the global DNA methylation. Prevention or reversal of hypermethylation-induced inactivation of tumour suppression genes or gene receptors by DNA Methyltransferase (DNMT) inhibitors could be an effective approach to cancer prevention [24].

Taking that into account, environmental exposures to nutritional, chemical and physical factors (such as smoke, diet and physical activity) could alter human and animal gene expression and modify the susceptibility to disease due to epigenomic changes [25]. Currently, biomedical research is focused on modifying the methylation pattern as a tool to understand cancer processes and others diseases. The ability of food compounds to influence the epigenome in cancer cells has been studied and has also been related to the individual's risk of developing cancer. Since epigenetic changes can be reversed in the human lifespan, the epigenetic focus is a good tool for the dietary prevention/treatment of cancers.

In vivo toxicity, antitoxicity, genotoxicity, antigenotoxiciy and lifespan assays using the model organism *Drosophila melanogaster*, and in vitro cytotoxicity, DNA fragmentation, methylation status and comet assays using HL-60 promyelocytic human cell line were carried out to evaluate some biological activities related to degenerative processes, food security and nutraceutic potential of CAR (caramel colour E150d-class IV).

2. Materials and Methods

2.1. Samples

A colour additive, caramel colour E150d-class IV (CAR), was selected for this study and was kindly provided by SANCOLOR (Barcelona, Spain).

2.2. In Vivo Fly Stocks

Two *Drosophila melanogaster* strains with genetic markers that affect the wing-hair phenotype were used: (i) *mwh/mwh*, carrying the recessive mutation *mwh* (multiple wing hairs) [26] and (ii) *flr3/In (3LR) TM3, rip^p se^p bx^{34e} e^s Bd^S*, where *flr^3* (*flare*) [27] marker is a homozygous recessive lethal mutation which is viable in homozygous somatic cells once larvae start developing and produce deformed trichomonas.

2.3. In Vitro Cell Culture Conditions

Promyelocytic human leukaemia (HL-60) cells were grown in RPMI-1640 medium (R5886, Sigma-Aldrich, St. Louis, MO, USA) supplemented with heat-inactivated foetal bovine serum (Linus, S01805, Madrid, Spain), L-glutamine 200 mM (G7513, Sigma-Aldrich, St. Louis, MO, USA) and 1× antibiotic–antimycotic solution (A5955, Sigma-Aldrich, St. Louis, MO, USA). Cells were incubated at 37 °C in a humidified atmosphere of 5% CO_2. Cultures were plated at 2.5×10^4 cells/mL density in 10 mL culture bottles and passed every 2 days.

2.4. In Vivo Safety Studies

2.4.1. Toxicity Assays

Toxicity was assayed according to our standard protocols. Before carrying out the assays, colour caramel was dissolved in distilled water in order to obtain the different concentrations as follows: 0.03, 0.125, 0.25, 1 and 4 mg/mL for CAR. The CAR concentrations range was calculated in order to assay the same amounts that are contained in the cola beverages [28]. A negative (H_2O) concurrent control was also assayed. Test groups consisted of larvae fed with *Drosophila* Instant Medium (Formula 4-24, Carolina Biological Supply, Burlington, NC, USA) supplemented with the tested compounds concentrations. Emerging adults of all groups were counted, and toxicity was determined as the percentage of hatched individuals in each treatment compared with the negative control.

Chi-square test was used to determine if the tested compounds significantly affected the survival of flies, as it was previously described by Mateo-Fernández, et al. [15], where an in parallel lethal dose 50 (LD_{50}) was performed to ascertain the toxicity of the compounds.

2.4.2. Genotoxicity Assay

Genotoxicity assays were carried out following the wing spot test standard procedure [11]. Briefly, transheterozygous larvae for *mwh* and *flr³* genes were obtained by crossing four-day-old virgin *flr³* females with *mwh* males in a 2:1 ratio. Four days after fertilization, females were allowed to lay eggs in fresh yeast medium (25 g yeast and 4mL sterile distilled water) for 8 h in order to obtain synchronised larvae. After 72 h, larvae were collected, washed with distilled water and clustered in groups of 100 individuals. Each group was fed with a mixture containing 0.85 g *Drosophila* Instant Medium (Formula 4-24, Carolina Biological Supply, Burlington, NC, USA) and 4 mL water, supplemented with the tested compounds at fixed concentrations (the highest and second lowest from the toxicity assays) and negative (H_2O) and positive (0.15 M H_2O_2) controls, until pupae hatching (10–12 days). Adult flies were collected and mounted on slides using Faure's solution. Mutant spots were scored in both dorsal and ventral surfaces of the wings in a bright light microscope at 400× magnification.

The frequencies of each type of mutant clone per wing (single, large or twin spot) were compared to the concurrent negative control and analysed by applying the Kastenbaum and Bowman binomial test [29]. Inconclusive results from this binomial test were resolved by using Mann–Whitney and Wilcoxon U-test [14,30].

2.5. In Vivo Evaluation of Nutraceutical Potential

2.5.1. Antitoxicity Assay

Antitoxicity was assessed using the same procedure and experimental concentrations as in toxicity assays, but in combined treatments with 0.15 M H_2O_2 and comparing the percentage of emerging adults with the positive toxicant control [31]. Chi-square test was also used for comparing treatments to the positive concurrent control [15].

2.5.2. Antigenotoxicity Assay

Antigenotoxicity tests were performed, following the method described by Anter, et al. [13]. The same concentrations used in genotoxicity assay were assayed in combined treatment with hydrogen peroxide (0.15 M) acting as concurrent genotoxicant. The inhibition percentages (IP) for the combined treatments were calculated when appropriate according to Abraham and Singh [32] formula:

$$\text{IP} = ((\text{genotoxin alone} - \text{combined treatment})/\text{genotoxin alone}) \times 100$$

Inconclusive results from Kastembaum–Bowman binomial test were also resolved using Mann–Whitney U-test [14,30].

2.5.3. Chronic Treatments: Lifespan and Healthspan Assays

In order to obtain comparable results in all the in vivo assays, we used an F_1 progeny from *mwh* and *flr³* parental strains produced in a 24 h egg-laying in yeast for all the longevity trials. We also tested the same concentrations as in the toxicity/antitoxicity experiments. Lifespan assays were carried out at 25 °C, according to the procedure described by Fernandez-Bedmar, et al. [33]. Briefly, synchronised 72 ± 12 h old transheterozygous larvae were washed in distilled water, collected and transferred in groups of 100 individuals into test vials containing 0.85 g *Drosophila* Instant Medium and 4 mL of the different concentrations of the compound to be assayed. Emerged adults from pupae were collected under CO_2 anaesthesia and placed in groups of 25 individuals of the same sex into sterile vials containing 0.21 g *Drosophila* Instant Medium and 1 mL of different concentrations of CAR. Flies were chronically treated during all their life. The number of survivors was determined twice a week. The statistical treatment of survival data was performed with the SPSS 19.0 (SPSS, Inc., Chicago, IL, USA) statistical software, using the Kaplan–Meier test. The significance of the curves was determined using the Log-Rank method (Mantel–Cox).

2.6. *In Vitro Evaluation of Nutraceutical Potential*

2.6.1. Cytotoxicity Assay

The effect of the assayed compounds on cell viability was determined by the trypan blue exclusion test, according to our standard procedures [13]. HL-60 cells were placed in 96 well plates (2×10^4 cells/mL) and cultured for 72 h and supplemented with the same concentrations of CAR from our toxicity assays. The wide range of tested concentrations was intended to estimate the cytotoxic inhibitory concentration 50 (IC_{50}). After culture, cells were stained with a 1:1 volume ratio of trypan blue dye (T8154, Sigma-Aldrich, St. Louis, MO, USA) and counted in a Neubauer chamber at $100\times$ magnification. The mean value of three independent assays of the alive treated cells was determined in order to obtain the tumoural growth inhibition curves. The standard error of the three replicas was calculated, and the curve given by the Excel Microsoft Office program was added. Finally, an estimation of inhibitory concentration 50 was calculated.

2.6.2. DNA Fragmentation Status

The ability of our compound to induce DNA fragmentation was determined as described by Anter, et al. [34]. Briefly, 10^6 HL-60 cells were co-cultured with five different concentrations of CAR (as selected in the toxicity assays) for 5 h. After treatment, genomic DNA was extracted using a commercial kit (Blood Genomic DNA Extraction Mini Spin Kit, Canvax Biotech, Cordoba, Spain). Subsequently, DNA was incubated overnight with RNAase at 37 °C and quantified in a spectrophotometer (Nanodrop_ND-1000, NanoDrop Technologies, Inc., Wilmington, DE, USA). Finally, 1200 ng DNA was electrophoresed in a 2% agarose gel for 40 min at 85 V, stained with GelRed and visualised under UV light. The apoptosis process can be detected by the appearance of internucleosomal DNA fragments that are multiple of 200 base pairs.

2.6.3. Clastogenicity: SCGE (Comet Assay)

DNA integrity was assayed by SCGE as described by Mateo-Fernández, et al. [15], with minor modifications. HL-60 cells (5×10^5) in exponential growing phase were incubated in 1.5 mL of culture medium supplemented with different CAR (0.03, 0.12 and 0.25 mg/mL) concentrations for 5 h. After treatment, cells were washed twice and adjusted to 6.25×10^5 cells/mL in PBS. Electrophoresis gels were prepared, pouring a 1:4 dilution (cells in liquid low-melting-point agarose at 40 °C, A4018, Sigma-Aldrich, St. Louis, MO, USA) into slides. Gels were covered with a coverslip and allowed to solidify at room temperature (RT) for 15 min. Once the slides solidified, the coverslips were carefully removed and slides were bathed in freshly prepared lysing solution (2.5 M NaCl, 100 mM Na-EDTA, 10 mM Tris, 250 mM NaOH, 10% DMSO and 1% Triton X-100; pH 13) for 1 h at 4 °C. Thereafter, slides were equilibrated in alkaline electrophoresis buffer (300 mM NaOH and 1 mM Na-EDTA, pH 13) for 20–30 min at 4 °C. Once equilibrated, the slides underwent electrophoresis (12 V, 400 mA for 8 min) in the dark and were immediately neutralised in cold neutral solution (0.4 M Tris-HCl buffer, pH 7.5) for 5 min. Finally, slides were dried overnight at RT in the dark. Gels were stained with 7 µL propidium iodide and photographed in a Leica DM2500 microscope at $400\times$ magnification (Leica Microsystems GmbH, Wetzlar, Germany). At least 100 single cells from each treatment were analysed, using the Open Comet software [35]. The Tail Moment (TM) data were analysed, applying a one-way ANOVA and post hoc Tukey's test with SPSS Statistics for Windows, Version 19.0 (IBM Corporation, Armonk, NY, USA), to determine the effect of the tested compound on HL-60 cell DNA integrity.

2.6.4. Methylation Status of HL-60 Cells

The methylation status assay was performed as it was described previously by Merinas-Amo, et al. [14]. Briefly, HL-60 cells were treated with different concentrations of CAR (0.12 mg/mL and 4 mg/mL) for 5 h. Then, DNA was extracted similarly to previously described DNA fragmentation assay. After that, the DNA was converted with bisulphite (EZ DNA Methylation-Gold

Kit, Zymo Research, Irvine, CA, USA). Bisulphite-modified DNA was used for fluorescence-based real-time quantitative Methylation-Specific PCR (qMSP), using 5 µM of each forward and reverse primer (Isogen Life Science B.V., Utrecht, The Netherlands), 2 µL of iTaq Universal SYBR Green Supermix (Bio-Rad Laboratories, Inc., Hercules, CA, USA, it contains antibody-mediated hot-start iTaq DNA polymerase, dNTPs, $MgCl_2$, SYBR Green Dye, enhancers, stabilizers, and a blend of passive reference dyes including ROX and fluorescein) and 25 ng of bisulphite converted genomic DNA. PCR conditions included initial denaturalisation at 95 °C for 3 min and amplification, which consisted of 45 cycles at 95 °C for 10 s, 60 °C for 15 s and 72 °C for 15 s, taking picture at the end of each elongation cycle.

After that, melting curve was determined, increasing 0.5 °C each 0.05 s from 60 °C to 95 °C and taking pictures. qMSP was carried out in 48 well plates in MiniOpticon Real-Time PCR System (MJ Mini Personal Thermal Cycler, Bio-Rad Laboratories, Inc., Hercules, CA, USA) and were analysed by Bio-Rad CFX Manager 3.1 software. The housekeeping Alu-C4 was used as a reference to correct for total DNA input. Alu-C4 and the target repetitive elements Alu M1, LINE-1 and Sat-α were obtained from Isogen Life Science B.V. (Utrecht, The Netherlands), and their sequences are shown in Table 1. Each sample was analysed in triplicate. The results of each CT (cycle number at which the amplification curves cross the threshold value) were obtained from each qMSP. Data were normalised with the housekeeping Alu-C4, using the Nikolaidis, et al. [36] and Liloglou, et al. [37] comparative CT method ($\Delta\Delta$CT). One-way ANOVA and post hoc Tukey's test were used to evaluate the differences between the tested compound, repetitive elements and concentrations.

Table 1. Primers information [38].

Primer	Forward Primer Sequence 5′ to 3′ (N)	Reverse Primer Sequence 5′ to 3′ (N)
ALU-C4	GGTTAGGTATAGTGGTTTATATTTGTAATTTTAGTA (-36)	ATTAACTAAACTAATCTTAAACTCCTAACCTCA (-33)
ALU-M1	ATTATGTTAGTTAGGATGGTTTCGATTTT (-29)	CAATCGACCGAACGCGA (-17)
LINE-1-M1	GGACGTATTTGGAAAATCGGG (-21)	AATCTCGCGATACGCCGTT (-19)
SAT-α-M1	TGATGGAGTATTTTTAAAATATACGTTTTGTAGT (-34)	AATTCTAAAAATATTCCTCTTCAATTACGTAAA (-33)

ALU (Short interspersed nuclear element –SINE- Alu-C4 sequence). LINE (Long Interspersed Nuclear Element M1). Sat-α (Satellite alpha DNA).

3. Results

3.1. In Vivo Assays

Table 2 shows the toxicity and antitoxicity results obtained in this study. These results revealed that although CAR was significantly toxic at every tested concentration, the LD_{50} was not reached in any concentration. According to antitoxicity assay, antioxidant properties were not found in any tested concentrations but, contrarily, the highest concentration of CAR, which was even more toxic than both the positive controls in *D. melanogaster*.

Genotoxicity and Antigenotoxicity assays of CAR are shown in Table 3. The concurrent positive control showed significant differences with respect to the negative control using the Kastembaum–Bowman statistical test, providing a mutation rate per wing of 0.425 against 0.195, respectively. This result proves the accuracy of the assay. As regards genotoxicity, CAR showed inconclusive results, and it was solved by applying Mann–Whitney test, which demonstrated CAR was not a genotoxic compound. According to antigenotoxicity assay, combined treatments of CAR and hydrogen peroxide showed positive (*) results, which means that there were significant differences between CAR and the positive control. The IP was calculated since positive results were found. The IPs obtained in lowest and highest concentrations of CAR were 61% and 79.5%.

Table 2. Toxicity and antitoxicity levels of CAR in *D. melanogaster*.

CAR	Survival (%)	
(mg/mL)	Simple Treatment [1]	Combined Treatment [2]
0	100	100
H_2O_2	-	62.64
0.03	61 *,[3]	62
0.125	65.7 *	54.02
0.25	64.3 *	53
1	61.32 *	54.02
4	63 *	23 *,[4]

[1] Data are expressed as percentage of survival adults with respect to 300 untreated 72 h old larvae from three independent experiments. [2] Combined treatments using standard medium and 0.15 M hydrogen peroxide. [3] Asterisks (*) indicate significant differences (one tail) with respect to the untreated control group and [4] the hydrogen peroxide control group: * Chi-square value higher than 5.02 [15]. CAR: caramel colour E150d-class IV.

Table 3. Genotoxicity and Antigenotoxicity assays of CAR in *D. melanogaster*.

Clones per Wings (Number of Spots)							
Compound	Wings Number	Small Single Spots (1–2 Cells) $m = 2$	Large Simple Spots (>2 Cells) $m = 5$	Twin Spots $m = 5$	Total Spots $m = 2$	Mann–Whitney Test	IP (%)
H2O	41	0.147 (6)	0.048 (2)	0	0.195 (8)		
H2O2 (0.15 M)	40	0.375 (15)	0.05 (2)	0	0.425 (17) +		
SIMPLE TREATMENT							
CAR (mg/mL)	40	0.25 (10)	0.125 (5)	0	0.375 (15) i	Λ	
[0.125] [4]	42	0.166 (7)	0.095 (4)	0.024 (1)	0.286 (12) i	Λ	
COMBINED TREATMENT							
CAR (mg/mL)	42	0.166 (7)	0	0	0.166 (7) *		61
[0.125] [4]	46	0.065 (3)	0.02 (1)	0	0.087 (4) *		79.5

Statistical diagnosis according to Frei and Wurgler [39]: + (positive), − (negative) and i (inconclusive) vs. negative control; * (positive), Δ (negative) and β (inconclusive) vs. respective positive control; m: multiplication factor. Kastenbaum–Bowman Test without Bonferroni correction, probability levels: $\alpha = \beta = 0.05$. No. of spots in parentheses. Mann-Whitney test was used when appropriate to resolve inconclusive results. Lambda (Λ) symbol mean that there were not significant differences with respect to the negative control when Mann-Whitney test is applied. Inhibition percentage values were included when appropriate.

Table 4 shows the lifespan and healthspan results, which reported that CAR does not exert any significant effect on *Drosophila* lifespan and healthspan, based on the survival curves as they are depicted in Figure 1.

Table 4. Effects of CAR treatments on the *Drosophila melanogaster* mean lifespan and healthspan.

CAR (mg/mL)	Mean Lifespan (Days)	Mean Lifespan Difference (%) [a]	Healthspan (80th Percentile) (Days)	Healthspan Difference (%) [a]
Control	64 ± 3.16	0	31.21 ± 2.37	0
0.125	59.65 ± 2.4	−6.8	31.03 ± 2.12	−0.5
0.25	60.83 ± 2.73	−4.9	33.68 ± 2.44	7.6
1	62.8 ± 2.78	−1.9	30.88 ± 2.1	−1.1
4	59 ± 3.35	−7.9	37.54 ± 4	20.28

[a] The difference was calculated by comparing treated flies with the concurrent water control. Positive numbers indicate lifespan increase, and negative numbers indicate lifespan decrease. Data are expressed as mean value ± SE.

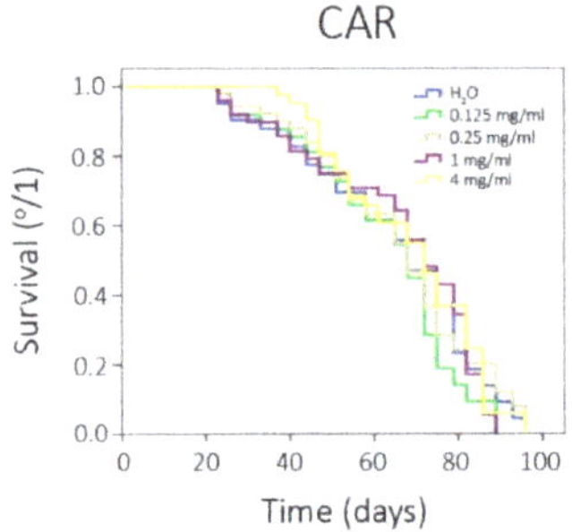

Figure 1. Survival curves obtained from log-rank test.

3.2. *In vitro Assays*

CAR showed cytotoxic effect against HL-60 leukaemia cell line, as it is shown in Figure 2. The IC_{50} was reached roughly at 1 mg/mL CAR.

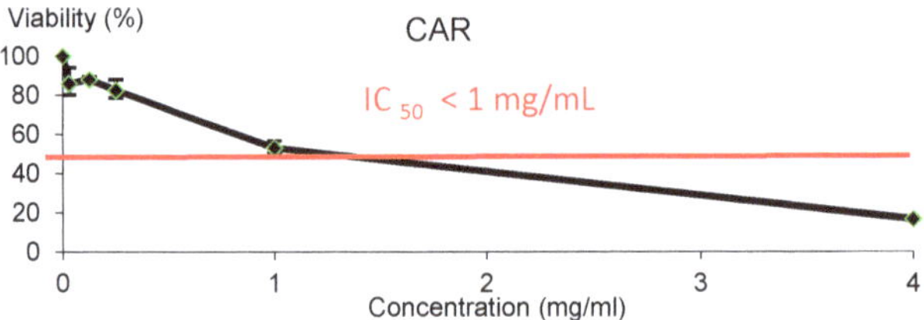

Figure 2. Viability of HL-60 cells treated with CAR for 72 h. Each point represents the percentage of viability with respect to the mean control ± SD of three independent experiments.

The electrophoresis of genomic DNA integrity of HL-60 cells treated with different concentrations of CAR is shown in Figure 3. No DNA damage was observed at any CAR concentrations.

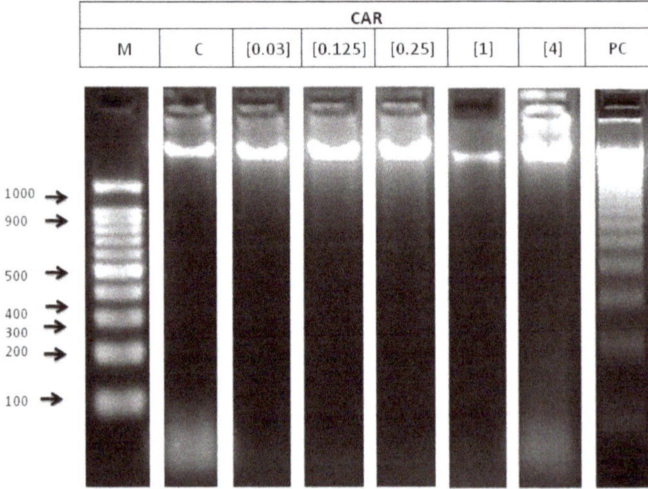

Figure 3. Internucleosomal DNA fragmentation after 5 h of HL-60 cells treated with CAR. Letters M and C mean weight size marker and negative control (RPMI), respectively, and lyophilised blond beer (62.5 mg/mL) has been used as a routine positive control (PC).

Figure 4 shows the results obtained in the single cell gel electrophoresis test or comet assay. According to this assay, CAR did not induce damage in human leukaemia HL-60 cell line at any tested concentration. The concentrations used in this SCGE assay were determined according to the results obtained in the previous cytotoxicity assay.

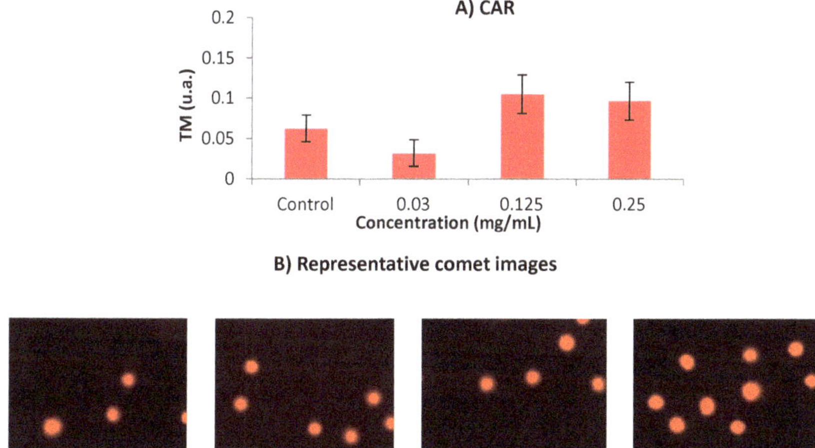

Figure 4. (**A**) Alkaline comet assay (pH > 13) of HL-60 cells after 5 h treatment with different concentrations of CAR. DNA migration is reported as mean TM. The plot shows mean TM values and standard errors. Statistical differences were analysed, applying one-way ANOVA and post hoc Tukey's test. (**B**) Representative images of each dose group from comet assay are shown following the order described in Figure 4A (control, 0.03 mg/mL, 0.125 mg/mL and 0.25 mg/mL, respectively). TM: Tail Moment.

Figure 5 shows the relative normalised methylation status (RMS) of the three repetitive sequences (LINE-1, Alu M1 and Sat-α) in HL-60 cell line treated with the tested compound. CAR hypomethylated Sat-α repetitive element of HL-60 cell line when these human leukaemia cells were treated with 4 mg/mL.

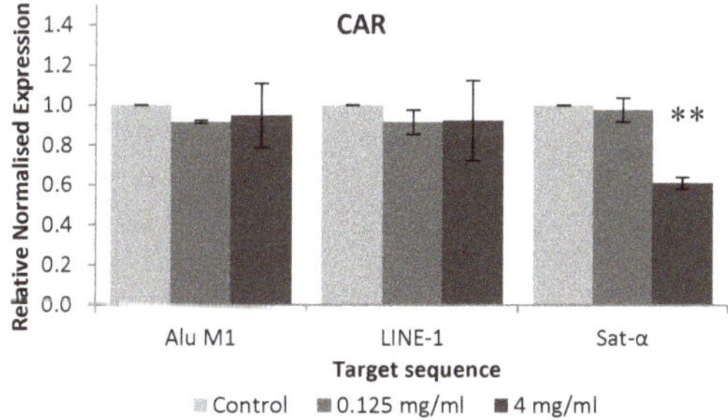

Figure 5. Relative normalised expression data of each repetitive element. Asterisk mark (*) is associated with different means, applying One-Way Anova test and post hoc Tuckey's test.

4. Discussion

4.1. Food Safety Assays: Toxicity and Genotoxicity

Food additives are still considered a big deal as many of their functions remain unknown. Precisely due to food colourants being massively used, an evaluation of their effect on public health should be needed [40]. Knowing that one third of human cancers are related to diet, much research is focused on small molecules added to food. This apparently easy concept became complicated as many compounds exert a dual, positive/negative effect that strongly depends on the dose [2]. The present work evaluates the safety of the caramel colour E150D (CAR) in in vivo toxicity and genotoxicity assays using the eukaryotic *Drosophila* model for the first time.

Our toxicity results revealed that although CAR was significantly toxic at every tested concentration, the LD_{50} was not reached in any concentrations. Therefore, CAR was not a toxic substance. This result did not agree with MacKenzie, et al. [41], who affirmed that caramel is not toxic. The toxicity of CAR could be caused by the 4-methyl imidazole (4-MeI) presence that has been described as a neurotoxic agent able to inhibit P450 cytochrome in the human liver [42,43] and even induced alveolar/bronchiolar adenoma and carcinoma in mice [44]. Anyway, The LD_{50} of CAR was not found at any concentrations, therefore both substances could be considered as nontoxic, being related to the report of MacKenzie, et al. [41].

Genotoxicity assays of tested compound showed inconclusive results, which were solved by applying Mann–Whitney test, which demonstrated that any of tested concentrations were genotoxic. This result agrees with the studies carried out by Brusick, et al. [45], who did not find evidence of genotoxicity in the *Salmonella* plate incorporation test using 5 standard strains or in the *Saccharomyces cerevisiae* gene conversion assay. The lack of genotoxicity was also demonstrated by Norizadeh Tazehkand, et al. [46], treating mice with 4-MeI. Colour caramel III was administrated to human males, and no significant differences were found in mean blood lymphocyte numbers compared to the respective control. The results supported the conclusion that this colourant does not pose a genotoxic hazard to humans [47] and CAR may provide nutraceutical potential.

The lack of genotoxicity observed in *Drosophila* for all CAR concentrations confirmed their safety. We hypothesised that the toxicity observed in our compound may either be induced by a different pathway than the genotoxic one, or it may be affecting different genes to those used in this assay.

4.2. Nutraceutical Potential Assays

Nutraceuticals and functional foods have become key issues in eating habits, nutrition and diets. The nutraceutical potential of food is recognised as an important domain of research [48]. The present study performed an evaluation of the nutraceutical potential of CAR by carrying out in vivo antitoxicity and lifespan assays as well as in vitro cytotoxicity, internucleosomal fragmentation, single and double DNA strands breaks and modulation of methylation patterns in the HL-60 leukaemia cells model.

4.3. In vivo Assays

In vivo antitoxicology assays have been performed through time in order to ascertain the health-promoting properties of the tested compounds. *D. melanogaster* model is increasingly used to study life extension, since there is a high homology between invertebrate and human genes involved in the aging process [17,49].

Our antitoxicity results demonstrated that CAR did not possess antioxidant effects in any tested concentrations since it was not able to revert the damage caused by hydrogen peroxide in *D. melanogaster*, except for the highest concentration of CAR, which was even more toxic than the positive control. According to antigenotoxicity assay, combined treatments of CAR and hydrogen peroxide showed positive (*) results, which means that there were significant differences between CAR and positive control, inhibiting the effect of the genotoxine in 61% and 79.5% for the lowest and the highest concentrations, respectively.

Tsai, et al. [50] concluded that CAR was overall antioxidant and this capacity depended on the colour of the caramel, and the brownest the additive the more antioxidant it is. Sengar and Sharma [10] reported a low antioxidant activity of CAR in a review. 4-MeI was demonstrated to be antigenotoxic in mice [46]. However, most of the literature regarding caramel focused on the identification of the caramelisation products. Therefore, more research is needed to evaluate the antioxidant and antigenotoxic properties of CAR as it is consumed.

The lifespan and healthspan results reported that CAR does not exert any significant effect on *Drosophila*'s lifespan and healthspan. As far as it is known, there is not any scientific information about CAR evaluating the aging and lifespan, only based on 4-MeI, which is one of the main components of CAR. 4-MeI was used in a chronic treatment conducted in rats. No observable adverse effects were found [41,47], being our results consistent with those obtained.

4.4. In vitro Assays

The in vitro evaluation of the anti-cancer properties of nutraceutical compounds is the first step of a large pathway to obtain suitable conclusions to be extrapolated to humans. The aim of the present trial was to determine the potential chemopreventive and genotoxic effect of CAR on a human cancer cell model (HL-60 cell line), performing cytotoxicity, DNA fragmentation, SCGE and epigenetic modulation assays.

CAR showed cytotoxic effect against HL-60 leukaemia cell line, and the IC_{50} was reached roughly at 1 mg/mL for CAR inducing cell death in HL-60 cell line. It seems to be the first attempt on ascertaining the chemopreventive potential of CAR. Our findings agreed in some extent with the cytotoxicity activity observed in lymphocytes induced by caramel colour additive [41]. Therefore, further research studies are needed to ascertain the chemopreventive potential of CAR.

Clastogenicity is involved in a process of DNA damage. DNA fragmentation test and comet assay were conducted in order to examine the clastogenic potential of CAR on HL-60 promyelocytic cell line. The degradation of genomic DNA into internucleosomal fragments was proposed as a major mechanism affecting cancer cell apoptosis. The typical ladder pattern has not been shown by any of the tested concentrations, thus they are not able to induce apoptosis (Figure 3). This result is not consistent with the reduction of the number of the mice follicles and oocytes observed by Suocheng, et al. [51], who concluded that this decrease could be due to apoptotic mechanisms. However, the cells used in this research differed from the used one in the present study. National Toxicology Progam (NTP) concluded there was no evidence of carcinogenic activity of 4-MeI in rats, but this compound should be better named as "some evidence", according the information found in previous studies [47].

Alkaline SCGE is performed in order to detect DNA damage [52], which is widely used to determine whether cells are undergoing apoptotic and/or necrotic pathways [53]. It is assumed that apoptosis occurs when treatments induce a TM > 30 (hedgehog pattern), whereas control cells remain lower than 2 (no tails). On the contrary, necrosis shows a short comet-tail pattern since the majority of the damaged DNA remains in the comet head [54]. Our result showed that CAR did not exhibit clastogenic activity, since TM values of all assayed concentrations remained in TM values lower than 1, as it is depicted in Figure 4. The concentrations used in this SCGE assay were determined according to the results obtained in the previous cytotoxicity assay. This finding means that CAR can be regarded as untreated cells (class 0) from the five TM classes proposed by Fabiani, et al. [55].

Clastogenic activity in CHO cells was induced when exposed to caramel colour [56]. These results are not in agreement with our findings, although the experimental models were different. The lack of in vitro genetic damage could be due to the fact that the assessed concentrations in comet assay were the three lowest ones, which are the less cytotoxic being the cell viability is roughly 80%. Furthermore, the results obtained in comet assay are congruent with those obtained in our DNA fragmentation test.

Despite the cytotoxic activity shown by CAR, internucleosomal DNA fragmentation and DNA damage at the assayed concentrations were not induced, thus the cell death was not due to proapoptotic mechanisms in our HL-60 model.

As for epigenetics, the genome is globally hypomethylated in cancer cells, inducing transposable element activity and thus triggering genome instability [57]. On the other hand, it is known that the silencing of tumour suppressor genes is closely associated with hypermethylation [58]. Repetitive elements are highly methylated in somatic normal cells, contributing to a global genomic hypermethylation [38], suppressing the transposable activity of repetitive elements. Three different repetitive elements: LINE-1, Alu-M4 and Sat-α were studied. Long interspersed nuclear elements (LINE) are abundant retrotransposons, and representing LINE-1 about 17% of the human genome, with a nonrandom distribution by accumulating primarily in G-positive bands, which are AT-rich regions of chromosomes [59]. LINE-1 elements are also accumulated in regions of low recombination rate, mainly in X-chromosome [60]. Alu elements belong to the SINE family (Short Interspersed Nuclear Elements), being the most abundant (accounting about 10% of the whole human genome [38] and predominantly present in noncoding and GC-rich regions [59,61]. Sat-α (Satellite alpha DNA) repeats are composed of tandem repeats of 170 bp DNA sequences, are AT-rich regions and represent the main DNA component of every human centromere, constituting about 5% of total human DNA [59,62]. Therefore, examination of the methylation status of LINE-1, Alu and Sat-α genomic regions has served as an approach for measuring global methylation levels since 32 % of the human genome has been evaluated [63].

To our knowledge, this is the first attempt at evaluating the ability of CAR for modulating the epigenome, thus there is not any information related to this assay using CAR on the scientific database. Our results showed that CAR hypomethylated Sat-α repetitive element of HL-60 cell line when these human leukaemia cells were treated with 4 mg/mL. In addition, it has been demonstrated that the expression of satellite sequences is associated with a hypomethylation, triggering cancer cells. Therefore, methylation process in satellite sequences is a potential mechanism for silencing its satellite expression in transformed cells [64], which is not induced by CAR.

Some recent human therapies against cancer are based on hypomethylating agents, since this activity is highly related to gene silencing, thus this fact could activate tumour suppressor genes and be a positive highlight. Despite this fact, it is not clear its benefit on human therapies, and many more research studies should be performed [65].

Further research using normal human cell line should be taken into account, to be compared to our carcinogenic cells once CAR is recommended to be monitored and reduced in soft drinks [66].

5. Conclusions

In conclusion, the food safety of CAR was demonstrated, since LD50 was not reached in toxicity assay and any of the tested concentrations induced mutation rates higher than that of the concurrent control in *D. melanogaster*. On the other hand, CAR protected DNA from oxidative stress provided by hydrogen peroxide in *D. melanogaster*, according to antigenotoxiciy assay. In addition, CAR was a first-step chemopreventive compound but it did not induce clastogenicity in human leukaemia cells. CAR modified the methylation status of HL-60 cell line, although much more information about the mechanisms of gene therapies related to epigenetic modulation by food is necessary.

Author Contributions: M.M.-F. wrote this manuscript and performed the genotoxicity tests, P.A.-M. performed the antigenotoxicity tests, M.R.-C., R.F., T.M.-A. performed all the in vitro assays, A.A.-M. designed this study and revised the manuscript. All the listed authors have read and approved the submitted manuscript.

Funding: This research received no external funding.

Conflicts of Interest: The authors declare no conflict of interest.

References

1. Fairweather, F.A.; Swann, C.A. Food additives and cancer. *Proc. Nutr. Soc.* **1981**, *40*, 21–30. [CrossRef]
2. Trewavas, A.; Stewart, D. Paradoxical effects of chemicals in the diet on health. *Curr. Opin. Plant Biol.* **2003**, *6*, 185–190. [CrossRef]
3. Redmon, T. Assessing the attitudes and bahaviors of pedestrians and drivers in traffic situations. *Inst. Transp. Eng.* **2003**, *73*, 26.
4. Willett, W.C. Diet and health: What should we eat? *Science (New York, NY)* **1994**, *264*, 532–537. [CrossRef] [PubMed]

5. Ishidate, M., Jr.; Sofuni, T.; Yoshikawa, K.; Hayashi, M.; Nohmi, T.; Sawada, M.; Matsuoka, A. Primary mutagenicity screening of food additives currently used in Japan. *Food Chem. Toxicol.* **1984**, *22*, 623–636. [CrossRef]

6. Tsubono, Y.; Ogawa, K.; Watanabe, Y.; Nishino, Y.; Tsuji, I.; Watanabe, T.; Nakatsuka, H.; Takahashi, N.; Kawamura, M.; Hisamichi, S. Food frequency questionnaire as a screening test. *Nutr. Cancer* **2001**, *39*, 78–84. [CrossRef] [PubMed]

7. Licht, B.; Shaw, K.; Smith, C.; Mendoza, M.; Orr, J.; Myers, D. Characterization of caramel colour iv. *Food Chem. Toxicol.* **1992**, *30*, 365–373. [CrossRef]

8. Golon, A.; Kuhnert, N. Unraveling the chemical composition of caramel. *J. Agric. Food Chem.* **2012**, *60*, 3266–3274. [CrossRef] [PubMed]

9. Delgado-Vargas, F.; Paredes-López, O. Chemicals and colorants as nutraceuticals. In *Natural Colorants for Food and Nutraceuticals Uses*; CRC Press: Boca Raton, FL, USA, 2003; pp. 257–305.

10. Sengar, G.; Sharma, H.K. Food caramels: A review. *J. Food Sci. Technol.* **2014**, *51*, 1686–1696. [CrossRef]

11. Graf, U.; Wurgler, F.F.; Katz, A.J.; Frei, H.; Juon, H.; Hall, C.B.; Kale, P.G. Somatic mutation and recombination test in drosophila melanogaster. *Environ. Mutagen.* **1984**, *6*, 153–188. [CrossRef]

12. Reiter, L.T.; Potocki, L.; Chien, S.; Gribskov, M.; Bier, E. A systematic analysis of human disease-associated gene sequences in drosophila melanogaster. *Genome Res.* **2001**, *11*, 1114–1125. [CrossRef]

13. Anter, J.; Fernandez-Bedmar, Z.; Villatoro-Pulido, M.; Demyda-Peyras, S.; Moreno-Millan, M.; Alonso-Moraga, A.; Munoz-Serrano, A.; Luque de Castro, M.D. A pilot study on the DNA-protective, cytotoxic, and apoptosis-inducing properties of olive-leaf extracts. *Mutat. Res.* **2011**, *723*, 165–170. [CrossRef] [PubMed]

14. Merinas-Amo, T.; Tasset-Cuevas, I.; Díaz-Carretero, A.M.; Alonso-Moraga, Á.; Calahorro, F. In vivo and in vitro studies of the role of lyophilised blond lager beer and some bioactive components in the modulation of degenerative processes. *J. Funct. Foods* **2016**, *27*, 274–294. [CrossRef]

15. Mateo-Fernández, M.; Merinas-Amo, T.; Moreno-Millán, M.; Alonso-Moraga, Á.; Demyda-Peyrás, S. In vivo and in vitro genotoxic and epigenetic effects of two types of cola beverages and caffeine: A multiassay approach. *BioMed Res. Int.* **2016**, *2016*, 7574843. [CrossRef]

16. Fleming, J.E.; Reveillaud, I.; Niedzwiecki, A. Role of oxidative stress in drosophila aging. *Mutat. Res.* **1992**, *275*, 267–279. [CrossRef]

17. Li, S.; Chen, K.; Li, X.; Zhang, X.; Liu, S.V. A new cultivation system for studying chemical effects on the lifespan of the fruit fly. *Exp. Gerontol.* **2010**, *45*, 158–162. [CrossRef] [PubMed]

18. Beckingham, K.M.; Armstrong, J.D.; Texada, M.J.; Munjaal, R.; Baker, D.A. Drosophila melanogaster-the model organism of choice for the complex biology of multi-cellular organisms. *Gravit. Space Res. Bull.* **2007**, *18*, 17–29.

19. Balls, M. Progressing toward the reduction, refinement and replacement of laboratory animal procedures: Thoughts on some encounters with dr iain purchase. *Toxicol. Vitr.* **2004**, *18*, 165–170. [CrossRef]

20. Leszczyniecka, M.; Roberts, T.; Dent, P.; Grant, S.; Fisher, P.B. Differentiation therapy of human cancer: Basic science and clinical applications. *Pharmacol. Ther.* **2001**, *90*, 105–156. [CrossRef]

21. Fesus, L.; Szondy, Z.; Uray, I. Probing the molecular program of apoptosis by cancer chemopreventive agents. *J. Cell. Biochem. Suppl.* **1995**, *22*, 151–161. [CrossRef]

22. Hong, W.K.; Sporn, M.B. Recent advances in chemoprevention of cancer. *Science (New York, NY)* **1997**, *278*, 1073–1077. [CrossRef]

23. Suzuki, M.M.; Bird, A. DNA methylation landscapes: Provocative insights from epigenomics. *Nat. Rev. Genet.* **2008**, *9*, 465. [CrossRef]

24. Roman-Gomez, J.; Jimenez-Velasco, A.; Agirre, X.; Castillejo, J.A.; Navarro, G.; San Jose-Eneriz, E.; Garate, L.; Cordeu, L.; Cervantes, F.; Prosper, F.; et al. Repetitive DNA hypomethylation in the advanced phase of chronic myeloid leukemia. *Leuk. Res.* **2008**, *32*, 487–490. [CrossRef]

25. Jirtle, R.L.; Skinner, M.K. Environmental epigenomics and disease susceptibility. *Nat. Rev. Genet.* **2007**, *8*, 253–262. [CrossRef]

26. Yan, J.; Huen, D.; Morely, T.; Johnson, G.; Gubb, D.; Roote, J.; Adler, P.N. The multiple-wing-hairs gene encodes a novel gbd-fh3 domain-containing protein that functions both prior to and after wing hair initiation. *Genetics* **2008**, *180*, 219–228. [CrossRef]

27. Ren, N.; Charlton, J.; Adler, P.N. The flare gene, which encodes the aip1 protein of drosophila, functions to regulate f-actin disassembly in pupal epidermal cells. *Genetics* **2007**, *176*, 2223–2234. [CrossRef]

28. Maes, P.; Monakhova, Y.B.; Kuballa, T.; Reusch, H.; Lachenmeier, D.W. Qualitative and quantitative control of carbonated cola beverages using 1h nmr spectroscopy. *J. Agric. Food Chem.* **2012**, *60*, 2778–2784. [CrossRef]

29. Frei, H.; Wurgler, F.E. Optimal experimental design and sample size for the statistical evaluation of data from somatic mutation and recombination tests (smart) in drosophila. *Mutat. Res.* **1995**, *334*, 247–258. [CrossRef]

30. Fernández-Bedmar, Z.; Arenas-Chaparro, R.; Merinas-Amo, T.; Mateo-Fernández, M.; Tasset-Cuevas, I.; Lozano-Baena, M.; de Haro-Bailón, A.; Campos-Sánchez, J. Modulator role of trilinolein/triolein and resveratrol on the health promoting effects of processed foods: Edible oils and red wine. *Toxicol. Lett.* **2016**, *258*, S159. [CrossRef]

31. Tasset-Cuevas, I.; Fernandez-Bedmar, Z.; Dolores Lozano-Baena, M.; Campos-Sanchez, J.; de Haro-Bailon, A.; Munoz-Serrano, A.; Alonso-Moraga, A. Protective effect of borage seed oil and gamma linolenic acid on DNA: In vivo and in vitro studies. *PLoS ONE* **2013**, *8*, e56986. [CrossRef]

32. Abraham, S.K.; Singh, S.P. Anti-genotoxicity and glutathione s-transferase activity in mice pretreated with caffeinated and decaffeinated coffee. *Food Chem. Toxicol.* **1999**, *37*, 733–739. [CrossRef]

33. Fernandez-Bedmar, Z.; Anter, J.; de La Cruz-Ares, S.; Munoz-Serrano, A.; Alonso-Moraga, A.; Perez-Guisado, J. Role of citrus juices and distinctive components in the modulation of degenerative processes: Genotoxicity, antigenotoxicity, cytotoxicity, and longevity in drosophila. *J. Toxicol. Environ. Health A* **2011**, *74*, 1052–1066. [CrossRef]

34. Anter, J.; Tasset, I.; Demyda-Peyrás, S.; Ranchal, I.; Moreno-Millán, M.; Romero-Jimenez, M.; Muntané, J.; Luque de Castro, M.D.; Muñoz-Serrano, A.; Alonso-Moraga, Á. Evaluation of potential antigenotoxic, cytotoxic and proapoptotic effects of the olive oil by-product "alperujo", hydroxytyrosol, tyrosol and verbascoside. *Mutat. Res. Genet. Toxicol. Environ. Mutagen.* **2014**, *772*, 25–33. [CrossRef] [PubMed]

35. Gyori, B.M.; Venkatachalam, G.; Thiagarajan, P.S.; Hsu, D.; Clement, M.V. Opencomet: An automated tool for comet assay image analysis. *Redox Biol.* **2014**, *2*, 457–465. [CrossRef]

36. Nikolaidis, G.; Raji, O.Y.; Markopoulou, S.; Gosney, J.R.; Bryan, J.; Warburton, C.; Walshaw, M.; Sheard, J.; Field, J.K.; Liloglou, T. DNA methylation biomarkers offer improved diagnostic efficiency in lung cancer. *Cancer Res.* **2012**, *72*, 5692–5701. [CrossRef]

37. Liloglou, T.; Bediaga, N.G.; Brown, B.R.; Field, J.K.; Davies, M.P. Epigenetic biomarkers in lung cancer. *Cancer Lett.* **2014**, *342*, 200–212. [CrossRef]

38. Weisenberger, D.J.; Campan, M.; Long, T.I.; Kim, M.; Woods, C.; Fiala, E.; Ehrlich, M.; Laird, P.W. Analysis of repetitive element DNA methylation by methylight. *Nucleic Acids Res.* **2005**, *33*, 6823–6836. [CrossRef] [PubMed]

39. Frei, H.; Wurgler, F.E. Statistical methods to decide whether mutagenicity test data from drosophila assays indicate a positive, negative, or inconclusive result. *Mutat. Res.* **1988**, *203*, 297–308. [CrossRef]

40. Vartanian, L.R.; Schwartz, M.B.; Brownell, K.D. Effects of soft drink consumption on nutrition and health: A systematic review and meta-analysis. *Am. J. Public Health* **2007**, *97*, 667–675. [CrossRef]

41. MacKenzie, K.; Boysen, B.; Field, W.; Petsel, S.; Chappel, C.; Emerson, J.; Stanley, J. Toxicity studies of caramel colour iii and 2-acetyl-4 (5)-tetrahydroxybutylimidazole in f344 rats. *Food Chem. Toxicol.* **1992**, *30*, 417–425. [CrossRef]

42. Hargreaves, M.B.; Jones, B.C.; Smith, D.A.; Gescher, A. Inhibition of p-nitrophenol hydroxylase in rat liver microsomes by small aromatic and heterocyclic molecules. *Drug Metab. Dispos.* **1994**, *22*, 806–810.

43. Moretton, C.; Crétier, G.; Nigay, H.; Rocca, J.L. Quantification of 4-methylimidazole in class iii and iv caramel colors: Validation of a new method based on heart-cutting two-dimensional liquid chromatography (lc-lc). *J. Agric. Food Chem.* **2011**, *59*, 3544–3550. [CrossRef]

44. Cunha, S.; Barrado, A.; Faria, M.; Fernandes, J. Assessment of 4-(5-) methylimidazole in soft drinks and dark beer. *J. Food Compos. Anal.* **2011**, *24*, 609–614. [CrossRef]

45. Brusick, D.; Jagannath, D.; Galloway, S.; Nestmann, E. Genotoxicity hazard assessment of caramel colours iii and iv. *Food Chem. Toxicol.* **1992**, *30*, 403–410. [CrossRef]

46. Norizadeh Tazehkand, M.; Topaktas, M.; Yilmaz, M.B. Assessment of chromosomal aberration in the bone marrow cells of swiss albino mice treated by 4-methylimidazole. *Drug Chem. Toxicol.* **2016**, *39*, 307–311. [CrossRef]

47. Vollmuth, T.A. Caramel color safety–an update. *Food Chem. Toxicol.* **2018**, *111*, 578–596. [CrossRef]

48. Batista, C.; Barros, L.; Carvalho, A.M.; Ferreira, I.C. Nutritional and nutraceutical potential of rape (brassica napus l. Var. Napus) and "tronchuda" cabbage (brassica oleraceae l. Var. Costata) inflorescences. *Food Chem. Toxicol.* **2011**, *49*, 1208–1214. [CrossRef]

49. Bell, R.; Hubbard, A.; Chettier, R.; Chen, D.; Miller, J.P.; Kapahi, P.; Tarnopolsky, M.; Sahasrabuhde, S.; Melov, S.; Hughes, R.E. A human protein interaction network shows conservation of aging processes between human and invertebrate species. *PLoS Genet.* **2009**, *5*, e1000414. [CrossRef]

50. Tsai, P.J.; Yu, T.Y.; Chen, S.H.; Liu, C.C.; Sun, Y.F. Interactive role of color and antioxidant capacity in caramels. *Food Res. Int.* **2009**, *42*, 380–386. [CrossRef]

51. Suocheng, W.; Huining, L.; Haoqin, L.; Luju, L.; Zhu, G. Coca-cola and pepsi-cola affect ovaries and follicles development. *Biomed. Res.* **2016**, *27*, 710–717.

52. Forchhammer, L.; Ersson, C.; Loft, S.; Möller, L.; Godschalk, R.W.; van Schooten, F.J.; Jones, G.D.; Higgins, J.A.; Cooke, M.; Mistry, V. Inter-laboratory variation in DNA damage using a standard comet assay protocol. *Mutagenesis* **2012**, *27*, 665–672. [CrossRef]

53. Olive, P.L.; Banáth, J.P. The comet assay: A method to measure DNA damage in individual cells. *Nat. Protoc.* **2006**, *1*, 23–29. [CrossRef]

54. Fairbairn, D.W.; O'Neill, K.L. Necrotic DNA degradation mimics apoptotic nucleosomal fragmentation comet tail length. *Vitr. Cell. Dev. Biol. Anim.* **1995**, *31*, 171–173. [CrossRef] [PubMed]

55. Fabiani, R.; Rosignoli, P.; De Bartolomeo, A.; Fuccelli, R.; Morozzi, G. Genotoxicity of alkene epoxides in human peripheral blood mononuclear cells and hl60 leukaemia cells evaluated with the comet assay. *Mutat. Res.* **2012**, *747*, 1–6. [CrossRef] [PubMed]

56. Kitts, D.D.; Wu, C.; Kopec, A.; Nagasawa, T. Chemistry and genotoxicity of caramelized sucrose. *Mol. Nutr. Food Res.* **2006**, *50*, 1180–1190. [CrossRef] [PubMed]

57. Lopez-Serra, L.; Esteller, M. Proteins that bind methylated DNA and human cancer: Reading the wrong words. *Br. J. Cancer* **2008**, *98*, 1881–1885. [CrossRef] [PubMed]

58. Qin, T.; Jelinek, J.; Si, J.; Shu, J.; Issa, J.P.J. Mechanisms of resistance to 5-aza-2'-deoxycytidine in human cancer cell lines. *Blood* **2009**, *113*, 659–667. [CrossRef] [PubMed]

59. Lander, E.S.; Linton, L.M.; Birren, B.; Nusbaum, C.; Zody, M.C.; Baldwin, J.; Devon, K.; Dewar, K.; Doyle, M.; FitzHugh, W. Initial sequencing and analysis of the human genome. *Nature* **2001**, *409*, 860–921.

60. Boissinot, S.; Entezam, A.; Furano, A.V. Selection against deleterious line-1-containing loci in the human lineage. *Mol. Biol. Evol.* **2001**, *18*, 926–935. [CrossRef]

61. Grover, D.; Majumder, P.P.; Rao, C.B.; Brahmachari, S.K.; Mukerji, M. Nonrandom distribution of alu elements in genes of various functional categories: Insight from analysis of human chromosomes 21 and 22. *Mol. Biol. Evol.* **2003**, *20*, 1420–1424. [CrossRef]

62. Waye, J.; Willard, H. Structure, organization, and sequence of alpha satellite DNA from human chromosome 17: Evidence for evolution by unequal crossing-over and an ancestral pentamer repeat shared with the human x chromosome. *Mol. Cell. Biol.* **1986**, *6*, 3156–3165. [CrossRef] [PubMed]

63. Martínez, J.G.; Pérez-Escuredo, J.; Castro-Santos, P.; Marcos, C.Á.; Pendás, J.L.L.; Fraga, M.F.; Hermsen, M.A. Hypomethylation of line-1, and not centromeric sat-α, is associated with centromeric instability in head and neck squamous cell carcinoma. *Cell. Oncol.* **2012**, *35*, 259–267. [CrossRef] [PubMed]

64. Ting, D.T.; Lipson, D.; Paul, S.; Brannigan, B.W.; Akhavanfard, S.; Coffman, E.J.; Contino, G.; Deshpande, V.; Iafrate, A.J.; Letovsky, S. Aberrant overexpression of satellite repeats in pancreatic and other epithelial cancers. *Science (New York, NY)* **2011**, *331*, 593–596. [CrossRef] [PubMed]

65. Wild, L.; Flanagan, J.M. Genome-wide hypomethylation in cancer may be a passive consequence of transformation. *BBA Rev. Cancer* **2010**, *1806*, 50–57. [CrossRef] [PubMed]

66. Lima, P.M.; Orsoli, P.C.; Araújo, T.G.; Brandao, D. Effects of a carbonated soft drink on epithelial tumor incidence in Drosophila melanogaster. *J. Pharm. Pharmacol.* **2018**, *6*, 240–2047.

© 2019 by the authors. Licensee MDPI, Basel, Switzerland. This article is an open access article distributed under the terms and conditions of the Creative Commons Attribution (CC BY) license (http://creativecommons.org/licenses/by/4.0/).

Article

Nutrient Properties and Nuclear Magnetic Resonance-Based Metabonomic Analysis of Macrofungi

Dan Liu [1], Yu-Qing Chen [1], Xiao-Wei Xiao [1], Ru-Ting Zhong [1], Cheng-Feng Yang [2], Bin Liu [1,3] and Chao Zhao [1,4,*]

1   College of Food Science, Fujian Agriculture and Forestry University, Fuzhou 350002, China
2   College of Food Science and Nutritional Engineering, China Agricultural University, Beijing 100083, China
3   China National Engineering Research Center of JUNCAO Technology, Fuzhou 350002, China
4   Key Laboratory of Marine Biotechnology of Fujian Province, Institute of Oceanology, Fujian Agriculture and Forestry University, Fuzhou 350002, China
*   Correspondence: zhchao@live.cn; Tel.: +86-591-8353-0197

Received: 25 July 2019; Accepted: 4 September 2019; Published: 7 September 2019

Abstract: Many delicious and nutritional macrofungi are widely distributed and used in East Asian regions, considered as edible and medicinal foods. In this study, 11 species of dried and fresh, edible and medicinal macrofungi, *Ganoderma amboinense*, *Agaricus subrufescens*, *Dictyophora indusiata*, *Pleurotus sajor-caju*, *Pleurotus ostreatus*, *Pleurotus geesteranu*, *Hericium erinaceus*, *Stropharia rugosoannulata*, *Pleurotus sapidus*, *Antrodia camphorata*, and *Lentinus edodes* (Berk.) Sing, were investigated to determine the content of their nutritional components, including proteins, fat, carbohydrates, trace minerals, coarse cellulose, vitamins, and amino acids. The amino acid patterns and similarity of macrofungi were distinguished through principal component analysis and hierarchical cluster analyses, respectively. A total of 103 metabolic small molecules of macrofungi were identified by nuclear magnetic resonance spectroscopy and were aggregated by heatmap. Moreover, the macrofungi were classified by principal component analysis based on these metabolites. The results show that carbohydrates and proteins are two main components, as well as the nutritional ingredients, that differ among various species and varied between fresh and dried macrofungi. The amino acid patterns in *L. edodes* and *A. subrufescens* were different compared with that of the other tested mushrooms.

Keywords: macrofungi; proximate compositions; small molecules; metabonomics; NMR

1. Introduction

More than 12,000 species of macrofungi (mushrooms) have been found, and there are at least 2200 edible species in the world [1]. Thousands of years ago, China, Japan, Korea, and parts of Africa have used macrofungi for nutritional and medical purposes. Edible mushrooms also are cultivated for consumption worldwide because of their special valued nutrients [2–4]. *Agaricus subrufescens*, *Dictyophora indusiata*, *Pleurotus sajor-caju*, *Lentinus edodes* (Berk.) Sing, *Pleurotus geesteranu*, *Stropharia rugosoannulata*, and *Pleurotus sapidus* offer special umami tastes, making them delicious table dishes. *A. subrufescens*, whose fruiting body is called *"Himematsutake"* in Japan, is regarded as a health food and is a significant component of traditional Chinese medicine. The so-called queen of mushrooms, *D. indusiata*, is famous for its good-looking appearance and delicious taste. Its polysaccharides have been verified to prevent tumor and oxidant stress [5,6]. Also, it has been demonstrated that glucan-rich polysaccharides from *P. sajor-caju* can prevent glucose intolerance and inflammation in high-fat diet mice [7]. The extracts of mycelium of *Pleurotus ostreatus* have been identified to be rich in selenium and to prevent cancer [8]. Moreover, the triterpene of *Ganoderma amboinense* has been documented to induce the senescence of HepG2 cells [9]. In addition, the extracts of *Hericium erinaceus* and *Antrodia camphorata* can manage neurodegenerative diseases, which has been utilized as Chinese medicine and in functional foods (Figure 1) [10]. In recent years, the nutritional values of edible mushrooms have attracted more and more attention globally [3,4,11]. There are many other biological activities, such as antifungal, antibacterial, immunomodulatory, antiviral [12], anti-inflammatory [13], cholesterol-reducing [14], anti-cytotoxic, and antihypoglycemic [3,11,15] effects that have been reported. The various functions of fungi are closely correlated with their abundant nutrients. The fruiting bodies of mushrooms are considered as the main sources of organic nutrients, which include abundant proteins, digestible and non-digestible carbohydrates, fiber, vitamins, mineral contents, and a low level of fat [16–18]. Edible mushrooms possess a large resource of proteins with abundant essential amino acids, with their ratio among total amino acids reaching from 30% to 40% [19]. The digestible carbohydrates contained mannitol (0.3%–5.5% dried sample weight (dw)) [20], glucose (0.5%–3.6% dw) [21], and glycogen. Non-digestible carbohydrates take up a large part of the total carbohydrates, and mainly include oligosaccharides and non-starch polysaccharides, such as chitin, crude fibre, β-glucans, and mannans [22].

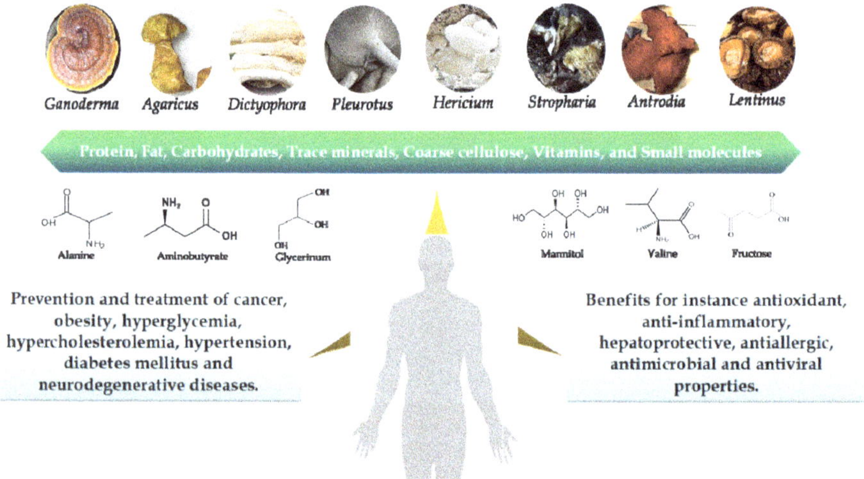

Figure 1. The beneficial roles of edible and medical mushrooms in human health.

In the last decade, the identification of mushrooms' proximate compositions and small molecules has mostly been investigated. Nonetheless, the broad, large-scale comparative metabolomics studies were lacking in terms of edible and medical mushrooms. In this study, the systematic broad-scale metabolomics of 11 species of mushrooms are investigated, including *G. amboinense*, *A. subrufescens*, *D. indusiata*, *P. sajor-caju*, *P. ostreatus*, *P. geesteranu*, *H. erinaceus*, *S. rugosoannulata*, *P. sapidus*, *A. camphorata*, and *L. edodes* (Berk.) Sing. Nuclear magnetic resonance (NMR) spectroscopy, a well-developed technique to perform metabolite profiling, was used to detect different metabolites and their assigned substances involved in metabolic pathways. Major metabolic and components differences between selected mushrooms and close relatives were also demonstrated. There is little relative research systematically focused on the metabolomic analysis of these mushrooms.

2. Materials and Methods

2.1. Cultivated Edible and Medical Mushrooms

The fruiting bodies of *G. amboinense*, *A. subrufescens*, *D. indusiata*, *P. sajor-caju*, *P. ostreatus*, *P. geesteranu*, *H. erinaceus*, *S. rugosoannulata*, *P. sapidus*, and *L. edodes* (Berk.) Sing were grown and harvested in the mushroom farm at the China National Engineering Research Center of JUNCAO Technology (Fuzhou, China). *A. camphorata* was obtained from the Institute of Food Science and Technology, National Taiwan University. The mushrooms were cut into pieces and were dried in an oven at 65 °C for 24 h, and then were grinded into powder by an ultrafine grinder (Hangzhou, China) for further detection. At the same time, the fresh mushrooms were smashed into particles with 2 mm diameter for comparison with the dried fungi. Then, the dried and fresh mushrooms were stored at −80 °C for further analysis.

2.2. Proximate Compositions and Trace Minerals

The chemical compositions (proteins, fat, and carbohydrates) of edible and medical mushrooms were analyzed by the Association of Official Analytical Chemists (AOAC) method [23]. The crude protein was estimated by Micro-Kjeldahl methods [24]. The crude fat was assessed through extracting a known weight of powdered sample with petroleum ether by Soxhlet apparatus. The carbohydrate contents of fungi were recalculated according to the glucose contents. The mineral elements of Cu, Fe, Zn, Mn, and Ca were detected by inductively coupled plasma-atomic emission spectrometry (ICP-AES) [24,25].

2.3. Coarse Fibers

The procedure of cellulose extraction was adjusted from the work done by Morán et al. [26]. The samples were washed several times by distilled water, dried by drying oven and cut to an approximate length of 6–12 mm. After that, the blocks were boiled with toluene/ethanol mixture (2:1, *v/v*) by Soxhlet apparatus for 6 h, then filtered and washed by ethanol for 30 min to be dried. The samples were incubated with 0.1 M NaOH (dissolved in 50% volume of ethanol) at 45 °C for 3 h under continuous agitation. The pre-treated samples were then treated with H_2O_2 at different concentrations (0.5%, 1.0%, 2.0%, and 3.0%, dissolved in buffer solution, pH 11.5) at 45 °C for 3 h under continuous agitation. Following that, they were incubated with 10% NaOH (*w/v*) and 1% $Na_2B_4O_7$ (*w/v*) at 28 °C for 15 h, and then with 70% HNO_3 and 80% HAc at 120 °C for 15 min. Finally, the extracts were washed by 95% ethanol, water, and 95% ethanol in turn, and dried at 60 °C in an oven until constant weight.

2.4. Vitamin A and C

Vitamin A content was analyzed by HPLC system with a fluorescence detector FP-2020 (Jasco, Tokyo, Japan) programmed at 320–330 nm [18,27]. The chromatography was used to identify the compounds by comparisons with standards (Sigma-Aldrich, Louis, MO, USA). On the basis of the fluorescence signal response of each standard, the internal standard method and calibration curves obtained were performed to quantify the content (µg) of vitamin A among 100 g dried sample weight (dw). The content of ascorbic acid (vitamin C) was measured by the 2,6-dichloroindophenol titrimetric method [28] on the basis of the calibration curve of L-ascorbic acid, in which the results were expressed as milligrams of ascorbic acid per 100 g of dw.

2.5. Amino Acids

Amino acid contents of samples were analyzed by gas chromatographic (GC) method as the previous study described [29]. Ten grams of each mushroom sample was defatted through extraction with 30 mL of the petroleum spirit three times by Soxhlet apparatus. The defatted samples were then hydrolyzed thoroughly at 112 °C for 24 h three times. The protein hydrolysates were extracted with 30 mL of dichloromethane three times until the final volume was 1.0 mL, then a GC system with a mass selective detector was used to test the amino acid content of the mushroom samples.

2.6. Sample Preparation and ^{1}H NMR Spectroscopic Measurement

The samples were defrosted at room temperature, and 250 µL aliquots were mixed with 250 µL of 1.5 M phosphate buffer (pH 7.4) to minimize variation in pH. The samples were centrifuged at 13,000 rpm for 10 min at 4 °C to separate any precipitate. Then, the filter liquor was added with 4, 4-dimethyl-4-silapentane-1-sulfonic acid (DSS) (50 µL) and vortexed for 10 s. Afterwards, the mixed liquor was centrifuged for 2 min at 13,000 rpm. Finally, the 480 µL of total supernatant was injected into the nuclear magnetic tube. ^{1}H NMR was measured on a DD2 600 MHz spectrometer (Agilent, CA, USA) operating at a 599.83 MHz magnet frequency and equipped with a triple-resonance cryoprobe. Therein, 256 scans were collected with a spectral width of 7225.434 Hz at 25 °C; the recycle delay time was set as 0.01 s and the water signals were suppressed during relaxation time. The free induction decay (FID) was transported into the Chenomx NMR suit (version 8.0, Edmonton, Alberta, Canada) software, and the ^{1}H NMR spectra were manually phased and baseline-corrected. The DSS at 0.0 ppm was used as reference for chemical shifts.

2.7. Statistical Analysis

For this study, data of amino acid derivatives and peptides, as well as metabolic profiling, were analyzed by unsupervised principle component analyses (PCA) and hierarchical cluster analyses (with complete linkage) using SIMCA-14.1 software (UMETRICS, Umea, Sweden) based on the Ward algorithm. For PCA, the data of mushrooms' components were normalized after filling the baseline value. Briefly, the logarithm of variance was calculated, and the Pareto scaling method was used for scaling. Then, a similar method was used in clustering analysis based on the results of PCA. The quantity and quality of metabolites were analyzed by Chenomx NMR suite (Version 8.0, Edmonton, Alberta, Canada). The relative abundances of small molecules from dried and fresh mushrooms were further visualized with heatmap by utilizing the R-3.2.2 software (Auckland University, Auckland, New Zealand) with a heatmap package.

3. Results and Discussion

3.1. Proximate Compositions and Trace Elements from Dried and Fresh Mushrooms

Edible mushrooms contain various kinds of nutrients, such as protein, fat, carbohydrates, and vitamins, which are the most essential macronutrients in human life with comprehensive nutritional values. Proximate compositions and trace elements from dried and fresh edible mushrooms were determined as shown in Figure 2 and Table 1. Eleven edible and medical mushrooms could be classified as follows: Agaricales (*A. subrufescens*, *P. sajor-caju*, *P. ostreatus*, *P. geesteranu*, *S. rugosoannulata*, *P. sapidus*, *L. edodes* (Berk.) Sing), Aphyllophorales (*G. amboinense*, *A. camphorata*), Phallales (*D. indusiata*), and Russulales (*H. erinaceus*). The content of nutrients varied in different species of fresh or dried mushrooms. Moreover, there was a vast difference in nutritional compositions between DI-F (*D. indusiata* (fresh)) and DI-D (*D. indusiata* (dried)), or PG-F (*P. geesteranu* (fresh)) and PG-D (*P. geesteranu* (dried)) (Table 1). The level of fat content was low in all mushrooms, except fresh *A. subrufescens* (15.7 g/100 g), which was similar to previous research, such as in the content of crude fat being 10.7 g/100 g in *Agricus* as reported by Liu et al. (2019) [30,31]. The protein and carbohydrates are the two main compounds of mushrooms, and the content of protein in Agricales mushrooms was generally higher than other mushrooms (Figure 2), with the highest content being found in fresh *A. subrufescens* (50.4 g/100 g) (Table 1). However, the levels of carbohydrates in Aphyllophorales, Phallales, and Russulales were higher than Agricales mushrooms. The total fibre of Aphyllophorales was also found to be higher than other, and the highest level of total fiber (40.8%) was presented in dried *G. subrufescens*. Dried *A. subrufescens* had the lowest (0.075%) content of total fiber, which was same as previous research results [32,33]. Above all, the rich amount of proteins, essential amino acids, carbohydrates, and essential minerals, in contrast to low fat levels, make many mushrooms a good choice for consumers. Vitamins and mineral elements play an indispensable role in satisfying bodily demand to promote health. *D. indusiata* and *H. erinaceus* showed the highest contents of vitamin A, with 770 μg/100 g and 550 μg/100 g, respectively. However, it was not detected in *A. camphorata*. Dried *D. indusiata* contained the highest level of vitamin C (111.4 mg/100g), and dried *A. subrufescens* (69.7 mg/100 g) followed. The trace elements, especially calcium, were abundantly found in Aphyllophorales and Phallales mushrooms (Figure 2). Dried *D. indusiata* had the highest contents of Zn (130 mg/kg), Fe (184 mg/kg), and Mn (75 mg/kg), which is consistent with previous research [34]. The highest levels of Cu and Ca were found in dried *A. subrufescens* and *A. camphorata*, respectively. Furthermore, dried and fresh samples had nearly the same composition. HE-F (*H. erinaceus* (fresh)) had the same composition with that in AS-F (*A. subrufescens* (fresh)), but the content of carbohydrate in HE-F was more than twice that of AS-F, while the protein content was twice more than that in GA-F (*G. amboinense* (fresh)).

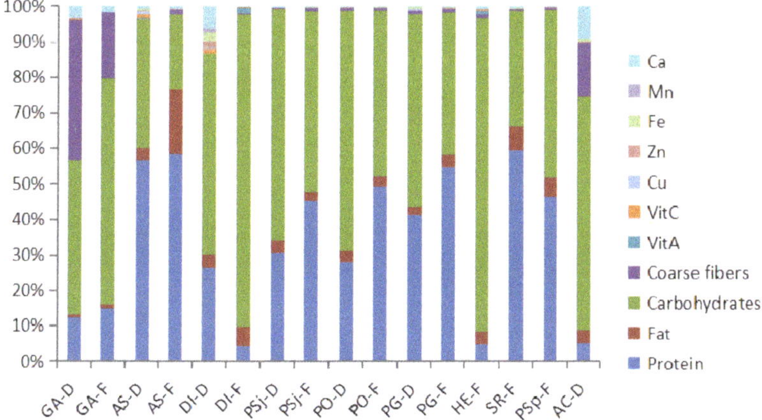

Figure 2. Proximate compositions and trace elements from the dried and fresh mushrooms. Note: GA-D, *G. amboinense* (dried); GA-F, *G. amboinense* (fresh); AS-D, *A. subrufescens* (dried); AS-F, *A. subrufescens* (fresh); DI-D, *D. indusiata* (dried); DI-F, *D. indusiata* (fresh); PSj-D, *P. sajorcaju* (dried); PSj-F, *P. sajorcaju* (fresh); PO-D, *P. ostreatus* (dried); PO-F, *P. ostreatus* (fresh); PG-D, *P. geesteranu* (dried); PG-F, *P. geesteranu* (fresh); HE-F, *H. erinaceus* (fresh); SR-F: *S. rugosoannulata* (fresh); PSp-F, *P. sapidus* (fresh); AC-D, *A. camphorata* (dried).

Table 1. Proximate compositions and trace elements from the dried and fresh edible mushrooms.

Macrofungi	Proximate Composition						Trace Elements (mg/kg)				
	Protein (g/100 g)	Fat (g/100 g)	Carbohydrates (g/100 g)	Coarse Fibers (g/100 g)	Vitamin A (μg/100 g)	Vitamin C (mg/100 g)	Cu	Zn	Fe	Mn	Ca
GA-D	12.7	0.8	45	40.8	15.2	10.5	4.7	19.2	13.4	11	321
GA-F	14.6	1	61	17.7	17.8	1.6	2.2	5.9	3.9	2.2	172
AS-D	37.2	2.2	23.9	0.075	<16	69.7	42.6	28	28.4	1.1	50.2
AS-F	50.4	15.7	18	1.2	7.38	1.1	14.6	13.5	14.4	0.9	53.8
DI-D	19.5	2.7	41.6	0.087	21	111.4	15	130	184	75	468
DI-F	1.9	2.6	40.2	0.085	770	16.7	0.83	0.39	1.7	-	4.1
PSj-D	23.2	2.7	49.4	0.4	1.46	0.6	0.6	4.8	5.4	0.7	14
PSj-F	35.5	1.98	40.1	0.8	4.05	0.6	1.9	9	0.2	1.2	18
PO-D	21.1	2.19	50.7	0.7	6.51	0.7	3.6	6.8	15	0.8	7
PO-F	36.4	2.06	34.7	0.5	4.42	0.5	2.5	7.7	14	1	11
PG-D	32.7	1.71	42.8	0.9	21.9	0.6	28	15	25	1.1	24
PG-F	39.8	2.26	29	0.8	2.98	0.8	6.6	13	15	1.4	22
HE-F	2.3	1.8	42.9	0.42	580	5.5	6.1	10.4	0.15	0.079	39
SR-F	41.1	4.5	22.4	0.4	1.5	1.6	2.1	3.4	13	1.4	32
PSp-F	33.8	4.1	34.6	0.5	2.61	0.5	1.2	8.4	10	1.2	8.5
AC-D	4.1	3	52.8	12.2	-	0.6	3.2	27	38	12	740

Note: GA-D, *G. amboinense* (dried); GA-F, *G. amboinense* (fresh); AS-D, *A. subrufescens* (dried); AS-F, *A. subrufescens* (fresh); DI-D, *D. indusiata* (dried); DI-F, *D. indusiata* (fresh); PSj-D, *P. sajorcaju* (dried); PSj-F, *P. sajorcaju* (fresh); PO-D, *P. ostreatus* (dried); PO-F, *P. ostreatus* (fresh); PG-D, *P. geesteranu* (dried); PG-F, *P.geesteranu* (fresh); HE-F, *H. erinaceus* (fresh); SR-F: *S. rugosoannulata* (fresh); PSp-F, *P. sapidus* (fresh); AC-D, *A. camphorata* (dried). Protein and fat were calculated by dried basis. "-": not detected.

3.2. Amino Acid Derivatives and Peptides

Amino acids can be divided into essential (eight kinds of amino acids for adults and nine kinds for infants), non-essential, and conditionally essential amino acids (cysteine (Cys) and tyrosine (Tyr)). The amino acids in these selected edible and medical mushrooms were of high quality and substantial similarity. The mushrooms mostly contained nine essential, six non-essential, and one kind of conditionally essential amino acids. To some extent, they can virtually be substituted for meat, eggs, and milk [19]. However, the contents of amino acids in fresh mushrooms were higher compared with dried samples, and different types of amino acids were varied in those selected mushrooms (Figure 3a and Table 2). The highest levels of total essential and conditionally essential amino acids were both found in fresh *A. blazei* at 133.7 and 5.9 g/kg, respectively. Total non-essential amino acids were found in *L. edodes* at 110.1 g/kg, which was slightly less than that of 144 g/kg tested by Li et al. (2018) [35]. *A. subrufescens* had the highest essential amino acid content, except for methionine (Met). Moreover, other mushrooms were also found to be rich in essential amino acids, especially *L. edodes* and *G. subrufescens*. Glutamate (Glu) is a major component of non-essential amino acids, which exists largely in selected mushrooms. In addition, tyrosine (Tyr), as a conditionally essential amino acid, was distributed in all selected mushrooms, especially in *L. edodes* (4.8 g/kg), *A. subrufescens* (5.9 g/kg), and *G. subrufescens* (2.8 g/kg).

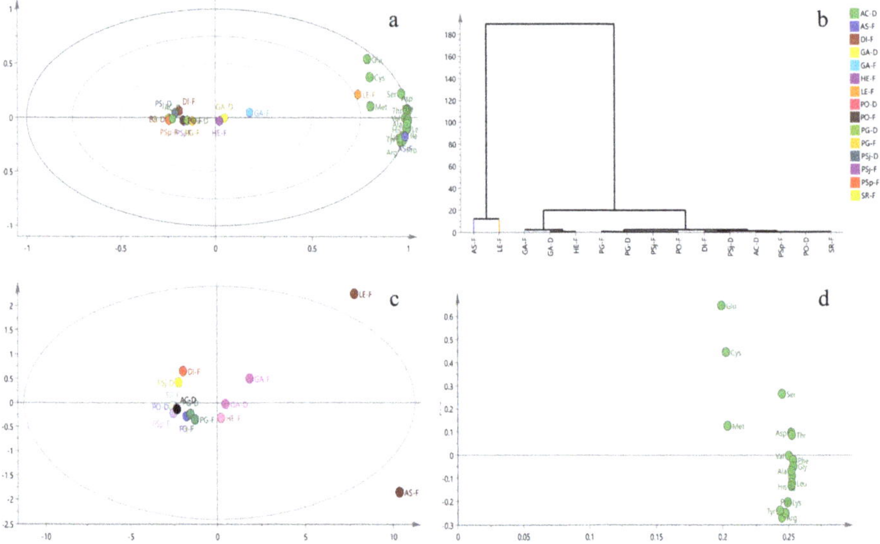

Figure 3. Biplot (**a**), clustering analysis (**b**), score plot (**c**), and loading plot (**d**) of amino acids in dried and fresh edible mushrooms. Note: GA-D, *G. amboinense* (dried); GA-F, *G. amboinense* (fresh); AS-D, *A. subrufescens* (dried); AS-F, *A. subrufescens* (fresh); DI-D, *D. indusiata* (dried); DI-F, *D. indusiata* (fresh); PSj-D, *P. sajorcaju* (dried); PSj-F, *P. sajorcaju* (fresh); PO-D, *P. ostreatus* (dried); PO-F, *P. ostreatus* (fresh); PG-D, *P. geesteranu* (dried); PG-F, *P. geesteranu* (fresh); HE-F, *H. erinaceus* (fresh); SR-F: *S. rugosoannulata* (fresh); PS-F, *P. sapidus* (fresh); AC-D, *A. camphorata* (dried); LE-F, *L. edodes* (Berk.) Sing (fresh).

Table 2. The amino acid composition of the dried and fresh edible mushrooms (dried weight: g/kg).

Amino Acid	GA-D	GA-F	AS-F	DI-F	PSj-D	PSj-F	PO-D	PO-F	PG-D	PG-F	HE-F	SR-F	PSp-F	AC-D	LE-F
TEAA	60.5	72	133.7	9.9	7.5	13.8	7.5	13.3	15.3	17.4	36.8	7.3	6.7	6	107.7
TNAA	23.5	30	91.2	6.4	6.5	10.3	6	10.6	12.4	14	18.1	5.4	4.3	7.4	110.1
CEAA	1.4	2.8	5.9	1.9	1.5	1.2	0.6	1.2	1.3	1.5	2.7	0.4	0.9	0.8	4.8
Asp	5.8	8.1	17.7	1.4	0.9	2.2	1.2	2.1	2.7	3.1	5.3	0.9	0.6	1.9	16.8
Thr	3.6	4.8	11	0.9	0.8	1.5	0.8	1.4	1.7	1.9	3.2	0.8	0.4	1.2	10.3
Met	33.2	36.3	33.5	0.7	0.4	0.9	0.5	0.8	0.8	0.8	11.6	0.2	0.2	-	29.7
Ile	2.7	3.4	9.8	0.7	0.7	1.3	0.6	1.2	1.5	1.6	2	0.6	0.8	1.1	7.1
Leu	4.2	5.3	16.8	1.2	1.1	2	1	1.9	2.1	2.6	3.8	1.3	1.4	1.6	11.7
Val	3.5	4.7	11.9	2.7	1.6	2.3	1.6	2.4	2.4	2.7	2.9	1.6	1.8	-	9.6
Phe	2.7	3.8	10.9	0.9	0.7	1	0.6	1	1.3	1.3	2.3	0.6	0.8	0.9	8.5
Lys	3.5	3.9	16.6	1	0.9	1.9	0.9	1.8	2	2.5	4.1	0.9	0.3	0.8	10.2
His	1.3	1.7	5.5	0.4	0.4	0.7	0.3	0.7	0.8	0.9	1.6	0.4	0.4	0.4	3.8
Ser	3.9	4.7	8.8	0.9	0.6	1.4	0.8	1.3	1.5	1.7	3	0.7	0.1	1.3	10.8
Clu	6.6	8.6	21.6	2.4	2.5	3.1	2.1	3.5	4.9	5.1	1.8	1.8	1.5	2.1	60.8
Gly	3.4	4.3	11.6	0.7	0.8	1.3	0.7	1.2	1.6	1.8	2.6	0.6	0.9	1.2	8.9
Ala	4	5.3	16	1.1	1.4	2.7	1.3	2.5	2.8	3.1	3.1	1.1	1	1.3	12.2
Arg	2.7	3.6	20.9	0.7	0.1	0.5	0.2	0.9	0.3	0.6	5.2	0.3	0.2	0.4	10.6
Pro	2.9	3.5	12.3	0.6	1.1	1.3	0.9	1.2	1.3	1.7	2.4	0.9	0.6	1.1	6.8
Cys	0.1	1	1.5	1.1	0.8	-	-	-	-	-	0.7	-	-	-	1.9
Tyr	1.3	1.8	4.4	0.8	0.7	1.2	0.6	1.2	1.3	1.5	2	0.4	0.9	0.8	2.9

Note: GA-D, *G. amboinense* (dried); GA-F, *G. amboinense* (fresh); AS-D, *A. subrufescens* (dried); AS-F, *A. subrufescens* (fresh); DI-D, *D. indusiata* (dried); DI-F, *D. indusiata* (fresh); PSj-D, *P. sajorcaju* (dried); PSj-F, *P. sajorcaju* (fresh); PO-D, *P. ostreatus* (dried); PO-F, *P. ostreatus* (fresh); PG-D, *P. geesteranu* (dried); PG-F, *P. geesteranu* (fresh); HE-F, *H. erinaceus* (fresh); SR-F: *S. rugosoannulata* (fresh); PS-F, *P. sapidus* (fresh); AC-D, *A. camphorata* (dried); LE-F, *L. edodes* (Berk.) Sing (fresh). TEAA: total essential amino acids; TNAA: total non-essential amino acids; CEAA: conditionally essential amino acids; "-": not detected.

Amino acids of mushrooms were classified by principal component analysis (PCA) as an unsupervised multidimensional statistical analysis method. Meanwhile, the principal components were acquired based on the content of metabolites which were measured by NMR. The model of all samples explained 95.03% of the principal components, with principal component 1 (PC1) interpreting 90.9% and principal component 2 (PC2) interpreting 4.13% of the total variance. The biplot indicated that the fresh *L. edodes* (Berk.) Sing and fresh *A. subrufescens* were remarkably separate from other mushrooms according to PC1 (Figure 3a). GA-D (*G. amboinense* (dried)), HE-F, GA-F, LE-F (*L. edodes* (Berk.) Sing (fresh)), and AS-F were located at PC1 with positive scores, nevertheless other mushrooms presented negative scores on PC1, suggesting that LE-F and AS-F were completely dissimilar in amino acid patterns compared with other mushrooms. LE-F and AS-F were further segregated on the direction of PC2, indicating the amino acid profiles were also different between them. The nutrients compositions of AS-F and AE-F were different with that of the other mushrooms (Figure 3c). The higher absolute value of data in loading plot meant the bigger contribution to the principle compounds. The loading plot of amino acids showed that Glu, Cys, and Met were the most contributive principles discriminative of LE-F, while Val, Phe, Gly, Ala, His, Ile, Pro, Lys, and Arg of AS-F were different compared with other mushrooms (Figure 3d). Hierarchical cluster analysis, a method to quantify the similarity of different mushrooms, was carried out based on amino acid profiles. All samples could be classified into three clusters if the phenon line was defined as the distance of 20. Cluster I was composed of AS-F and LE-F. Cluster II was made up with GA-D, GA-F, and HE-F. Most mushrooms were gathered at Cluster III, which included PG-F, PG-D, PSj-F (*P. sajorcaju* (fresh)), PO-F (*P. ostreatus* (fresh)), DI-F, PSj-D (*P. sajorcaju* (dried)), AC-D (*A. camphorata* (dried)), PSp-F (*P. sapidus* (fresh)), PO-D (*P. ostreatus* (dried)), and SR-F (*S. rugosoannulata* (fresh)) (Figure 3b).

3.3. Metabolic Profiling of Selected Mushrooms

A total of 103 different kinds of small molecules, including organic acids, amino acids, polyols, amines, sugars, vitamins, esters, and others were determined in the dried and fresh mushrooms by NMR (Table S1 and Figure S1). It was identified that the PG-D had the most content of metabolites, which was followed by fresh *P. sapidus* and dried *H. erinaceus*. 4-Aminobutyrate, formate, fumarate, glucose, glutamine, isoleucine, methanol, serine, sn-glycero-3-phosphocholine, uridine, and valine were found to be existing in all tested fungi, and the content of them in different fungi was diverse. The total content of trehalose in all mushrooms was the most prevalent, especially present in fresh *P. sajorcaju*, followed by mannitol (Table S1). Total glucose also made up a relatively significant share in all water-soluble small molecules, especially in DI-D, and the glucose content of *D. indusiata* was similar to the former study [36]. Mannitol existed in all selected mushrooms, which can support and expand the mushroom fruiting bodies [30].

Metabolites of edible and medical mushrooms were classified by PCA (Figure 4a). Meanwhile, the principal components were acquired based on the content of metabolites using NMR (Table S1). The model of all samples explained 41.4% of the principal components, with principal component 1 (PC1) interpreting 24.9% and principal component 2 (PC2) interpreting 16.5% of the total variance. The mushrooms were categorized into three clusters based on metabolites using the ward algorithm: Cluster I (HE-F, AS-D (*A. subrufescens* (dried)), and AS-F), Cluster II (PSj-F, PG-D, PSj-D, PO-D, and PO-F), and Cluster III (SR-F, LE-F, GA-D, GA-F, HE-D (*H. erinaceus* (dried)) and DI-D) (Figure 4b). The loading plot of metabolites showed that different metabolites were distributed in different quadrants (Figure 4c,d). For example, uridine (C98), mannitol (C61), propylene glycol (C79), proline (C78), inosine (C50), and valine (C100) were the main important discriminative factors contributing to HE-F from other mushrooms. Hierarchical cluster analyses of different mushrooms based on metabolites showed that samples could be sorted into four clusters if the phenon line was defined as the distance of 100. Cluster I was composed of HE-F; Cluster II was made up with AS-F and AS-D; Cluster III contained DI-D, HE-D, GA-F, and GA-D. Most mushrooms were sorted on Cluster IV, which included PSj-F, PG-F, LE-F, SR-F, PO-F, PO-D, PG-D, PSj-D, and PSp-F. The compositions of metabolites were different between fresh and dried mushrooms, especially in *P. sajorcaju* and *H. erinaceus* (Figure 4a,b). Primary metabolites, especially saccharides, lipids, and amino acids, could affect plants' own growth and involvement in the biosynthesis of necessary materials [37,38]. The content of metabolites from the same genus may have great differences, such as *Agricale* with different species. The metabolites of mushrooms contained abundant sugars. Mannitol, trehalose, and glucose were the main constituents in the tested mushrooms, which coincided with previous studies [39]. Sugars had great effects, not only in cellular energy metabolism, but also on the formation of structural polysaccharides [40,41]. The various alcohol derivatives, especially mannitol and arabinitol transformed from sugars, were reported to support the growth of the fruiting bodies of mushrooms [30]. Except for sugars and amino acids, 2-octenoate and 1-octen-3-ol, which belong to the "fungi alcohol", were important elements for the special fragrance present in mushrooms [41]. Aspartic and glutanmic acids in mushrooms could give the most typical mushroom taste, umami, or palatable taste [42]. The fresh mushrooms were immensely distinguished from dried ones in compositions and metabolites.

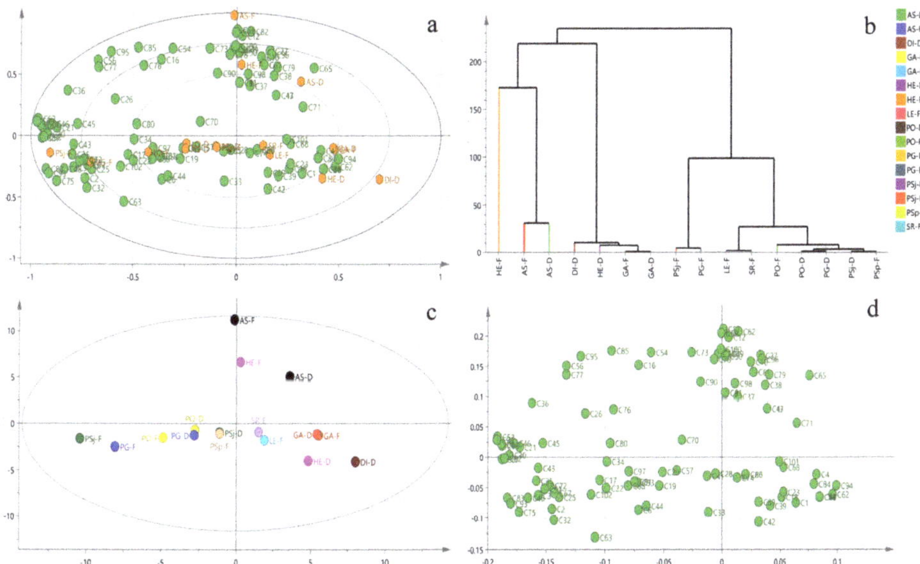

Figure 4. Biplot (**a**), clustering analysis (**b**), score plot (**c**), and loading plot (**d**) of metabolites.

The same results could be obtained by heatmap analysis (Figure 5). HE-F was rich in lysine, acetate, methionine, ethanolamine, trimethylamine, uracia, choline, proline, ornithine, arginine, succinate, sarcosing, urocanate, 3-methyl-2-oxovalerate, 2-hydroxyisovutyrate, 2-oxoglutarate, oxypurinol, malonate, and isobutyrate. In contrast, HE-D had more abundant malate, glutamine, arabinitol, and mannitol. Some compositions of metabolites existed in all mushrooms, but others were only found in one or few mushrooms. The metabolites with glucose, methanol, fumarate, uridine, glutamine, serine, alanine, isoleucine, valine, 4-aminobutyrate, formate, and sn-glycero-3-phosphocholine were found in all selected mushrooms (Figure 5). The metabolites, such as 3-methyl-2-oxovalerate, 2-oxoisocaproate, 2-hydroxyisobutyrate, and 2-oxoglutarate, were only found in HE-F, while 4-hydroxybenzoate and guanosine only existed in AS-D, and pantothenate was only found in PSj-D, PO-F, and DI-D. The metabolites of HE-F were also quite different from HE-D.

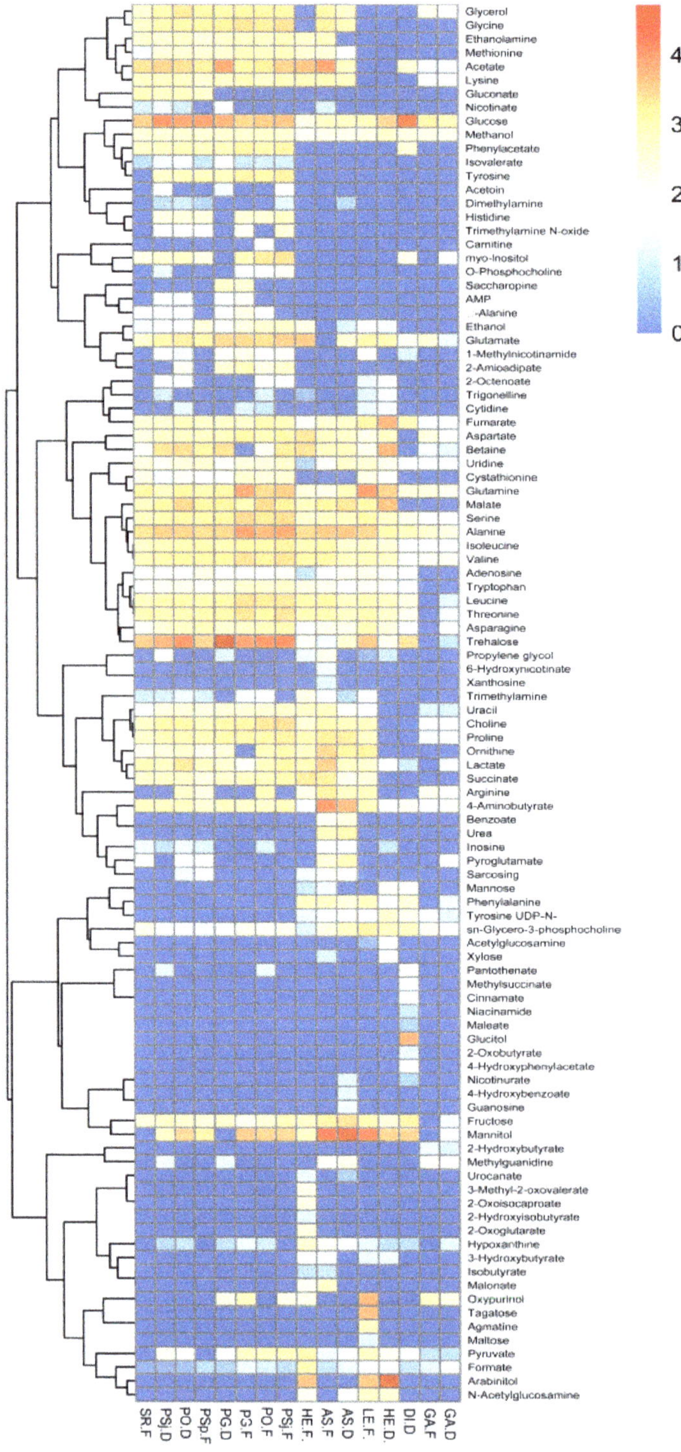

Figure 5. Heatmap of small molecules from the dried and fresh mushrooms. The red color indicates the content more than 2.5 mM, and the blue color indicates the content less than 2.5 mM.

Clustering analysis of compositions from dried and fresh edible and medical mushrooms showed that they could be divided into three kinds of mushrooms with the distance for 1000 (Figure 6). Cluster I was composed of HE-F, GA-D, GA-F, AC-D, and DI-F. Cluster II was made up of AS-F, AS-D, PG-F, and SR-F. Cluster III contained DI-D, PSj-D, PO-D, PSj-F, PG-D, PO-F, and PSp-F. The metabolic compositions of mushrooms showed great differences, even in the same order. Cluster II and Cluster III were species of *Agaricales*. However, these two clusters had much difference. Interestingly, there was a difference in terms of the mushrooms' composition. For instance, the dried and fresh *D. indusiata* were divided into two clusters, which may be caused by the big difference between them. Moreover, the fresh *G. amboinense* were similar in composition with dried *A. camphorata*. It was demonstrated that the drying process may influence the composition of mushrooms. The different functions of the mushrooms may be closely associated with the nutrient (protein, fat, carbohydrate, vitamins, trace minerals, and amino acids) composition of mushrooms. These data would be a desirable choice for analyzing the functions of mushrooms.

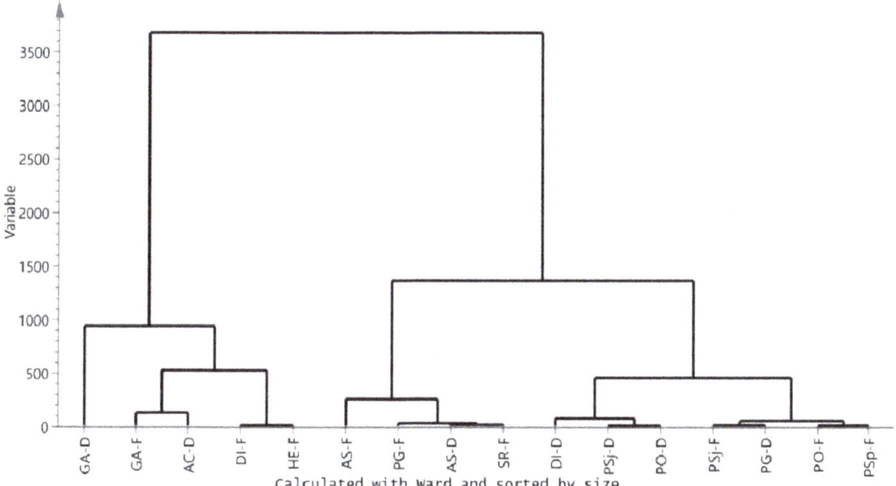

Figure 6. Clustering analysis of compositions from the dried and fresh edible mushrooms.

4. Conclusions

Mushrooms are traditionally regarded as nutritional and delicious foods worldwide. Because of their abundant bioactive phytochemicals, they are also generally present in traditional Chinese medicine. In this study, we presented a systematic broad-scale metabolomic investigation of 11 species of dried and fresh edible and medicinal mushrooms. The nutritional component analysis of these selected 11 species suggested that mushrooms contained a wide range of proteins, carbohydrates, amino acids, vitamins, and small molecules. The results showing the chemical components of the selected mushrooms provide fundamental data for the development of functional foods from mushrooms.

Supplementary Materials: The following are available online at http://www.mdpi.com/2304-8158/8/9/397/s1, Table S1: The concentrations of small molecules from dried and fresh edible mushrooms measured by nuclear magnetic resonance spectroscopy. Figure S1: The nuclear magnetic resonance (NMR) spectra of mushrooms.

Author Contributions: Methodology, C.-F.Y.; software, D.L. and Y.-Q.C.; formal analysis, C.-F.Y.; writing—original draft preparation, D.L. and Y.-Q.C; writing—review and editing, X.-W.X. and R.-T.Z.; supervision, C.Z. and B.L.; project administration, C.Z.; funding acquisition, C.Z.

Funding: This work was supported by the Science and Technology Innovation Projects (CXZX2017291 and CXZX2017024) of the Fujian Agriculture and Forestry University, the Key Project of the Fuzhou Municipal Bureau of Science and Technology (2017-N-36), the China National Engineering Research Center of JUNCAO Technology (JCGG1406), and the Key Laboratory of the Marine Biotechnology of Fujian Province.

Conflicts of Interest: The authors declare no conflict of interest.

References

1. Wu, S.R.; Luo, X.L.; Liu, B.; Gui, M.Y. Analyse and advise to research and development of wild edible fungi. *Food Sci. Technol.* **2010**, *35*, 100–103.
2. Zhao, C.; Fan, J.; Liu, Y.; Guo, W.; Cao, H.; Xiao, J.; Wang, Y.; Liu, B. Hepatoprotective activity of *Ganoderma lucidum* triterpenoids in alcohol-induced liver injury in mice, an iTRAQ-based proteomic analysis. *Food Chem.* **2019**, *271*, 148–156. [CrossRef] [PubMed]
3. Chen, Y.Q.; Liu, D.; Wang, D.Y.; Lai, S.S.; Zhong, R.T.; Liu, Y.Y.; Yang, C.F.; Liu, B.; Sarker, M.M.R.; Zhao, C. Hypoglycemic activity and gut microbiota regulation of a novel polysaccharide from *Grifola frondosa* in type 2 diabetic mice. *Food Chem. Toxicol.* **2019**, *126*, 295–302. [CrossRef] [PubMed]
4. Zhao, C.; Yang, C.F.; Wai, S.T.C.; Zhang, Y.B.; Portillo, M.Y.; Paoli, P.; Wu, Y.J.; Cheang, W.S.; Liu, B.; Carpéné, C.; et al. Regulation of glucose metabolism by bioactive phytochemicals for the management of type 2 diabetes mellitus. *Crit. Rev. Food Sci. Nutr.* **2019**, *59*, 830–847. [CrossRef] [PubMed]
5. Deng, C.; Fu, H.; Teng, L.; Hu, Z.; Xu, X.; Chen, J. Anti-tumor activity of the regenerated triple-helical polysaccharide from *dictyophora indusiata*. *Int. J. Biol. Macromol.* **2013**, *61*, 453–458. [CrossRef] [PubMed]
6. Hua, Y.; Yang, B.; Tang, J.; Ma, Z.; Gao, Q.; Zhao, M. Structural analysis of water-soluble polysaccharides in the fruiting body of *Dictyophora indusiata* and their in vivo antioxidant activities. *Carbohydr. Polym.* **2012**, *87*, 343–347. [CrossRef]
7. Kanagasabapathy, G.; Kuppusamy, U.R.; Malek, S.N.A.; Abdulla, M.A.; Chua, K.H.; Sabaratnam, V. Glucan-rich polysaccharides from *Pleurotus sajor-caju*(fr.) singer prevents glucose intolerance, insulin resistance and inflammation in c57bl/6j mice fed a high-fat diet. *BMC Complementary Altern. Med.* **2012**, *12*, 261. [CrossRef]
8. Cao, X.Y.; Liu, J.L.; Yang, W.; Hou, X.; Li, Q.J. Antitumor activity of polysaccharide extracted from *Pleurotus ostreatus* mycelia against gastric cancer in vitro and in vivo. *Mol. Med. Rep.* **2015**, *12*, 2383. [CrossRef]
9. Chang, U.M.; Li, C.H.; Lin, L.I.; Huang, C.P.; Kan, L.S.; Lin, S.B. Ganoderiol F, a *Ganoderma* triterpene, induces senescence in hepatoma HepG2 cells. *Life Sci.* **2006**, *79*, 1129–1139. [CrossRef]
10. Phan, C.W.; David, P.; Naidu, M.; Wong, K.H.; Sabaratnam, V. Therapeutic potential of culinary-medicinal mushrooms for the management of neurodegenerative diseases: Diversity, metabolite, and mechanism. *Crit. Rev. Biotechnol.* **2015**, *35*, 355–368. [CrossRef]
11. Chen, Y.Q.; Liu, Y.Y.; Sarker, M.M.; Yan, X.; Yang, C.F.; Zhao, L.N.; Lv, X.C.; Liu, B.; Zhao, C. Structural characterization and antidiabetic potential of a novel heteropolysaccharide from *Grifola frondosa* via IRS1/PI3K-JNK signaling pathways. *Carbohydr. Polym.* **2018**, *198*, 452–461. [CrossRef] [PubMed]
12. Zhao, C.; Gao, L.; Wang, C.; Liu, B.; Jin, Y.; Xing, Z. Structural characterization and antiviral activity of a novel heteropolysaccharide isolated from *Grifola frondosa* against enterovirus 71. *Carbohydr. Polym.* **2016**, *144*, 382–389. [CrossRef] [PubMed]
13. Komura, D.L.; Carbonero, E.R.; Gracher, A.H.P.; Baggio, C.H.; Freitas, C.S.; Marcon, R.; Santos, A.R.S.; Gorin, P.A.J.; Iacomini, M. Structure of *Agaricus* spp. fucogalactans and their anti-inflammatory and antinociceptive properties. *Bioresour. Technol.* **2010**, *101*, 6192–6199. [CrossRef] [PubMed]
14. Jeong, S.C.; Jeong, Y.T.; Yang, B.K.; Islam, R.; Koyyalamudi, S.R.; Pang, G.; Cho, K.Y.; Song, C.H. White button mushroom (*Agaricus bisporus*) lowers blood glucose and cholesterol levels in diabetic and hypercholesterolemic rats. *Nutr. Res.* **2010**, *30*, 49–56. [CrossRef] [PubMed]
15. Dey, B.; Bhunia, S.K.; Maity, K.K.; Patra, S.; Mandal, S.; Maiti, S.; Maiti, T.K.; Sikdar, S.R.; Islam, S.S. Glucans of *Pleurotus florida* blue variant: Isolation, purification, characterization and immunological studies. *Int. J. Biol. Macromol.* **2012**, *50*, 591–597. [CrossRef]
16. Barros, L.; Cruz, T.; Baptista, P.; Estevinho, L.M.; Ferreira, I.C. Wild and commercial mushrooms as source of nutrients and nutraceuticals. *Food Chem. Toxicol.* **2007**, *46*, 2742–2747. [CrossRef]
17. Zhao, C.; Liu, Y.; Lai, S.; Cao, H.; Guan, Y.; San Cheang, W.; Liu, B.; Zhao, K.; Miao, S.; Riviere, C.; et al. Effects of domestic cooking process on the chemical and biological properties of dietary phytochemicals. *Trends Food Sci. Technol.* **2019**, *85*, 55–66. [CrossRef]

18. Pereira, E.; Barros, L.; Martins, A.; Ferreira, I.C.F.R. Towards chemical and nutritional inventory of Portuguese wild edible mushrooms in different habitats. *Food Chem.* **2012**, *130*, 394–403. [CrossRef]

19. Wang, X.M.; Zhang, J.; Wu, L.H.; Zhao, Y.L.; Li, T.; Li, J.Q.; Wang, Y.Z.; Liu, H.G. A mini-review of chemical composition and nutritional value of edible wild-grown mushroom from China. *Food Chem.* **2014**, *151*, 279–285. [CrossRef]

20. Vaz, J.A.; Barros, L.; Martins, A.; Santos-Buelga, C.; Vasconcelos, M.H.; Ferreira, I.C.F.R. Chemical composition of wild edible mushrooms and antioxidant properties of their water soluble polysaccharidic and ethanolic fractions. *Food Chem.* **2011**, *126*, 610–616. [CrossRef]

21. Minyoung, K.; Lllmin, C.; Sunjoo, L.; Joungkuk, A.; Eunhye, K.; Mijung, K.; Sunlim, K.; Hyungin, M.; Heemyong, R.; Kang, E.Y. Comparison of free amino acid, carbohydrates concentrations in Korean edible and medicinal mushrooms. *Food Chem.* **2009**, *113*, 386–393.

22. Pck, C. The nutritional and health benefits of mushrooms. *Nutr. Bull.* **2010**, *35*, 292–299.

23. AOAC. *Official Methods of Analysis of AOAC International*, 16th ed.; Cunniff, P.A., Ed.; AOAC International: Arlington, VA, USA, 1995; Volume 1.

24. Allen, W.F. A micro-Kjeldahl method for nitrogen determination. *Oil Fat Ind.* **1931**, *10*, 391–397. [CrossRef]

25. Brzostowski, A.; Falandysz, J.; Jarzyńska, G.; Zhang, D. Bioconcentration potential of metallic elements by Poison Pax (*Paxillus involutus*) mushroom. *J. Environ. Sci. Health Part A* **2011**, *46*, 378–393. [CrossRef] [PubMed]

26. Brzostowski, A.; Jarzyńska, G.; Kojta, A.K.; Wydmańska, D.; Falandysz, J. Variations in metal levels accumulated in Poison Pax (*Paxillus involutus*) mushroom collected at one site over four years. *J. Environ. Sci. Health Part A* **2011**, *46*, 581–588. [CrossRef] [PubMed]

27. Morán, J.I.; Alvarez, V.A.; Cyras, V.P.; Vázquez, A. Extraction of cellulose and preparation of nanocellulose from sisal fibers. *Cellulose* **2008**, *15*, 149–159. [CrossRef]

28. Leal, A.R.; Barros, L.; Barreira, J.C.M.; Sousa, M.J.; Martins, A.; Santos-Buelga, C.; Ferreira, I.C.F.R. Portuguese wild mushrooms at the "pharma–nutrition" interface: Nutritional characterization and antioxidant properties. *Food Res. Int.* **2013**, *50*, 1–9. [CrossRef]

29. Grangeia, C.; Heleno, S.A.; Barros, L.; Martins, A.; Ferreira, I.C.F.R. Effects of trophism on nutritional and nutraceutical potential of wild edible mushrooms. *Food Res. Int.* **2011**, *44*, 1029–1035. [CrossRef]

30. Barros, L.; Venturini, B.A.; Baptista, P.; Estevinho, L.M.; Ferreira, I.C.F.R. Chemical composition and biological properties of portuguese wild mushrooms: A comprehensive study. *J. Agric. Food Chem.* **2008**, *56*, 3856–3862. [CrossRef]

31. Liu, P.H.; Yuan, J.; Jiang, Z.H.; Wang, Y.X.; Weng, B.Q.; Li, G.X. A lower cadmium accumulating strain of *Agaricus brasiliensis* produced by ^{60}Co-γ-irradiation. *LWT Food Sci. Technol.* **2019**, *114*, 108370. [CrossRef]

32. Kalač, P. Chemical composition and nutritional value of European species of wild growing mushrooms: A review. *Food Chem.* **2009**, *113*, 9–16. [CrossRef]

33. Ouzouni, P.K.; Petridis, D.; Koller, W.D.; Riganakos, K.A. Nutritional value and metal content of wild edible mushrooms collected from West Macedonia and Epirus, Greece. *Food Chem.* **2009**, *115*, 1575–1580. [CrossRef]

34. Jiang, Z.H. An analysis of amino acids in *A. blazei* fruitbodies cultured by three different composts. *Acta Edulis Fungi* **1999**, *6*, 30–34. (In Chinese)

35. Li, S.; Wang, A.; Liu, L.; Tian, G.; Wei, S.; Xu, F. Evaluation of nutritional values of shiitake mushroom (*Lentinus edodes*) stipes. *J. Food Meas. Charact.* **2018**, *12*, 2012–2019. [CrossRef]

36. Neilan, B.A.; Pearson, L.A.; Muenchhoff, J.; Moffitt, M.C.; Dittmann, E. Environmental conditions that influence toxin biosynthesis in cyanobacteria. *Environ. Micrbiol.* **2013**, *15*, 1239–1253. [CrossRef] [PubMed]

37. Zhou, S.; Lou, Y.R.; Tzi, V.; Jander, G. Alteration of plant primary metabolism in response to insect herbivory. *Plant Physiol.* **2015**, *169*, 1488–1498. [CrossRef]

38. Beluhan, S.; Ranogajec, A. Chemical composition and non-volatile components of Croatian wild edible mushrooms. *Food Chem.* **2011**, *124*, 1076–1082. [CrossRef]

39. Zhao, C.; Wu, Y.J.; Liu, X.Y.; Liu, B.; Cao, H.; Yu, H.; Sarker, S.D.; Nahar, L.; Xiao, J.B. Functional properties, structural studies and chemo-enzymatic synthesis of oligosaccharides. *Trends Food Sci. Technol.* **2017**, *66*, 135–145. [CrossRef]

40. Yang, C.F.; Lai, S.S.; Chen, Y.H.; Liu, D.; Liu, B.; Ai, C.; Wan, X.Z.; Gao, L.Y.; Chen, X.H.; Zhao, C. Anti-diabetic effect of oligosaccharides from seaweed *Sargassum confusum* via JNK-IRS1/PI3K signalling pathways and regulation of gut microbiota. *Food Chem. Toxicol.* **2019**, *131*, 110562. [CrossRef]

Foods **2019**, *8*, 397

41. Maga, J.A. Mushroom flavor. *J. Agric. Food Chem.* **1981**, *29*, 1–4. [CrossRef]
42. Tsai, S.Y.; Tsai, H.L.; Mau, J.L. Non-volatile taste components of *Agaricus blazei, Agrocybe cylindracea* and *Boletus edulis. Food Chem.* **2008**, *107*, 977–983. [CrossRef]

 © 2019 by the authors. Licensee MDPI, Basel, Switzerland. This article is an open access article distributed under the terms and conditions of the Creative Commons Attribution (CC BY) license (http://creativecommons.org/licenses/by/4.0/).

Article

Analysis of the Acid Detergent Fibre Content in Turnip Greens and Turnip Tops (*Brassica rapa* L. Subsp. *rapa*) by Means of Near-Infrared Reflectance [†]

Sara Obregón-Cano [1], Rafael Moreno-Rojas [2], Ana María Jurado-Millán [1], María Elena Cartea-González [3] and Antonio De Haro-Bailón [1,*]

[1]   Department of Plant Breeding, Institute for Sustainable Agriculture (CSIC), 14004 Córdoba, Spain
[2]   Department of Food Science and Technology, University of Córdoba, Rabanales Campus, 14014 Córdoba, Spain
[3]   Department of Plant Genetics, Biological Mission of Galicia (CSIC), 36143 Pontevedra, Spain
*   Correspondence: adeharobailon@ias.csic.es
†   This work is a chapter of Dr. Sara Obregón-Cano's PhD thesis.

Received: 30 June 2019; Accepted: 21 August 2019; Published: 26 August 2019

Abstract: Standard wet chemistry analytical techniques currently used to determine plant fibre constituents are costly, time-consuming and destructive. In this paper the potential of near-infrared reflectance spectroscopy (NIRS) to analyse the contents of acid detergent fibre (ADF) in turnip greens and turnip tops has been assessed. Three calibration equations were developed: in the equation without mathematical treatment the coefficient of determination (R^2) was 0.91, in the first-derivative treatment equation $R^2 = 0.95$ and in the second-derivative treatment $R^2 = 0.96$. The estimation accuracy was based on RPD (the ratio between the standard deviation and the standard error of validation) and RER (the ratio between the range of ADF of the validation as a whole and the standard error of prediction) of the external validation. RPD and RER values were of 2.75 and 9.00 for the treatment without derivative, 3.41 and 11.79 with first-derivative, and 3.10 and 11.03 with second-derivative. With the acid detergent residue spectrum the wavelengths were identified and associated with the ADF contained in the sample. The results showed a great potential of NIRS for predicting ADF content in turnip greens and turnip tops.

Keywords: *Brassica rapa*; turnip greens; turnip tops; acid detergent fibre; NIRS

---

1. Introduction

The plants of the genus *Brassica* constitute one of the economically most important plant groups in the world. They are valuable sources of roots, stems, leaves, shoots and inflorescences, as well as of oils, condiments and forage for nutrition or industrial use [1]. Depending on the part of the plant used, these crops are classified as being oleaginous, forage, horticultural products and condiments. The growing scientific interest in this botanical group has increased in parallel to its economic importance and recent achievements in investigation. The consumption of vegetables of the genus *Brassica* has been related to human health with regard to the reduction in the risk of suffering from certain types of chronic diseases, such as cardiovascular problems and different types of cancer [2,3]. Within the genus *Brassica*, four species, *Brassica oleracea*, *Brassica rapa*, *Brassica napus* and *Brassica juncea*, are the crops with a horticultural use. *Brassica rapa* L. subsp. *rapa*, commonly known as turnip, is one of the oldest crops used for human consumption. It was the first species of *Brassica* domesticated by humans thousands of years ago, and it was already cited in Sanskrit literature under the name of Siddharta, which proves the antiqueness of its cultivation [4]. In the north of Spain and Portugal turnip greens and turnip tops are rising in value and they occupy a prominent place in traditional Galician and Portuguese agriculture. Turnip greens are the young leaves of turnips harvested in their vegetative period, whereas turnip tops are the floral stalks collected just before the flower opens. In the case of turnip tops, the diversification of this product is acquiring special importance, and the number of firms packing and freezing it is increasing, not only in Galicia but also in other parts of Spain. An important factor to be taken into account in the nutritional composition of both turnip tops and turnip greens is their fibre content, in addition to the presence of other components like some vitamins and minerals which partly complement the daily dietary demands. The fibre content in vegetables is essential to the digestibility of the food. It has been recognized that the ingestion of fibre is of great benefit to human health, contributing to the prevention of cancer of the colon and reducing the risk of developing cardiovascular diseases, cerebral infarction, hypertension, diabetes, obesity and certain gastrointestinal complaints [5]. Traditionally, the structural carbohydrates of foodstuffs have been estimated via the analysis of their crude fibre content. Crude fibre can be defined as being the residue resulting from submitting the food to a double hydrolysis: acid (with sulphuric acid) and alkaline (with potassium hydroxide), using the protocol developed by the Weende method [6]. One drawback of double hydrolysis is that it solubilizes part of the hemicellulose and of the lignin of the cell wall, so that the result obtained of the crude fibre content is lower than the real content in structural carbohydrates. This problem is avoided by using detergent solutions for the fibre analysis, following the method proposed by Goering and Van Soest [7]. Neutral detergent fibre (NDF) estimates the content in cellulose, hemicellulose, lignin, cutine and insoluble minerals in the cell wall, and is determined as being the residue remaining after extraction with the neutral detergent solution (made up of sodium lauryl sulphate and EDTA). Acid detergent fibre (ADF) is an estimator of the content in cellulose, lignin, cutine and insoluble minerals in the cell wall and it is determined as the residue remaining after the digestion of the sample with an acid detergent solution (made up of diluted sulphuric acid and cetyl-trimethyl-ammonium bromide). The difference between NDF and ADF is the fraction of hemicellulose. With the ADF method the hemicellulose is hydrolysed so that the determination of ADF is more closely associated with degradability and digestibility, whereas the NDF content is only related to ingestion or to a fraction of fibre still highly usable by the organism [8]. Several authors have documented the negative correlation existing between the content of NDF and ADF with the digestibility of vegetable products [9,10]. In the same sense, the high negative correlation between the ADF content and digestibility in vitro has been demonstrated, therefore, the ADF content in a vegetable could be considered as being a good indicator of its digestibility and quality [11–13].

Standard wet chemistry analytical techniques currently used to determine plant fibre constituents (as those described above) are costly, time-consuming and destructive. Additionally, they need specialized workers for their application. During the last 40 years technology based on near-infrared reflectance spectroscopy (NIRS) has become one of the most attractive analytical techniques that is routinely used to estimate numerous quality components in agriculture and food research, since analysis can be carried out at a low cost, with an important saving of time, and without using hazardous chemicals. Moreover, NIRS is a non-destructive technique which requires minimal or zero sample preparation [14–20]. Nowadays, NIRS technology is applied routinely in plant breeding programs for many vegetable species to determine their content in fibre, moisture, oil, protein, minerals, glucosinolates and fatty acid composition of their edible parts [21–24]. The first calibrations for the crude fibre content in seeds in the genus *Brassica* were carried out by Panford, Williams and Man [25] and Michalski, Ochodzki and Cicha [26]. More recently, calibrations for ADF in seeds of different *Brassica* species have been performed by Font, Del Río, Fernández and De Haro-Bailón [27], Font et al. [16], Dimov, Suprianto, Hermann and Möllers [28] and Wittkop, Snowdon and Friedt [29]. Lately, the NIRS technique has been used for the rapid determination of the quality of crude matter starting from the study of fibre as a component of biomass [30], in order to determine the digestibility of cane sugar [31], or to study the fibre content in food for ruminants [32].

This work has aimed to develop and validate NIRS calibration equations for the determination of acid detergent fibre (ADF) in aerial edible parts of *Brassica rapa* (turnip greens and turnip tops), in order to employ them as a tool for a fast and non-destructive analysis in the screening of germplasm and in the selection of genotypes of the highest quality with respect to this component.

2. Materials and Methods

2.1. Plant Material

During the seasons 2013–2014 and 2014–2015, a set of five varieties of *Brassica rapa* L. subsp. *rapa* were grown on the Institute for Sustainable Agriculture experimental farm in Córdoba, (37°51′ N, 4°48′ W, Spain) in a random block design with three replications. The climate is a typical Mediterranean one with a mean rainfall of 650 mm and deep loamy-sandy soil classified as Typic Xerofluent.

The *Brassica rapa* L. subsp. *rapa* varieties came from the *Brassica* Germplasm Bank at the Biological Mission of Galicia (Pontevedra), where they had been characterized by their agronomic characteristics and their aptitude for the production of turnip greens and turnip tops. During each agricultural season, and at the optimal consumption moment, samples of turnip greens (4–5 leaves per plant) and of turnip tops (3–4 flower stalks per plant) were harvested from the plants selected for each of the varieties studied (Figure 1). In total, 134 samples were harvested, 78 in the 2013–2014 season (34 turnip greens and 44 turnip tops) and 56 samples in the 2014–2015 season (29 turnip greens and 27 turnip tops). All the vegetable material was thoroughly washed with tap water to remove dirt and dust from its surface and, finally, it was rinsed with deionized water. Next, it was stored at −80 °C until its lyophilisation, which was done in Telstar® model Cryodos-50 (Telstar, Terrasa, Spain) equipment. The lyophilized samples were ground in an IKA-Labortechnik® (Staurfen, Germany) model A10 mill for 20 s and stored in desiccators up to the moment of being analysed by the reference method or scanned in the NIRS equipment.

(a) (b)

Figure 1. (**a**) Turnip greens; (**b**) turnip tops.

2.2. Analysis of Acid Detergent Fibre

The ADF content was determined following the procedures described by Goering and Van Soest [7] in a Dosi-Fibre (Selecta®, Barcelona, Spain) machine. 0.5 g of lyophilized sample was weighed in glass filtering crucibles (porosity 2). This was digested for one hour in 100 mL of hot cetyl-methyl-ammonium bromide in an acid medium (sulphuric acid) and then filtered to obtain the residue considered as being the acid detergent fibre of the sample. Next, the residue was washed with hot water and acetone and dried in a stove at 110 °C for 90 min. Then it was stored in desiccators for 30 min to temper the crucibles and prevent the sample from becoming moist, after which the sample was weighed. The acid detergent residue (ADR) remaining after digestion was removed from the crucibles and stored to obtain the NIRS spectrum from the pure residue.

The acid detergent fibre of the sample was calculated according to Equation (1):

$$\text{ADF }(\%) = \frac{P3 - P1}{P2 - P1} \times 100, \tag{1}$$

where P1 is the crucible weight, P2 is the weight of the crucible with the sample and P3 is the weight of the crucible with the acid detergent residue after digestion. Each sample was analysed in duplicate.

2.3. Development of NIRS Equations

Sample spectra were recorded with a Model 6500 (Foss-NIRSystems®, Inc., Silver Spring, MD, USA) near-infrared spectrophotometer in the reflectance mode. One spectrum was recorded for each sample. The samples were placed in a round capsule 3 cm in diameter made of quartz glass and anodized aluminium to prevent interferences in their absorption. From each sample, reflectance spectra in the wavelength range of 400–2500 nm, at 2 nm intervals, were obtained. Collection of spectral data and their chemometric analysis was conducted with the WinISI II v1,50 software (Infrasoft International, Port Matilda, PA, USA).

The spectral outliers were detected by a principal component analysis (PCA) applied to the whole set of the population based on the calculation of the Mahalanobis distance (H) [33,34]. In addition to being a tool for the selection of samples from the calibration set, this is a highly useful technique in the analysis for converting original spectra data (absorbance values) into new orthogonal variables (principal components) thus eliminating collinearity (redundant information) [35]. The CENTER algorithm included in the WinISI II software (version 1.50, Infrasoft International, Port Matilda, PA, USA) was used to calculate the H distances between the spectra of the different samples with respect to the mean spectrum. In agreement with the work of Shenk and Westerhaus [33], samples with a statistical H value of over three units were defined as being atypical spectra and they were eliminated for the establishment of the equations. A total of four spectra found were eliminated from the set of samples employed in the work. The final number of samples selected was of 130, the calibration set was composed of 104 samples and was used for the development of the different calibration equations; the external validation set was formed by 20% of the total samples (n = 26) and was used to evaluate the prediction capacity of each of the equations developed. The external validation set samples were selected by taking the list of samples ordered on the basis of their H values, choosing 1 of each of the five samples on the list [33]. In this way, the samples selected represented all the variability in the whole of the population [36]. To develop the calibration equations, the method of regression by modified minimum partial least squares (MPLS) was applied. The usefulness of this method has been demonstrated for the evaluation of fibre content, using the whole spectrum range (400–2500 nm) [17,27].

The spectrum correction procedure SNV-DT was applied. The latter provides the WinISI software for the elimination of dispersion due to the effects caused by the differences in particle size or the variation in length, halfway between the dispersion of the samples and fitting the baseline [37]. The treatment selected for one parameter in a dataset is not always the best option for the same parameter in any other set of samples [24]; this confirms the importance of optimizing the treatment for each parameter and dataset. In this sense, the mathematical treatments selected and applied to the spectra in our work were (0, 0, 1, 1), (1, 4, 4, 1) and (2, 5, 5, 2), in which: the first number indicates the order of the derivative (first or second derivative of the logarithm of 1/R); the second number is the amplitude or distance between the segments to be subtracted; the third number is the length of the segment to be smoothed; and the fourth number indicates a second smoothing [38]. The statistics defining the calibration equations obtained are the coefficient of determination (R^2) which shows the percentage of the variability in the ADF concentrations explained by the regression equation, and the standard error of calibration (SEC), which is the standard error in the residuals for the calibration set. It should be noted that the standard error in the calibration only advises one of the fitting of the reference values to the regression line, so that it cannot be considered as being an adequate statistic for assessing the validity of the calibration equation obtained [34].

2.4. Equation Validation

To evaluate the prediction capacity of the calibration equations, two validation models were used, permitting the establishment of a comparison (through different statistical criteria) between the true value (obtained by the reference method) and the estimated one (obtained by NIRS).

2.4.1. Cross Validation

A cross validation was made based solely on the data employed at the calibration stage, in order to calculate the optimal number of terms in the regression. The algorithm selects different calibration and validation sets within the whole population considered, making with each selection a simulation of the regression algorithm [33,35]. Finally, the calculation software chose the equation which made the minimum standard error of cross validation (SECV). The statistics resulting from the cross validation were: the coefficient of determination of cross validation (r^2_{cv}), the standard error of cross validation (SECV), which represents the standard error of the residuals for the cross validation set; and the statistic (RPD) [2] which is the ratio between the standard deviation and the standard error of cross validation (SD/SECV). The RPD_{cv} is a statistic which permits the evaluation of the SECV in terms of the standard deviation of the reference data for the population being studied [39]:

$$\text{RPD} = \text{SD}\left\{\left[\left(\sum_{i=1}^{n}\left(y_i - \hat{y}_i\right)^2\right)(n-k-1)^{-1}\right]^{0.5}\right\}^{-1}, \tag{2}$$

where y_i = laboratory reference value for the sample; $\hat{y}_i$ = NIR mean value; n = number of samples, k = number of wavelengths used in an equation; SD = standard deviation of the chemical data.

2.4.2. External Validation

The calibration equations selected with samples which did not intervene in the calibration (validation set, $n = 26$ in our work) were evaluated. The external validation statistics include: the coefficient of determination of validation (r^2_{ev}), the standard error of prediction (SEP), the RPD_{ev} (which is the ratio SD/SEP), and the RER [3], which is the ratio between the range of ADF of the validation as a whole and the standard error of prediction:

$$\text{RER} = \text{range}\left\{\left[\left(\sum_{i=1}^{n}\left(y_i - \hat{y}_i\right)^2\right)(n-k-1)^{-1}\right]^{0.5}\right\}^{-1}, \tag{3}$$

where y_i = laboratory reference value for the sample; $\hat{y}_i$ = NIR mean value; n = number of samples, k = number of wavelengths used in an equation; SD = standard deviation of the chemical data.

The RPD and RER statistics permitted a comparison of the performance of the model through populations with different standard deviations [18]. The best calibration equations for the ADF analysis were selected by considering the optimal combination of the following external validation statistics: high values of coefficients of determination (r^2_{ev}) and high RPD_{ev} and RER values. Those equations in which RPD is higher than 3 were considered to have an excellent prediction ability, those with RPDs of between 2 and 3 allowed approximate predictions to be made, and those whose RPD was between 1.5 and 2 could only be used for classification purposes in groups with a high-medium-low content. Similarly, the RER values obtained with the different calibration equations with a good prediction capacity should be over 10 [39,40].

The standard error of laboratory (SEL) for the ADF analysis was determined and compared with the SEP for all the equations. To obtain the total error of the reference method (SEL), 10 samples of turnip tops and turnip greens were selected and analysed in duplicate at different times and by different analysts. The statistical ratio SEP/SEL permitted the NIRS error to be related to the error in the reference method.

3. Results and Discussion

3.1. ADF Reference Analysis in Samples of Turnip Tops and Turnip Greens

A collection of 134 samples of *Brassica rapa* were analysed (63 turnip greens and 71 turnip tops) by the Goering and Van Soest method [7]. The mean ADF content in turnip greens and turnip tops was 11.53% and 15.98%, respectively (Table 1). A *t*-test, showed significant differences between the means. ($p < 0.001$).

Table 1. Fibre content in samples of turnip greens and turnip tops of *Brassica rapa*, analysed in the laboratory.

Plant Material	ADF (%)		
	Range	Mean	SD [1]
Turnip greens ($n = 63$)	8.55–15.27	11.53	1.54
Turnip tops ($n = 71$)	10.41-21.91	15.98	2.54

[1] SD = Standard deviation; n = number of samples.

The differences in the ADF content between turnip greens and turnip tops samples can be explained by the fact the turnip greens are formed by young leaves and the turnip tops by flower stems with a higher content of fibre. Therefore, we can conclude that the maturity of plants and the increase in structural carbohydrates lead to higher accumulation of fibre amounts in turnip tops when compared to turnip greens. These results highlight that *Brassica rapa* was a good source of fibre with high concentrations in some samples (21.91%) and lower concentrations in others (8.55%). Figure 2 shows the distribution of the frequency of fibre content in turnip greens and in turnip tops from the samples studied.

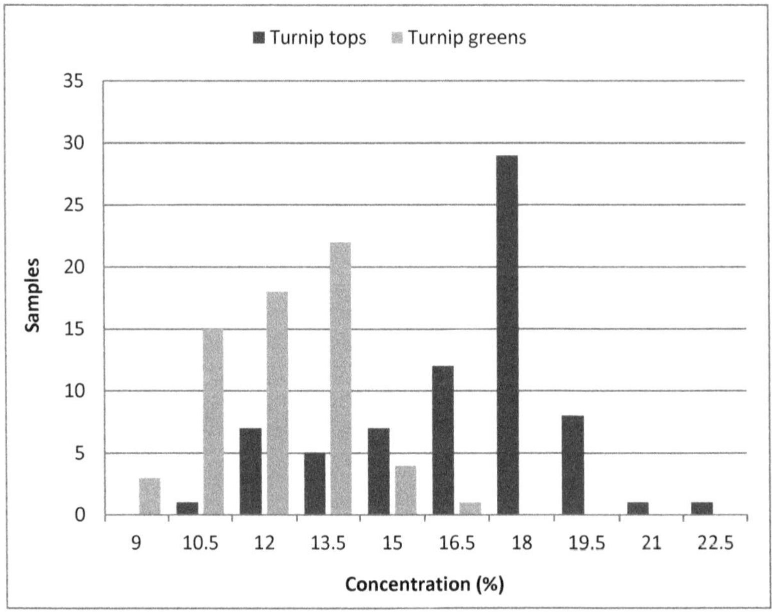

Figure 2. Distribution plot for ADF content in turnip greens and turnip tops.

The variability in the ADF content in the samples analysed in this work was similar to that published in others studies, in which ADF values present in leaves of *Brassica rapa* were 23.50% [41]; in crude fibre 12.9% and in ADF 23.5% [42]. Previous works on the fibre content in turnip (the thickened hypocotyls of *B. rapa* widely used in human nutrition) have found values of 11.20% [41] and 14.68% [43]. In rapeseed flour the ADF content values were comprised between 9.5% and 15.2% [44]; and in seeds of other *Brassica* their values ranged from 5.33% (*B. carinata*) to 16.31% (*B. juncea*) [27].

The statistical data describing the calibration and validation sets are shown in Table 2. The range of values of the set of validation samples were included within the range of the values of the calibration samples, which were required to generate a calibration model with a reliable predictive ability [45].

Table 2. ADF content in the turnip greens and turnip tops samples from the calibration and validation set of *Brassica rapa* analysed following the reference method.

Sample Groups	ADF (%)		
	Range	Mean	SD [1]
Calibration set (*n* = 104)	8.75–20.02	13.87	2.98
Validation set (*n* = 26)	8.55–18.81	13.67	3.01

[1] SD = Standard deviation; *n* = number of samples.

3.2. Calibration and Validation

The principal component analysis was carried out to locate any possible spectral outliers from the calibration set [33]. Figure 3a shows the mean spectrum of the *Brassica rapa* samples in the range of 400 to 2500 nm; Figure 3b depicts the first derivative spectrum (1, 4, 4, 1; SNV-DT) and Figure 3c the second derivative spectrum (2, 5, 5, 2; SNV-DT), both derived with the application of a spectra correction treatment.

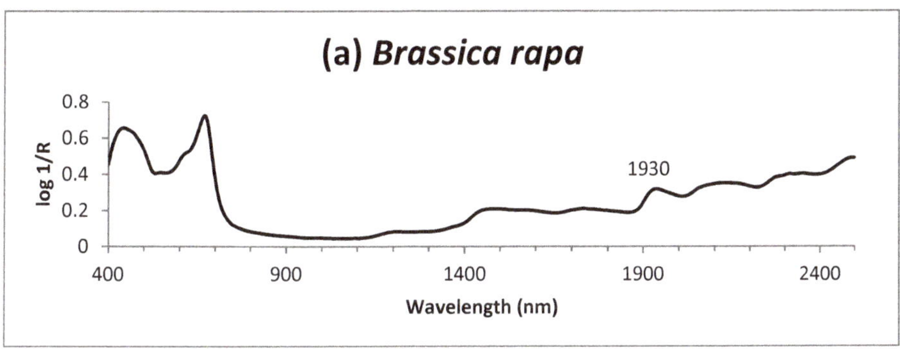

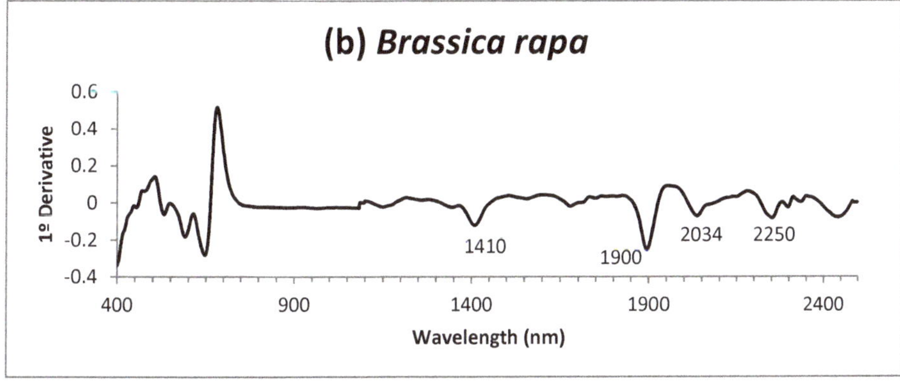

Figure 3. *Cont.*

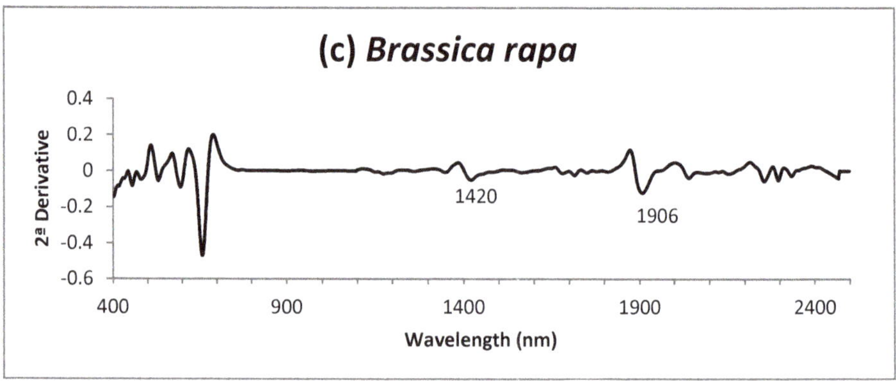

Figure 3. (**a**) Mean spectrum of the lyophilized green parts of *Brassica rapa*; (**b**) first derivative (SNV-DT) of the mean spectrum of the lyophilized green parts of *Brassica rapa*; (**c**) second derivative (SNV-DT) of the mean spectrum of the lyophilized green parts of *Brassica rapa*.

With the aim of identifying wavelengths and associate spectrum bands with the ADF contained in the sample, the acid detergent residue spectrum (ADR), Figure 4, was compared with the spectrum of the green parts of *Brassica rapa*, Figure 3, in order to identify the NIRS spectrum regions which might be more related to the ADF content in the sample. In the spectra of *B. rapa* and ADR, Figures 3 and 4, absorption similarities were found in certain wavelengths. It is worth noting that the wavelengths of 1420 nm related to aromatic groups, 1906 nm related to groups OH, C = O and CO_2H and 2278 nm related to groups CH and CH_2 associated with the structural polysaccharides of the plants, 2468 nm related to groups CH, CH_2 and C-N-C associated with proteins [46] (WinISI II v1,50 software). Those wavelengths would participate more highly in the development of robust calibrations for the ADF content.

The results of the calibration equations obtained by MPLS regression with the three mathematical treatments is shown in Table 3. In the evaluation of the treatments applied in the development of those equations, a clear difference was found between the statistics values obtained in the equations without treatment (0, 0, 1, 1) (R^2 = 0.91) and the equations with treatments with derivative the value of R^2 = 0.95 in the first derivative (1, 4, 4, 1; SNV + DT) and a value of R^2 = 0.96 in the second one (2, 5, 5, 2; SNV + DT), with both values being very similar to each other (Table 3).

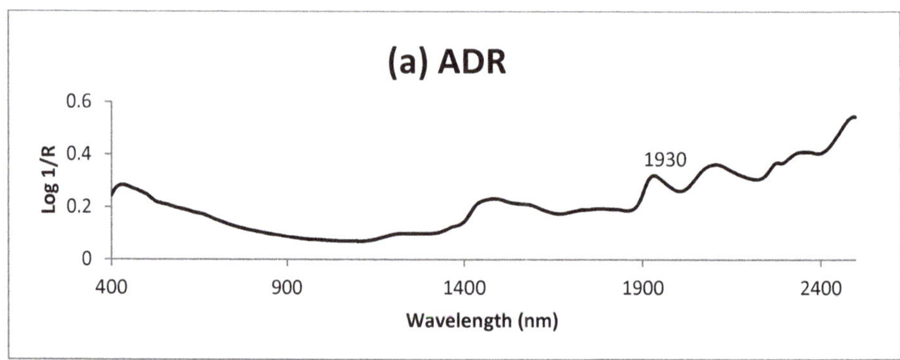

Figure 4. *Cont.*

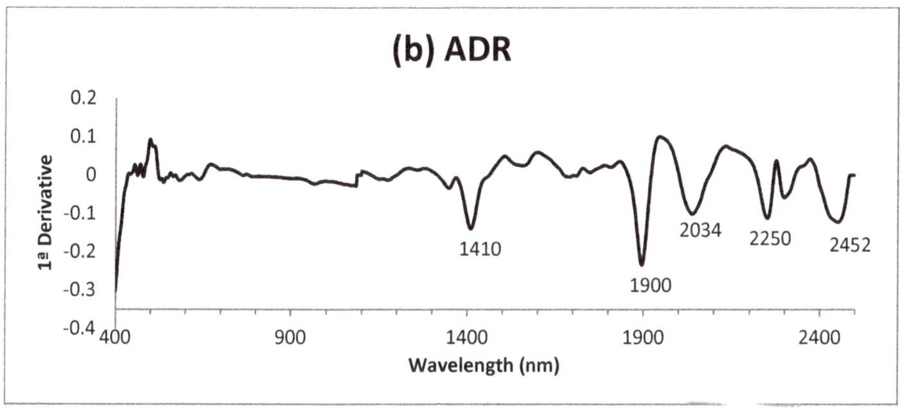

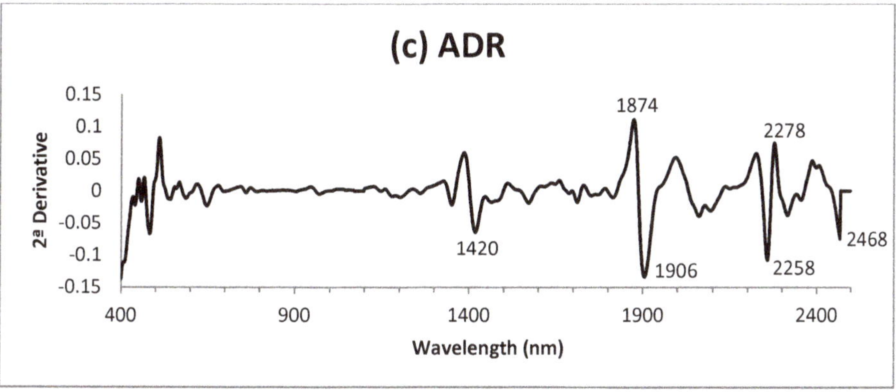

Figure 4. (**a**) Mean spectrum of the acid detergent residue of ADF; (**b**) first derivative (SNV-DT) of the mean spectrum of ADR; (**c**) second derivative (SNV-DT) of the mean spectrum of ADR.

Table 3. Calibration and cross validation statistics for ADF content in *Brassica rapa*.

			Calibration				Cross Validation		
TM [1]	Range	Samples	Mean	SD [2]	SEC [3]	R^2 [4]	SECV [5]	RPD_{cv} [6]	R^2_{cv} [7]
0, 0, 1, 1	8.75–20.02	101	13.81	2.95	0.86	0.91	1.07	2.77	0.87
1, 4, 4, 1	8.75–20.02	104	13.80	2.96	0.65	0.95	0.88	3.36	0.91
2, 5, 5, 2	8.75–20.02	103	13.82	2.95	0.56	0.96	0.89	3.33	0.91

[1] Mathematical treatment of the spectra. [2] Standard deviation. [3] Standard error of the calibration. [4] Coefficient of determination of the calibration. [5] Standard error of the cross validation. [6] Relation between the standard deviation and the standard error of the cross validation. [7] Coefficient of determination of the cross validation.

3.2.1. Cross Validation

On the basis of the statistics obtained in the cross validation, the final calibration equations for the ADF content were selected on the premise of maximizing the r^2_{vc} and minimizing the SECV. The values of RPD_{cv} of the cross validation for ADF obtained were 3.36 (for the treatment 1, 4, 4, 1) and 3.33 (for the treatment 2, 5, 5, 2). In both equations the RPD_{cv} values were higher than 3, proving the ability of the calibration equations to be used for diagnosis and investigation purposes [40].

The two derivatization treatments (1, 4, 4, 1 and 2, 5, 5, 2) successfully optimized the model getting some optimal results in the statistics values. Both two models were valid for calibration. In the study of the profile of fatty acids in milk calibration equations, the first derivative and the second derivative were developed, (1, 5, 5, 1) and (2, 5, 5, 1), and both treatments were valid to be used in the characterization of the fat content in milk [24]. Other authors have evaluated the prediction of protein and amylose in brown rice and rice bran, where five treatments were tested including first and second derivatives, and stating that two of the treatments (1, 6, 6, 1 and 1, 4, 4, 1) were equally valid for the development of calibration equations to predict amylose content [47].

Figure 5 depicts the laboratory values compared to the NIRS prediction ones of the cross validation as a whole for ADF content.

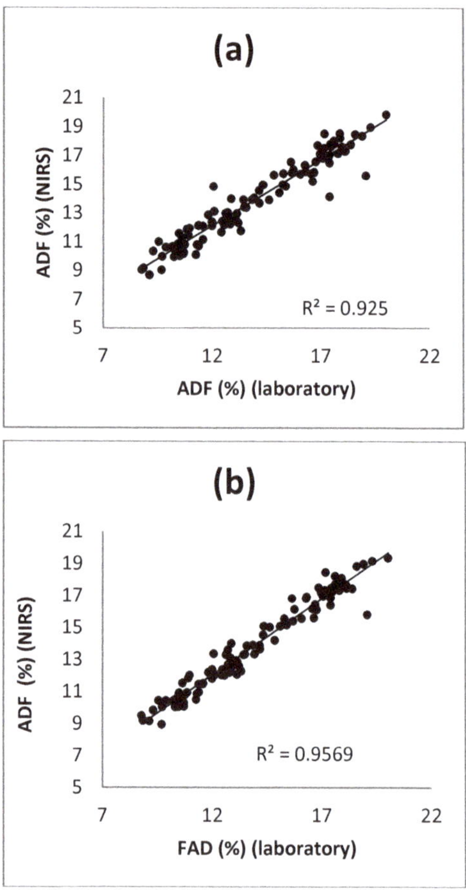

Figure 5. Scatter plot of the reference values against the predicted values in cross validation with respect to the ADF content applying the equations 1, 4, 4, 1 (**a**) and 2, 5, 5, 2 (**b**).

3.2.2. External Validation

Once the equations were obtained and the cross validation was performed, a second evaluation of the equations was made by using the samples not included in the calibration (external validation set) for the prediction of the ADF content. Table 4 presents the statistics of the external validation obtained for the ADF equations developed with the three mathematical treatments. In treatment 0, 0, 1, 1, a value of $r^2_{ev} = 0.87$ was obtained, which led to lower RPD_{ev} and RER values than those of the treatments with derivatives. In treatments (1, 4, 4, 1) and (2, 5, 5, 2) the same high value in the coefficients of determination of the prediction, ($r^2_{ev} = 0.91$) was obtained. The RPD_{ev} values were very similar to each other, 3.41 in the first derivative and 3.10 in the second one. Those results were very similar to those of RPD_{cv} obtained in the cross validation (values of over 3) and they confirmed the excellent ability to predict ADF content by using both the equations developed with treatments (1, 4, 4, 1) and (2, 5, 5, 1) [39,40]. Finally, the RER statistic was calculated and values greater than 10 were obtained in both cases: 11.79 (1, 4, 4, 1) and 11.03 (2, 5, 5, 2). This was an additional proof of the high predictive ability of the calibration models developed for ADF [39,40].

Table 4. Statistics of the external validation ($n = 26$) applied to the calibration equations of the fibre content in *Brassica rapa*.

TM [1]	Range	Samples	Mean	SD [2]	SEP [3]	r^2_{ev} [4]	RPD$_{ev}$ [5]	RER [6]
0, 0, 1, 1	8.55–18.81	26	13.67	3.13	1.14	0.87	2.75	9.00
1, 4, 4, 1	8.55–18.81	25	13.55	2.96	0.87	0.91	3.41	11.79
2, 5, 5, 2	8.55–18.81	25	13.55	2.89	0.93	0.91	3.10	11.03

[1] Mathematical treatment. [2] Standard deviation of the reference data of the external validation set. [3] Standard error in the prediction. [4] Coefficient of determination of the external validation. [5] Relation between the standard deviation and the standard error in the prediction. [6] Relation between the data range and the standard error in the prediction.

No calibration equations of the ADF content in green parts of *Brassica rapa* have been described in the bibliography up to now. However, some equations developed for the ADF content in leaves of woody species have been reported with higher values than those described (RPD$_{ev}$ = 5.3), possibly due to the heterogeneity in the samples selected for the development of the calibration, in which different woody species collected on different dates were included [48]. ADF calibration in corn plants gave RPD$_{ve}$ values of 2.9 [49], and other works investigating grasses leaves and red clover presented RPD = 3.4 values in NDF, which were similar to those found in this work with ADF [50].

Regarding calibrations within the genus *Brassica*, the values obtained in the calibration equations in our work are the highest described to the moment for ADF content, compared to the equations developed for ADF content in seeds found in the literature. To summarized, calibrations in intact *Brassica napus* seeds were described, with RPD$_{cv}$ values of: 2.13 and 2.20 (in a volume of 10 mL of seed) and values of 1.91 and 2.34 (in a volume of 1 mL of seed) [16,29]; these results coincide with those of other authors who also obtained values of 1.92 in seeds of the same species [28]. As for the external validation results, those obtained in our work were also higher than those found in the literature in *B. napus* seeds, with RPD$_{ev}$ of 2.2 and RER of 10.03 [28].

To evaluate the precision of the equations the reference method error (SEL) was calculated and was related to the SEP. The SEL value obtained was of 0.25. The SEP/SEL ratio shown in the ADF was of 4.56 in the treatment 0, 0, 1, 1, which indicates a poor precision, and values of 3.48 and 3.72 were obtained for the treatments of 1, 4, 4, 1 and 2, 5, 5, 2, respectively, which reveal a good precision; those values are similar to the ones obtained by other authors in other *Brassica* species [27].

3.3. Modified Partial Least Squares Loadings of the Lyophilized Green Parts Model

Panels a, b, and c of Figure 6 represent MPLS loading spectra for factors 1, 2, and 3, respectively. These plots show the regression coefficients of each wavelength to ADF for each factor. Wavelengths represented here as participating more highly in the development of each factor are those of a greater variation and with a higher correlation with the ADF in the calibration set. In the second derivative, peaks pointing downwards indicate the positive influence of absorbers on the development of the equations, while peaks pointing upwards evidence negative correlations. Factors 1 and 3 of the lyophilized green parts model showed those most highly correlated with ADF, presenting a loading with major positive correlations at 1404, 2308 and 2348 nm, associated with the absorbance of C–H and C–O groups of lipids (Figure 6a) [27,29,46]. Factor 1 was also influenced by groups N–H at 1996 nm. Factor 2 was the one most highly correlated with amide groups in the protein region at 2052 and 2300 nm. Factor 3 was also influenced by water, as indicated by the band at 1932 nm. Wavelengths for specific absorbance of oil functional groups are known as being major contributors to NIRS calibrations for ADF in *Brassica* species and for dietary fibre in high-fat cereal products [29].

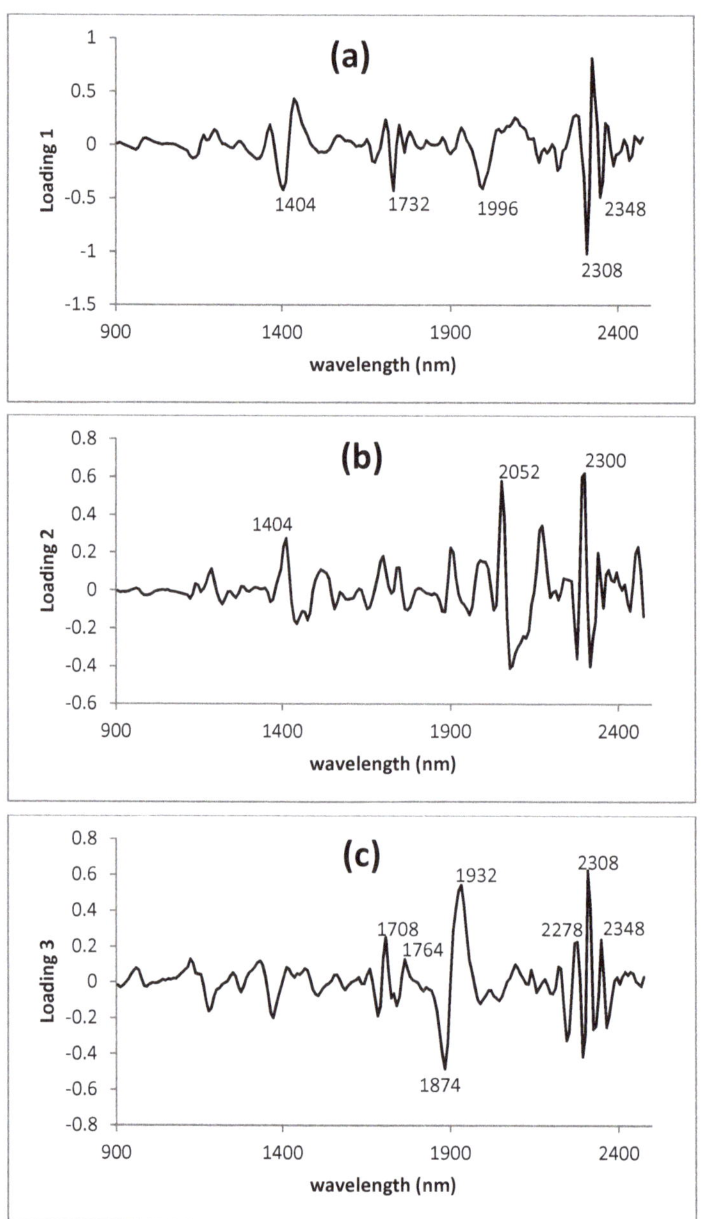

Figure 6. MPLS loading spectra for ADF in *Brassica rapa* in the second derivative (2, 5, 5, 2) transformations. Panels (**a**), (**b**) and (**c**) represent loadings for factors 1, 2 and 3, respectively.

On the basis of the similarities between the second-derivative transformation of the ADR spectrum (Figure 4c) and the third MPLS loading for *Brassica rapa* (Figure 6), it seems that absorbers of the ADR participated directly in the modelling this factor, specifically, 1874 and 2278 nm related to groups CH and CH_2 associated with the structural polysaccharides of the plants.

The study of the MPLS loadings of the ADF equation developed in this study suggests that OH groups of water, CH and CH_2 group of structural polysaccharides, CO groups of lipids and also NH groups of amides (proteins) were the molecular associations most frequently used in modelling the equation. Shape and positioning the bands presented by the different loadings very closely resembled those reported by Font et al. [27] for oilseed *Brassicas*, in which effects due to CH groups of lipids and OH groups of water were the most important in the model. Recent NIRS calibrations for fibre fractions in intact seeds of *Brassica napus* also showed a significant contribution to the model of the CH, OH and NH groups in aromatic and protein regions [29].

The results obtained in the present work, both in the cross-validation and in the external validation confirm the reliability and potential of the calibration equations developed with treatments (1, 4, 4, 1) and (2, 5, 5, 2) to predict accurately and precisely the ADF content in turnip greens and turnip tops. In addition, both calibration equations (with treatments 1, 4, 4, 1 and 2, 5, 5, 2) displayed the same ability prediction of the ADF content in samples of turnip greens and turnip tops.

As a conclusion, the accurate predictions provided by the NIR equations developed in this work confirm that NIR technology could be very useful for the rapid evaluation of the ADF content in turnip greens and turnip tops. Furthermore, this technique allows us to save considerable time and money in comparison to the standard methods of analysis, making it possible to conduct large numbers of analyses for ADF content in a short time.

Author Contributions: S.O.-C. and A.M.J.-M. performed all physicochemical analyses. R.M.-R., E.C.G. and A.D.H.-B. designed this study and revised the manuscript. S.O.-C. wrote this manuscript. All the listed authors have read and approved the submitted manuscript.

Funding: This research was funded by the Project "Metabolitos secundarios en Brassicaceae. Implicaciones en la mejora genética y defensa a estreses" Ref. RTI2018-096591-B-I00 of the Spanish Government, which was co-financed by the European Regional Development Fund (ERDF).

Conflicts of Interest: The authors declare no conflict of interest.

References

1. Gómez-Campo, C. Brassica Crops and Wild Allies. In *Morphology and Morpho-Taxonomy of the Tribe Brassiceae*; Scientific Societies Press: Tokyo, Japan, 1980.
2. Cartea, M.E.; Velasco, P. Glucosinolates in Brassica foods:bioavailability in food and significance for human health. *Phytochem. Rev.* **2008**, *7*, 213. [CrossRef]
3. Traka, M.; Mithen, R. Glucosinolates, isothiocyanates and human health. *Phytochem. Rev.* **2009**, *8*, 269–282. [CrossRef]
4. Prakash, O. *Food and Drinks in Ancient India*; Munshi Ram Manohar Lal: Delhi, India, 1961; pp. 165–168.
5. Anderson, J.W.; Baird, P.; Davis, R.H., Jr.; Ferreri, S.; Knudtson, M.; Koraym, A.; Waters, V.; Williams, C.L. Health benefits of dietary fiber. *Nutr. Rev.* **2009**, *67*, 188–205. [CrossRef] [PubMed]
6. AOAC. Official Method 978.10. Fiber (Crude) in animal feed Fritted glass crucible method. In *AOAC Official Methods of Analysis*, 16th ed.; Association of Official Analytical Chemists, Inc.: Arlington, VA, USA, 1995; Volume 1, pp. 20–21.
7. Goering, H.K.; Van Soest, P.J. Forage fiber analysis; apparatus, reagents, procedures and some applications. In *USDA-ARS Agricultrual*; Handbook no 379; U.S. Agricultural Research Service: Washington, DC, USA, 1970.
8. Van Soest, P.J.; Wine, R.H. Use of detergents in the analysis of fibrous feeds. IV. Determination of plant cell-wall constituents. *J. Assoc. Off. Anal. Chem.* **1967**, *50*, 50–55.
9. Oh, H.K.; Baumgardt, B.R.; Scholl, J.M. Evaluation of forages in the laboratory. V. Comparison of chemical analysis, solubility tests, and in vitro fermentation. *J. Dairy Sci.* **1966**, *49*, 850–855. [CrossRef]
10. Archer, K.A.; Decker, A.M. Relationship between fibrous components and in vitro dry matter digestibility of autumn-saved grasses. *Agron. J.* **1997**, *69*, 610–612. [CrossRef]
11. Hill, R.R.; Barnes, R.F. Genetic Variability for Chemical Composition of Alfalfa. II. Yield and Traits Associated with Digestibility. *Crop Sci.* **1977**, *17*, 948–952. [CrossRef]

12. Soh, A.C.; Frakes, R.V.; Chilcote, D.O.; Sleper, D.A. Genetic variation in acid detergent fiber, neutral detergent fiber, hemicellulose, crude protein, and their relationship with in vitro dry matter digestibility in tall fescue. *Crop Sci.* **1984**, *24*, 721–727. [CrossRef]

13. Van Soest, P.J.; Robertson, J.B. *Analysis of Forages and Fibrous Foods*; Cornell University Publication: Ithaca, NY, USA, 1985; Volume 165.

14. Scotter, C. Use of near infrared spectroscopy in the food industry with particular reference to its applications to on/in-line food processes. *Food Control* **1990**, *1*, 142–149. [CrossRef]

15. Bochereau, L.; Bourgine, P.; Palagos, B. A method for prediction by combining data analysis and neural networks: Application to prediction of apple quality using near infra-red spectra. *J. Agric. Eng. Res.* **1992**, *51*, 207–216. [CrossRef]

16. Font, R.; Wittkop, B.; Badani, A.G.; Del Río-Celestino, M.; Friedt, W.; Lühs, W.; De Haro-Bailón, A. The measurements of acid detergent fiber in rapeseed by visible and near-infrared spectroscopy. *Plant Breed.* **2005**, *124*, 410–412. [CrossRef]

17. Font, R.; Del Río, M.; De Haro-Bailón, A. The use of near-infrared spectroscopy (NIRS) in the study of seed quality components in plant breeding programs. *Ind. Crops Prod.* **2006**, *24*, 307–313. [CrossRef]

18. Cozzolino, D.; Moron, A. Exploring the use of near infrared reflectance spectroscopy (NIRS) to predict trace minerals in legumes. *Anim. Feed Sci. Technol.* **2004**, *111*, 161–173. [CrossRef]

19. Sinelli, N.; Spinardi, A.; Di Egidio, V.; Mignani, I.; Casiraghi, E. Evaluation of quality and nutraceutical content of blueberries (*Vaccinium corymbosum* L.) by near and mid-infrared spectroscopy. *Postharv. Biol. Technol.* **2008**, *50*, 31. [CrossRef]

20. Davey, M.W.; Saeys, W.; Hof, E.; Ramon, H.; Swennen, R.L.; Keulemans, J. Application of visible and nearinfrared reflectance spectroscopy (Vis/NIRS) to determine carotenoid contents in banana (*Musa* spp.) fruit pulp. *J. Agric. Food Chem.* **2009**, *57*, 1742. [CrossRef] [PubMed]

21. Helgerud, T.; Wold, J.P.; Pedersen, M.B.; Liland, K.H.; Ballance, S.; Knutsen, S.H.; Rukke, E.O.; Afseth, N.K. Towards on-line prediction of dry matter content in whole unpeeled potatoes using near-infrared spectroscopy. *Talanta* **2015**, *143*, 138–144. [CrossRef] [PubMed]

22. Hell, J.; Prückler, M.; Danner, L.; Henniges, U.; Apprich, S.; Rosenau, T.; Kneifel, W.; Böhmdorfer, S. A comparison between near-infrared (NIR) and mid-infrared (ATR-FTIR) spectroscopy for the multivariate determination of compositional properties in wheat bran samples. *Food Control* **2016**, *60*, 365–369. [CrossRef]

23. Guo, Y.; Ni, Y.; Kokot, S. Evaluation of chemical components and properties of the jujube fruit using near infrared spectroscopy and chemometrics. *Spectrochim. Acta Part A Mol. Biomol. Spectrosc.* **2016**, *153*, 79–86. [CrossRef]

24. Núñez-Sánchez, N.; Martínez-Marín, A.L.; Polvillo, O.; Fernández-Cabanás, V.M.; Carrizosa, J.; Urrutia, B.; Serradilla, J.M. Near Infrared Spectroscopy (NIRS) for the determination of the milk fat fatty acid profile of goats. *Food Chem.* **2016**, *190*, 244–252. [CrossRef]

25. Panford, A.; Williams, P.C.; Man, J.M. Analysis of oilseeds for protein, oil, fiber and moisture by near-infrared reflectance spectroscopy. *J. Am. Oil Chem. Soc.* **1988**, *65*, 1627–1634. [CrossRef]

26. Michalski, K.; Ochodzki, P.; Cicha, B. Determination of fiber, sulphur amino acids and lysine in oilseed rape by NIT. In *Making Light Work: Advances in Near Infrared Spectroscopy*; Murray, I., Cowe, I.A., Eds.; VCH Weinheim: New York, NY, USA, 1992; pp. 333–335.

27. Font, R.; Del Río, M.; Fernández, J.M.; De Haro-Bailón, A. Acid Detergent Fiber Analysis in Oilseed *Brassicas* by Near-Infrared Spectroscopy. *J. Agric. Food Chem.* **2003**, *51*, 2917–2922. [CrossRef] [PubMed]

28. Dimov, Z.; Suprianto, E.; Hermann, F. Möllers. Genetic variation for seed hull and fiber content in a collection of European winter oilseed rape material (*Brassica napus* L.) and development of NIRS calibrations. *Plant Breed.* **2012**, *131*, 361–368. [CrossRef]

29. Wittkop, B.; Snowdon, R.; Friedt, W. New NIRS Calibrations for Fiber Fractions Reveal Broad Genetic Variation in *Brassica napus* Seed Quality. *J. Agric. Food Chem.* **2012**, *60*, 2248–2256. [CrossRef] [PubMed]

30. Foster, A.J.; Kakani, V.G.; Ge, J.; Mosali, J. Rapid assessment of bioenergy feedstock quality by near infrared reflectance spectroscopy. *Agron. J.* **2013**, *105*, 1487–1497. [CrossRef]

31. Daniel, J.L.P.; Capelesso, A.; Cabezas-Garcia, E.H.; Zopollatto, M.; Santos, M.C.; Huhtanen, P.; Nussio, L.G. Fiber digestion potential in sugarcane across the harvesting window. *Grass Forage Sci.* **2014**, *69*, 176–181. [CrossRef]

32. Krizsan, S.J.; Rinne, M.; Nyholm, L.; Huhtanen, P. New recommendations for the ruminal in situ determination of indigestible neutral detergent fiber. *Anim. Feed Sci. Technol.* **2015**, *205*, 31–41. [CrossRef]
33. Shenk, J.S.; Westerhaus, M.O. Population definition, sample selection, and calibration procedures for near infrared reflectance spectroscopy. *Crop. Sci.* **1991**, *31*, 469–474. [CrossRef]
34. Shenk, J.S.; Westerhaus, M.O. Calibration the ISI way. In *Near Infrared Spectroscopy: The Future Waves*; Davies, A.M.C., Williams, P., Eds.; NIR Publications: Chichester, UK, 1996; pp. 198–202.
35. Martens, H.; Naes, T. *Multivariate Calibration*; John Wiley and Sons: Chichester, UK, 1989.
36. Hruschka, W.R. Data Analysis Wavelength Selection Methods. In *Near Infrared Technology in the Agricultural, Food Industries*; Williams, P.C., Norris, K.H., Eds.; American Association of Cereal Chemist: St. Paul, MN, USA, 2001; pp. 35–55.
37. Barnes, R.J.; Dhanoa, M.S.; Lister, S.J. Standard normal variate transformation and de-trending of near-infrared diffuse reflectance spectra. *Appl. Spectrosc.* **1989**, *43*, 772–777. [CrossRef]
38. Shenk, J.S.; Workman, J.J.; Westerhaus, M.O. Application of NIR spectroscopy to agricultural products. In *Handbook of Near-Infrared Analysis*; Burns, D.A., Ciurczak, E., Eds.; Dekker Inc.: New York, NY, USA, 1992; pp. 383–431.
39. Williams, P.C. Implementation of Near-Infrared technology. In *Near Infrared Technology in the Agricultural, Food Industries*; Williams, P.C., Norris, K.H., Eds.; American Association of Cereal Chemist: St. Paul, MN, USA, 2001; pp. 145–169.
40. Williams, P.C.; Sobering, D.C. How do we do it: A brief summary of the methods we use in developing near infrared calibrations. In *Near Infrared Spectroscopy: The Future Waves*; Davies, A.M.C., Williams, P.C., Eds.; NIR Publications: Chichester, UK, 1996; pp. 185–188.
41. Türk, M.; Albayrak, S.; Balabanli, C.; Yüksel, O. Effects of fertilization on root and leaf yields and quality of forage turnip (*Brassica rapa* L.). *J. Food Agric. Environ.* **2009**, *7*, 339–342.
42. Francisco, M.; Velasco, P.; Lema, M.; Cartea, M.E. Genotypic and Environmental Effects on Agronomic and Nutritional Value of *Brassica rapa*. *Agron. J.* **2011**, *103*, 735–742. [CrossRef]
43. Azam, A.; Khan, I.; Mahmood, A.; Hameed, A. Yield, chemical composition and nutritional quality responses of carrot, radish and turnip to elevated atmospheric carbon dioxide. *J. Sci. Food Agric.* **2013**, *93*, 3237–3244. [CrossRef] [PubMed]
44. Daszykowski, M.; Wrobel, M.S.; Czarnik-Matusewicz, H.; Walczak, B. Near-infrared reflectance spectroscopy and multivariate calibration techniques applied to modelling the crude protein, fiber and fat content in rapeseed meal. *Analyst* **2008**, *133*, 1523–1531. [CrossRef] [PubMed]
45. Petisco, C.; García-Criado, B.; Vázquez de Aldana, B.R.; Zabalgogeazcoa, I.; Mediavilla, S.; García-Ciudad, A. Use of near-infrared reflectance spectroscopy in predicting nitrogen, phosphorus and calcium contents in heterogeneous woody plant species. *Anal. Bioanal. Chem.* **2005**, *382*, 458–465. [CrossRef] [PubMed]
46. Osborne, B.G. *Near Infrared Spectroscopy in Food Analysis*; Longman Scientific and Technical: New York, NY, USA, 1986.
47. Bagchi, T.B.; Sharma, S.; Chattopadhyay, K. Development of NIRS models to predict protein and amylose content of brown rice and proximate compositions of rice bran. *Food Chem.* **2016**, *191*, 21–27. [CrossRef]
48. Petisco, C.; García-Criado, B.; Mediavilla, S.; Vázquez de Aldana, B.R.; Zabalgogeazcoa, I.; García-Ciudad, A. Near-infrared reflectance spectroscopy as a fast and non-destructive tool to predict foliar organic constituents of several woody species. *Anal. Bioanal Chem.* **2006**, *386*, 1823–1833. [CrossRef]
49. Campo, L.; Monteagudo, A.B.; Salleres, B.; Castro, P.; Moreno-Gonzalez, J. NIRS determination of non-structural carbohydrates, water soluble carbohydrates and other nutritive quality traits in whole plant maize with wide range variability. *Span. J. Agric. Res.* **2013**, *11*, 463–471. [CrossRef]
50. Nordheim, H.; Volden, H.; Fystro, G.; Lunnan, T. Prediction of in situ degradation characteristics of neutral detergent fiber (aNDF) in temperate grasses and red clover using near-infrared reflectance spectroscopy (NIRS). *Anim. Feed Sci. Technol.* **2007**, *139*, 92–108. [CrossRef]

© 2019 by the authors. Licensee MDPI, Basel, Switzerland. This article is an open access article distributed under the terms and conditions of the Creative Commons Attribution (CC BY) license (http://creativecommons.org/licenses/by/4.0/).

Article

Seed Oil Quality of *Brassica napus* and *Brassica rapa* Germplasm from Northwestern Spain

Elena Cartea [1], Antonio De Haro-Bailón [2,*], Guillermo Padilla [3], Sara Obregón-Cano [2], Mercedes del Rio-Celestino [4] and Amando Ordás [1]

[1] Plant Genetic and Breeding Department, Biological Mission of Galicia (CSIC), Apartado 28, E-36080 Pontevedra, Spain

[2] Plant Breeding Department, Institute for Sustainable Agriculture (CSIC), Alameda del Obispo s/n, 14080 Córdoba, Spain

[3] Bioinformatics and Biostatistics Service, Biological Reseach Center (CSIC), Ramiro de Maeztu 9, 28040 Madrid, Spain

[4] Agri-food Laboratory, (CAPDER), Avda Menéndez Pidal, s/n, 14080 Córdoba, Spain

[*] Correspondence: adeharobailon@ias.csic.es

Received: 30 June 2019; Accepted: 25 July 2019; Published: 27 July 2019

Abstract: The seed oil content and the fatty acid composition of a germplasm collection of *Brassica napus* and *Brassica rapa* currently grown in Galicia (northwestern Spain) were evaluated in order to identify potentially interesting genotypes and to assess their suitability as oilseed crops for either edible or industrial purposes. The seeds of the *B. rapa* landraces had higher oil content (mean 47.3%) than those of *B. napus* (mean 42.8%). The landraces of both species showed a similar fatty acid profile (12% oleic acid, 13% linoleic acid, 8–9% linolenic acid, 8–9% eicosenoic acid, and 50–51% erucic acid). They were very high in erucic acid content, which is nutritionally undesirable in a vegetable oil, and very low in oleic and linoleic acid contents. Therefore, they could be used for industrial purposes but not as edible oil. The erucic acid content ranged from 42% to 54% of the total fatty acid composition with an average value of 50% in the *B. napus* landraces whereas in *B. rapa*, it ranged from 43% to 57%, with an average value of 51%. Considering the seed oil and the erucic acid content together, three varieties within the *B. napus* collection and two varieties within the *B. rapa* one seem to be the most promising genotypes for industrial purposes.

Keywords: *Brassica napus*; *Brassica rapa*; fatty acid composition; germplasm; oil content

---

1. Introduction

Brassica oilseed crops have become the third most important source of edible vegetable oils in the world [1]. Although edible oils currently represent the largest market for *Brassica* oilseed crops, the prevalence of agricultural surpluses in many developed countries has focused attention toward the possible industrial use of *Brassica* seed oils. The usefulness and quality characteristics of seed oils are determined by the proportion of its main constituent fatty acids [2–4]. Consequently, one of the most important objectives in *Brassica* breeding is the genetic modification of seed oil by maximizing the proportion of specific fatty acids [5–8].

Brassica oil is considered beneficial from a health point of view. It contains linoleic acid, which is desirable for nutritional purposes, and oleic acid, whose thermostability makes it desirable for cooking oil [9]. High oleic acid oil tastes better and may also have health benefits. The oxidative stability of this fatty acid also makes it suitable for some industrial applications [10]. Nevertheless, *Brassica* oil is characterised by significant amounts of erucic acid (about 50% of the total fatty acids), which is absent in any other commercial plant oils [11,12]. Erucic acid (cis-13-docosenoic acid, 22:1) has 22 carbon atoms with one double bond at the cis-13 position of the carbon chain. Oil with high erucic acid content has anti-nutritional properties but is suitable for some industrial applications, such as anti-blocking agents in polyethylene films, adhesives in printing, and anticorrosive materials in the steel sheet metal industry [9,13,14]. They may also be used in the manufacture of cosmetics products through the synthesis of waxes that could be used as a jojoba oil substitute [15,16]. The oleochemical industry demands oils with high levels of erucic, behenic, and arachidic fatty acids. In recent decades, oilseed *Brassica* crops have also gained attention not only as a source of edible oils but also as a source of bio-fuel and industrial feed-stock. These genera have regained interest for use in cosmetics, in the emollient industry for lubricant, and for adhesive and biodegradable plastic products [4,17]. A medicinal application has also been found for erucic acid, administrated in therapeutic doses, to treat adrenoleukodystrophy (X-ALD), a genetic disorder that damages the nervous system and is associated with the accumulation of very long chain fatty acids [18,19]. Therefore, the fatty acid compositions of rapeseed oils have been modified according to specific objectives through conventional and molecular breeding [8]. The production of biodiesel has offered new opportunities and also lead to changes in the orientation of rapeseed consumption and utilization. Moreover, the emerging emphasis on renewable energy, chemical feed stocks, industrial oils, and the steadily growing bioeconomy will provide significant growth opportunities for industrial *Brassica* oils.

The development of commercial varieties free of erucic acid and with very high erucic acid content are breeding objectives in *Brassica* oilseed crops [7,13,17,20]. Agronomically acceptable cultivars producing low erucic acid oils were first available in the *B. napus* cultivar 'Oro' in 1968 and in the *B. rapa* cultivar 'Span' in 1971 [17,21]. The term 'canola oil' describes the oil profile of the current *B. napus* and *B. rapa* cultivars used for the production of edible oil with very low erucic acid content.

The genus *Brassica* encompasses very diverse types of plants grown as vegetables, fodder, and sources of oils and condiments. The species *B. napus*, *B. rapa*, *B. juncea*, and *B. carinata*, generally known as rapeseed, form the oilseed group [4,17]. Within the *B. rapa* and *B. napus* species there are also vegetable crops used for human nutrition, such as turnip, turnip tops or turnip greens (*B. rapa* ssp. *rapa*) and leaf rape (*B. napus* var. *pabularia*), which are widely grown in Galicia (northwestern Spain). *B. napus* var. *pabularia* crops grown in Galicia are known as 'nabicol' [22]. These populations are the result of mass selection carried out by growers who have been using them as leafy greens for many years, since the use of commercial varieties in this area is not common yet. The agronomic performance, morphological attributes, and leaf nutritional value of the *Brassica* germplasm grown in northwestern Spain have been extensively studied in *B. napus* [23,24]) and *B. rapa* [25–27]. The potential use of the genetic diversity existing in the *B. napus* landraces was described by Cartea et al. [21]. De Haro et al. [28] reported a preliminary work about the seed oil composition for a set of *Brassica* landraces from northwestern Spain and found that its accessions had very high erucic and very low oleic and linolenic acid contents. The suitability of other *Brassica* species as sources of new potential oilseed crops has been reported for *B. carinata* [29–32] and *B. juncea* [33,34].

The objectives of the present work were to evaluate the seed quality (the seed oil content and fatty acid composition) of a germplasm collection of *B. napus* and *B. rapa*, to identify potentially interesting genotypes, and to assess its suitability as oilseed crops for edible or industrial purposes.

2. Materials and Methods

2.1. Plant Material

A set of 41 accessions of 'nabicol' (*B. napus* ssp. *pabularia*) comprising 38 landraces and 3 commercial varieties, and a set of 169 accessions of turnip, turnip tops, and turnip greens (*B. rapa* ssp. *rapa*), including 162 landraces and 7 commercial varieties from the germplasm collection of the Biological Mission of Galicia (Misión Biológica de Galicia, MBG, Pontevedra), in Spain were analysed for total seed oil content and fatty acid composition. This material is stored and maintained as an active collection at the MBG.

These accessions represent the genetic variability of the *B. rapa* and *B. napus* germplasm currently grown in Galicia. The landraces were collected directly from growers at different sites throughout northwestern Spain from the eighties to present, and some of them have been propagated under isolation at the MBG in different years. Seed samples from different genotypes were taken from accessions kept at the germplasm bank at the MBG under the same low temperature and seed moisture conditions. Varieties were multiplied over several years, but always in the same location, with the same experimental plot, and under the same growing conditions. Due to the high number of genotypes evaluated in this study, it would have been impossible to multiply all the varieties in the same year. The geographical distribution of the *B. rapa* and *B. napus* landraces is shown in Figure 1. The number of landraces comprising the *B. napus* collection was lower than the *B. rapa* ones since the growing region of *B. napus* is restricted to the coastal area of southern Galicia and to areas near the Portuguese border (Figure 1). Here, the crop is well adapted and common in the human diet [21]. Five accessions of *B. napus* were collected in inland areas, probably due to both human migration and the sale of seeds in local markets. All the *B. napus* landraces come from Galicia, Spain whereas the 3 *B. napus* commercial varieties were bought in Vila Real (North of Portugal), since commercial varieties of *B. napus* are not found in Galicia. All the *B. rapa* entries come from Galicia, Spain.

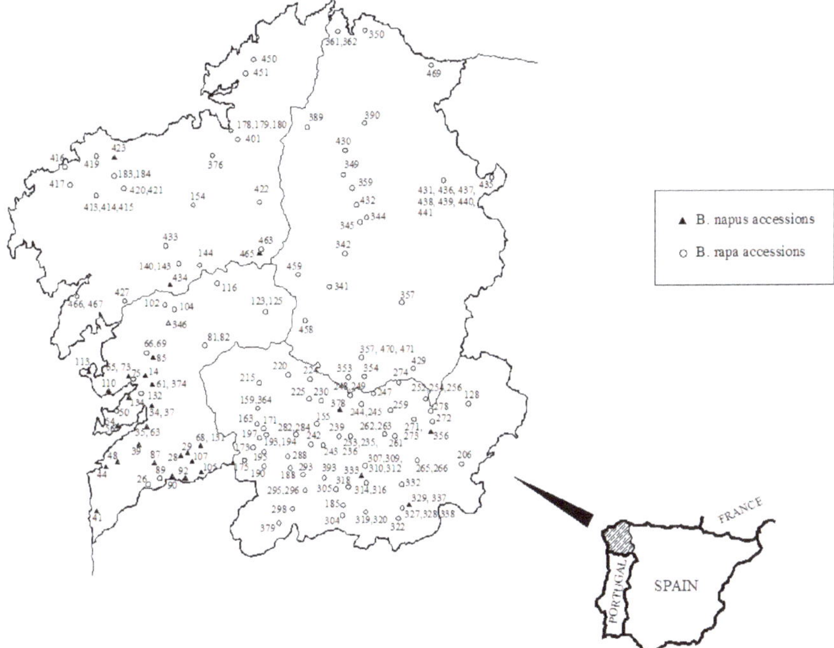

Figure 1. Map of Galicia (northwestern Spain) with the geographical distribution of the *B. rapa* and *B. napus* landraces evaluated in this study.

2.2. Lipid Analysis

Bulk samples of seeds of each accession (landraces and commercial varieties) were screened for oil content. The oil content of the seeds was determined by nuclear magnetic resonance (NMR) with an Oxford 4000 Analyzer (Oxford Analytical Instruments Ltd., Abingdon, OX, UK), following desiccation at 50 °C for 72 h. For fatty acid composition, ten seeds were randomly selected and individually analysed for each local and commercial variety. The content of seven major fatty acids present in the oil extracted from *Brassica* crops (palmitic, stearic, oleic, linoleic, linolenic, eicosenoic, and erucic), as well as the content of other minor fatty acids (arachidic, arachidonic, and behenic), was determined. In order to evaluate the fatty acid composition of seed samples, the lipids were extracted, transmethylated, and purified using the one-step method of Garcés and Mancha [35] with some modifications. Individual seed samples were heated at 80 °C for 2 h in MeOH: toluene: dimethoxypropane: H_2SO_4: heptane (33:14:2:1:50; by vol.), and, after cooling, the fatty acid methyl esters were recovered in the upper phase. The analysis of the fatty acid methyl esters composition was developed in a Perkin Elmer Autosystem gas-liquid chromatograph (Perkin-Elmer Corporation, Norwalk, USA) equipped with a flame ionization detector (FID) and a 2 m long column packed with 3% SP-2310/2% SP-2300 on a Chromosorb WAW (Supelco Incorporated, Bellefonte, USA). The gas chromatograph was programmed for an initial temperature of 190 °C for 10 min followed by an increase of 2 °C per min to 220 °C; this final temperature was maintained for a further 5 min. The injector and flame-ionization detector were held at 275 and 250 °C, respectively. The fatty acids were identified by comparison with known fatty acid methyl esters standards (F.A.M.E. Mix, CRM18920 Supelco and ME14-1KT Supelco, Sigma-Aldrich). The analyses were performed at the 'Institute for Sustainable Agriculture (*Instituto de Agricultura Sostenible*, IAS), Spain.

Individual and combined analyses of variance were performed for each trait of seed composition, using the general lineal model (GLM) procedure of the SAS statistical package [36]. The accessions and the species were considered as fixed effects. Comparisons of means among populations and species were performed for each trait using Fisher's protected least significant difference (LSD) at $p = 0.05$ [37].

3. Results and Discussion

3.1. Oil Content in the Brassica Collection

The oil content in the seeds of the *B. napus* landraces ranged from 29.1% (for MBG-BRS0423) to 50.1% (for MBG-BRS0044), with an average value of 42.5%. The oil content of the *B. rapa* landraces ranged from 31.4% (for MBG-BRS0285) to 56.3% (for MBG-BRS0245), with an average value of 47.3% (Table 1). The *Brassica rapa* genotypes were significantly higher in oil content than the *B. napus* ones (Table 1). This result agrees with Mandal et al. [38], who found that seed oil content was higher in a collection of *B. rapa* (about 42%) than in a collection of *B. napus* (about 35%). The seed oil content from the *B. napus* landraces was significantly lower than that from the commercial varieties, whereas no significant differences were found between the landraces and the commercial varieties in *B. rapa*. The genotypes evaluated in this work showed values for oil content similar to those found on cultivars of the major *Brassica* oilseed crops (*B. napus*, *B. rapa*, *B. carinata*, and *B. juncea*), with an average oil content between 45% and 50% [30,39], even though the *Brassica* germplasm from northwestern Spain is not grown as an oilseed crop. Since only one datum per population was obtained, mean oil content comparisons among populations are not reported. Despite this fact, the highest seed oil content (more than 47%) in the *B. napus* collection was found for landraces MBG-BRS0044, MBG-BRS0087, MBG-BRS0329, MBG-BRS0434, and MBG-BRS0105, along with the commercial variety, MBG-BRS0373. For the *B. rapa* collection, landraces MBG-BRS0245, MBG-BRS0236, MBG-BRS0125, MBG-BRS0249, and MBG-BRS0139 had the highest levels.

Table 1. Mean, range (minimum and maximum values), and standard deviation of oil content in the seeds of *B. napus* and *B. rapa* varieties from northwestern Spain.

	Oil Content (%)			
Accessions	N°	Mean	Range	Standard Deviation
Brassica napus collection	41	42.80		
Landraces	38	42.46	(29.06–50.11)	3.88
Commercial varieties	3	47.13	(46.35–48.50)	1.19
LSD (5%)		4.60		
Brassica rapa collection	169	47.27		
Landraces	162	47.22	(31.38–56.34)	4.31
Commercial varieties	7	48.59	(46.04–52.85)	2.09
LSD (5%)		3.24		
LSD (5%) between landraces [1]		1.50		

N° = Number of accessions studied; LSD = least significative differenc; [1] Comparison between 38 *B. napus* and 162 *B. rapa* landraces.

3.2. Fatty Acid Composition in the Brassica Collection

Significant differences among the *B. napus* landraces were found for all the fatty acids analysed, whereas the commercial varieties of this species were not significantly different for linoleic and erucic acid contents (Table 2). The fatty acid profile observed between the landraces and the commercial varieties of *B. napus* was different. The average erucic acid content was considerably higher in the *B. napus* landraces compared to the commercial seeds (Table 3). The commercial varieties had zero erucic acid and followed the typical profile of canola varieties even though they are commonly used as vegetable crops but not as oilseed crops. The collection of the *B. napus* evaluated includes all the germplasm currently grown in Galicia, which means that in this region, only the *pabularia* type is grown, and *B. napus* is not used for oil production. The significant differences between the *B. napus* landraces and the commercial varieties do not mean that the commercial *pabularia* breeding is based on modern zero erucic varieties of *B. napus*. Since commercial varieties of *B. napus* are not common in Galicia, the seeds used in this study came from Portugal, where they were bought as 'couve-nabiça' (Portuguese *B. napus* landrace), but they are probably zero erucic rapeseed varieties.

Table 2. Mean squares of the analysis of variance for fatty acids in the *B. napus* and *B. rapa* varieties studied.

Sources of Variation	df	C16:0	C18:0	C18:1	C18:2	C18:3	C20:1	C22:1
Brassica napus	40	2.35 **	0.31 **	1754.28 **	56.91 **	11.24 **	49.73 **	1803.20 **
Landraces	37	0.83 **	0.08 **	70.15 **	15.50 **	9.41 **	15.11 **	85.53 **
Commercial var.	2	0.97 *	0.01 *	37.79 *	0.85	40.54 **	0.31 **	0.005
Between types	1	61.37 **	9.26 **	67,501.8 **	1700.9 **	20.20 **	1409.30 **	68,961.78 **
Brassica rapa	168	2.21 **	0.41 **	42.34 **	8.73 **	8.64 **	13.53 **	67.21 **
Landraces	161	2.27 **	0.42 **	40.37 **	8.66 **	8.12 **	13.55 **	67.22 **
Commercial var.	6	0.72 **	0.11 **	32.68 **	11.05 **	12.47 **	9.35 **	35.36 **
Between types	1	2.64 *	0.32 *	418.70 **	6.32	70.32 **	37.41 **	257.05 **
Between landraces [a]	1	43.67 **	0.66 **	7.78	11.34 **	4.57 *	6.78 *	16.21

*, ** significant at $p < 0.05$ and at $p < 0.01$, respectively. [a] Comparison between 38 *B. napus* and 162 *B. rapa* landraces. C16:0 = palmitic acid, C18:0 = stearic acid, C18:1 = oleic acid, C18:2 = linoleic acid, C18:3 = linolenic acid, C20:1 = eicosenoic acid, and C22:1 = erucic acid. df = degrees of freedom.

Table 3. Fatty acid composition [a] of the seed oil (mean, minimum and maximum values) of *B. napus* and *B. rapa* varieties from northwestern Spain.

Accessions	No.	C16:0	C18:0	C18:1	C18:2	C18:3	C20:1	C22:1
Brassica napus	41							
Landraces	38	2.95	0.58	12.37	12.72	8.26	8.27	49.83
		(2.30–3.50)	(0.40–0.77)	(7.85–18.73)	(7.94–15.74)	(6.51–10.87)	(6.23–10.58)	(42.35–54.09)
Commercial varieties	3	4.43	1.16	61.64	20.54	9.12	1.10	0.02
		(4.25–4.79)	(1.09–1.26)	(59.91–63.74)	(20.21–20.71)	(6.86–10.72)	(0.90–1.24)	(0.004–0.05)
LSD (5%) [b]		0.172	0.054	1.265	0.610	0.510	0.646	1.463
Brassica rapa	169							
Landraces	162	1.75	0.72	11.86	13.33	8.65	8.73	50.56
		(0.66–2.57)	(0.25–1.28)	(7.56–18.53)	(10.91–16.56)	(5.25–10.84)	(4.72–11.95)	(42.75–56.96)
Commercial varieties	7	1.95	0.79	14.36	13.02	7.63	9.48	48.60
		(1.57–2.37)	(0.59–0.93)	(11.75–16.64)	(11.33–14.27)	(6.23–9.33)	(8.30–11.07)	(46.03–51.39)
LSD (5%) [b]		0.181	0.067	0.709	-	0.334	0.442	0.951
LSD (5%) between landraces [c]		0.159	0.070	-	0.355	0.325	0.419	-

[a] These values are the means of ten single seeds, expressed as % of the total fatty acids. [b] Comparison between the landraces and the commercial varieties of each species. [c] Comparison between 38 *B. napus* and 162 *B. rapa* landraces. C16:0 = palmitic acid, C18:0 = stearic acid, C18:1 = oleic acid, C18:2 = linoleic acid, C18:3 = linolenic acid, C20:1 = eicosenoic acid, and C22:1 = erucic acid. N° = Number of accessions studied; LSD = least significative difference.

Within the *B. rapa* collection, the landraces and the commercial varieties were significantly different for all the fatty acids. The seeds of the *B. rapa* commercial varieties, which included crops of turnips, turnip greens, and turnip tops, were significantly different in erucic acid content (Table 2), although their values were significantly lower than those found in most landraces (Table 3). Despite this fact, the genotypes of both types of germplasm (landraces and commercial varieties) showed a similar fatty acid profile (Table 3).

Significant differences were found between the *B. napus* and *B. rapa* landraces for all the fatty acids, except for oleic and erucic acids (Table 2). The seeds of the *B. rapa* landraces were higher than the *B. napus* ones for stearic, linoleic, linolenic, and eicosenoic acids, while the *B. napus* landraces were only higher for palmitic acid content (Table 3).

The landraces of both species were high in erucic acid and low in oleic, linoleic, linolenic, and eicosenoic acids; palmitic and stearic acids were minor, and arachidic, arachidonic, and behenic acid were negligible (Table 3). The fatty acid profile of the oil contained in the *B. napus* seeds (12% oleic acid, 13% linoleic acid, 8% linolenic acid, 8% eicosenoic acid, and 50% erucic acid) was very similar to that found in the seeds of *B. rapa* (12% oleic acid, 13% linoleic acid, 9% linolenic acid, 9% eicosenoic acid, and 51% erucic acid) (Table 3). Both fatty acid profiles contrast with the typical profile of canola oil, which can be represented as 61% oleic acid, 21% linoleic acid, 11% linolenic acid, and no erucic acid [4,8,10].

Since erucic acid was the major fatty acid found, and because it is a trait of large interest for plant breeding, most discussion will be focused on this fatty acid. The erucic acid content ranged from 42% to 54% of the total fatty acid composition, with an average value of 50% in the *B. napus* landraces and from 43% to 57% of the total fatty acid composition in the *B. rapa* landraces, with an average value of 51% (Table 3). Similar values for erucic acid content were found by Sharafi et al. (2015) [12] for three rapeseed varieties and for four entries of *B. rapa*, although crops were different from those evaluated in this study. The lowest content of erucic acid was found within the *B. napus* species in seeds of MBG-BRS0333 (about 42% of the total fatty acids) and within the *B. rapa* species in seeds of MBG-BRS0379 (about 43% of the total fatty acids). Both values are still very high, and, therefore, vegetable oil of the genotypes evaluated should be considered unsuitable for edible oil production. On the other hand, oils with high erucic acid content are desirable for industrial purposes. For both species, the landraces with high erucic acid content are included in Table 4. The highest erucic acid composition, more than 53% of the total fatty acids, was found in *B. napus* landraces MBG-BRS0329, MBG-BRS0041, MBG-BRS0048, and MBG-BRS0105 (Table 4). *B. rapa* landraces

MBG-BRS0235, MBG-BRS0416, MBG-BRS0202, MBG-BRS0139, MBG-BRS0190, and MBG-BRS0239 showed the highest erucic acid composition—more than 55% of the total fatty acids (Table 4). Considering their high oil content and high erucic acid content together, genotypes MBG-BRS0329, MBG-BRS0434, and MBG-BRS0105 within the *B. napus* collection, and MBG-BRS0139 and MBG-BRS0101 within the *B. rapa* collection, offer interesting prospects for future industrial applications. The *Brassica* landraces grown in Galicia have been traditionally improved by growers for vegetable use but not for their use as oil sources. Nowadays, efforts to develop low erucic acid genotypes of both *B. napus* var. *pabularia* and *B. rapa* ssp. *rapa* have not been undertaken because the main usage of these crops is for leaf consumption.

Table 4. Seed oil content and fatty acid composition [a] (major fatty acids) of *B. napus* and *B. rapa* landraces with the highest seed oil content and the highest erucic acid content.

Landraces	Oil Content	C16:0	C18:0	C18:1	C18:2	C18:3	C20:1	C22:1
Brassica napus								
MBG-BRS0329	47.47	2.3 ± 0.1	0.6 ± 0.1	16.5 ± 2.0	7.9 ± 1.0	6.6 ± 0.7	8.1 ± 1.6	54.1 ± 2.9
MBG-BRS0041	45.31	2.6 ± 0.1	0.5 ± 0.1	9.7 ± 1.4	12.0 ± 0.6	9.7 ± 0.7	6.4 ± 0.8	53.8 ± 1.2
MBG-BRS0048	42.88	2.9 ± 0.3	0.6 ± 0.1	9.9 ± 1.5	12.6 ± 0.7	8.1 ± 1.0	6.6 ± 1.1	53.8 ± 1.8
MBG-BRS0105	46.61	2.6 ± 0.2	0.5 ± 0.1	10.0 ± 1.4	12.2 ± 0.7	8.9 ± 0.5	7.2 ± 1.6	53.8 ± 2.6
MBG-BRS0465	42.55	2.7 ± 0.2	0.6 ± 0.1	9.7 ± 0.9	11.9 ± 0.5	9.8 ± 0.4	6.8 ± 0.9	52.9 ± 1.8
MBG-BRS0065	39.46	2.9 ± 0.3	0.6 ± 0.1	9.0 ± 1.2	13.9 ± 0.5	8.4 ± 0.5	7.0 ± 1.1	52.6 ± 1.9
MBG-BRS0028	40.77	2.8 ± 0.3	0.6 ± 0.1	9.7 ± 2.2	13.8 ± 0.9	13.8 ± 0.9	7.1 ± 1.7	52.5 ± 2.1
MBG-BRS0037	43.09	3.3 ± 1.5	0.5 ± 0.2	11.0 ± 2.1	12.5 ± 1.3	8.8 ± 1.2	6.3 ± 1.0	52.4 ± 2.5
MBG-BRS0063	40.01	3.1 ± 0.2	0.5 ± 0.1	10.1 ± 1.0	12.3 ± 0.8	8.3 ± 0.3	7.1 ± 0.6	52.2 ± 2.4
MBG-BRS0014	45.16	2.7 ± 0.2	0.6 ± 0.2	11.1 ± 2.0	12.1 ± 0.6	9.2 ± 0.9	7.7 ± 1.5	51.8 ± 2.5
MBG-BRS0434	47.30	2.7 ± 0.2	0.5 ± 0.1	10.8 ± 1.8	11.9 ± 0.7	9.4 ± 0.9	8.0 ± 1.5	51.8 ± 3.1
MBG-BRS0054	43.94	3.0 ± 0.1	0.5 ± 0.1	7.9 ± 1.1	13.9 ± 0.8	10.9 ± 0.6	6.2 ± 0.7	51.7 ± 1.9
MBG-BRS0374	45.48	3.0 ± 0.2	0.6 ± 0.1	11.8 ± 1.3	12.6 ± 0.6	8.1 ± 0.5	8.1 ± 1.1	51.5 ± 2.4
MBG-BRS0068	46.18	3.1 ± 0.3	0.6 ± 0.1	10.0 ± 1.6	13.1 ± 0.7	8.7 ± 0.5	7.7 ± 1.0	51.4 ± 2.9
Brassica rapa								
MBG-BRS0235	46.35	0 [b]	0.1 ± 0.2	10.0 ± 1.4	14.6 ± 1.5	9.5 ± 0.8	6.0 ± 0.9	57.0 ± 1.6
MBG-BRS0416	46.63	1.6 ± 0.6	0.7 ± 0.2	8.8 ± 1.6	13.7 ± 0.8	9.0 ± 1.1	6.3 ± 1.8	55.7 ± 3.0
MBG-BRS0202	42.92	2.3 ± 0.5	0.9 ± 0.2	7.6 ± 1.7	12.7 ± 0.9	10.8 ± 1.1	4.7 ± 1.0	55.3 ± 2.6
MBG-BRS0139	54.34	1.3 ± 0.5	0.8 ± 0.1	9.9 ± 2.5	11.8 ± 0.8	9.1 ± 1.3	7.5 ± 1.8	54.9 ± 3.0
MBG-BRS0190	42.14	1.2 ± 0.5	0.6 ± 0.2	8.2 ± 1.1	13.8 ± 1.3	9.7 ± 1.3	6.7 ± 0.8	54.7 ± 1.5
MBG-BRS0239	47.97	0 [b]	0.1 ± 0.1	11.4 ± 1.2	13.7 ± 1.1	9.9 ± 1.0	7.7 ± 1.0	54.5 ± 1.4
MBG-BRS0231	48.80	0.4 ± 0.8	0.2 ± 0.4	10.0 ± 3.5	15.6 ± 1.9	7.4 ± 1.1	8.2 ± 2.6	54.5 ± 6.8
MBG-BRS0342	51.29	1.5 ± 0.4	0.7 ± 0.1	8.9 ± 1.4	12.7 ± 1.0	10.0 ± 1.3	6.8 ± 1.7	54.4 ± 2.8
MBG-BRS0228	40.38	0.2 ± 0.7	0 [b]	11.2 ± 2.2	15.4 ± 1.1	8.9 ± 0.8	6.6 ± 0.9	54.3 ± 2.3
MBG-BRS0224	50.26	1.7 ± 1.6	0.5 ± 0.3	9.6 ± 1.0	13.3 ± 1.2	9.5 ± 1.4	6.6 ± 1.3	54.2 ± 4.6
MBG-BRS0181	47.96	1.4 ± 0.5	0.7 ± 0.2	8.9 ± 1.6	14.3 ± 0.9	8.4 ± 1.1	7.5 ± 1.9	54.0 ± 3.3
MBG-BRS0237	47.38	0 [b]	0 [b]	10.7 ± 2.4	14.2 ± 1.4	9.8 ± 0.8	8.4 ± 2.5	54.0 ± 3.7
MBG-BRS0244	47.24	2.0 ± 0.2	0.6 ± 0.2	10.4 ± 1.9	11.6 ± 1.0	8.6 ± 0.9	8.2 ± 2.2	54.0 ± 3.2
MBG-BRS0467	50.09	2.3 ± 0.6	0.7 ± 0.2	0.0 ± 1.6	12.5 ± 0.7	8.4 ± 0.5	8.0 ± 1.0	53.8 ± 2.9
MBG-BRS0252	50.89	1.9 ± 0.2	0.6 ± 0.1	10.7 ± 1.3	11.2 ± 0.6	9.7 ± 0.5	8.2 ± 1.1	53.7 ± 2.4
MBG-BRS0349	44.61	2.2 ± 0.3	0.7 ± 0.1	9.3 ± 1.6	13.2 ± 0.7	9.2 ± 0.5	6.7 ± 1.0	53.7 ± 2.9
MBG-BRS0304	44.26	2.0 ± 0.3	0.8 ± 0.3	10.0 ± 1.6	13.0 ± 0.7	8.2 ± 0.5	7.5 ± 1.0	53.6 ± 2.9
MBG-BRS0435	49.13	1.6 ± 0.3	0.8 ± 0.2	9.3 ± 1.6	12.4 ± 0.7	9.8 ± 0.5	7.5 ± 1.0	53.5 ± 2.9
MBG-BRS0101	53.27	2.1 ± 0.3	0.7 ± 0.1	10.4 ± 1.6	11.9 ± 0.7	8.4 ± 0.5	8.5 ± 1.0	53.5 ± 2.9

[a] Fatty acids, given as mean ± standard deviation, expressed in % seed oil. [b] Values were negligible (only traces). C16:0 = palmitic acid, C18:0 = stearic acid, C18:1= oleic acid, C18:2 = linoleic acid, C18:3= linolenic acid, C20:1 = eicosenoic acid, and C22:1 = erucic acid.

High erucic acid contents in the seed oil of different *Brassica* crops have been previously reported by several authors. De Haro et al. [30] found high values of erucic acid in the seed oil of some accessions analysed from these same collections, between 43.3 and 57.2% in *B. napus*, and between 42.7 and 53.2% in *B. rapa*. Velasco et al. [40] reported high levels of erucic acid (about 40% of the total fatty acids) in a *B. napus* collection comprising 25 accessions, where four were *B. napus* crops. High erucic acid levels were also found by Velasco et al. [40] in a set of 72 genotypes of *B. rapa* (about 45% of the total fatty acids), where five were *B. rapa* ssp. *rapa*, although the average erucic acid content for these five accessions was lower (37.6%) than the values found in the present work. Mandal et al. [38] evaluated the fatty acid content of several cruciferous species and reported erucic acid contents of 41% in genotypes of *B. napus*, 47.4% in *B. rapa* ssp. *dichotoma*, and 51.6% in *B. rapa* ssp. *trilocularis*. Rapeseed oil genotypes with higher proportions of erucic acid than the levels found in traditional cultivars (about 50%) are sought by breeders for use in well-known industrial products [4,7,13,20]. In this way, the above-mentioned populations of *B. napus* and *B. rapa* could be used in industrial processes but they are not appropriate for edible uses.

In general, low intra-population variability was observed for the fatty acid composition of the seed oil. However, some landraces displayed an important variation among the ten seeds individually analysed for erucic acid content. Some of them had seeds with either decreased or increased values of that fatty acid. In the *B. napus* collection, three individual seeds of MBG-BRS0337 were found to have less than 35% erucic acid content (27.2%, 31.7%, and 33.1%). On the other hand, one individual seed of both MBG-BRS0105 and MBG-BRS0329 was found to have more than 56% of erucic acid content. In the *B. rapa* collection one individual seed of MBG-BRS0431 and two seeds of MBG-BRS0463 had low values for erucic acid content, (32.3%, 33.9%, and 34.1%, respectively), whereas three landraces, MBG-BRS0195, MBG-BRS0224, and MBG-BRS0231 had individual seeds with values higher than 60% of erucic acid content. The decreased content of erucic acid could be due to an unintended cross with a zero or low erucic commercial variety. However, in order to avoid the problem of pollen contamination, the seeds analyzed were directly obtained from growers or from multiplications made in isolation cages at the MBG. The decreased or increased levels of erucic acid content found in some seeds could presumably correspond to heterozygous seeds for that fatty acid. It would be interesting to analyse more seeds from the above-mentioned populations using the half-seed method described for other oilseed crops [41] and to select genotypes in segregating populations with lower or higher erucic acid content than that observed in their respective landraces.

There are numerous studies on the nutritional and industrial value of the seed oil of Brassica oilseed crops, but there are no previous studies on the potential of seed oil in these vegetable Brassica crops. Therefore, the present work provides relevant information and discussion on the potential of the oil of the seeds of two horticultural crops of Brassica, nabicol and turnip greens, for food or industrial purposes. The reported results offer the first insights into the variability of the current gene pool of the *B. rapa* and *B. napus* varieties grown in Galicia. These valuable genetic resources will certainly be studied regarding for important traits. The content of glucosinolates should be taken into account, as glucosinolates are important in modern rapeseed breeding and are probably very high in the landraces of both species. As a conclusion, the germplasm evaluated in this work displayed variability in the fatty acid composition of its seed oil. Some accessions of both species could be further used as sources of oil for industrial purposes because their seeds were high in erucic acid content and low in oleic, linoleic, and linolenic acid content. Further research will be needed for some accessions having seeds with reduced or increased values of erucic acid content, in order to select valuable genotypes that could be used for both nutritional and industrial applications.

Author Contributions: E.C., A.O., and A.D.H.-B. designed the work, G.P. collected plant samples and acquired the data, S.O.-C. and M.d.R.-C performed all the chemical analyses, E.C. and G.P. wrote this manuscript, E.C., A.D.H.-B., and A.O. revised the manuscript. All the listed authors have read and approved the submitted manuscript.

Funding: This research was funded by the Project "Metabolitos secundarios en Brassicaceae. Implicaciones en la mejora genética y defensa a estreses" Ref. RTI2018-096591-B-I00 of the Spanish Government.

Acknowledgments: Padilla acknowledges a fellowship from the La Palma Island Council.

Conflicts of Interest: The authors declare no conflict of interest.

References

1. Friedt, W.; Lühs, W. Recent developments and perspectives of industrial rapeseed breeding. *Fett Lipid* **1998**, *100*, 219–226. [CrossRef]
2. Ohlrogge, J.B. Design of new plant products: Engineering of fatty acid metabolism. *Plant Physiol.* **1994**, *104*, 821–826. [CrossRef]
3. Singh, S.P.; Jeena, A.S.; Kumar, R.; Sacan, J.N. Variation for quality parameters and fatty acids in *Brassica* and related species. *Agric. Sci. Dig.* **2002**, *22*, 205–206.
4. McVetty, P.B.E.; Mietkiewska, E.; Omonov, T.; Curtis, J.; Taylor, D.C.; Weselake, R.J. Brassica spp. Oils. In *Industrial Oil Crops*; McKeon, T.A., Hayes, D.G., Hildebrand, D.F., Weselake, R.J., Eds.; AOCS Press: London, UK, 2016; pp. 113–156.
5. Buzza, G.C. Plant Breeding. In *Brassica Oilseeds. Production and Utilization*; Kimber, D.S., McGregor, D.I., Eds.; CAB International: Wallingford, Oxon, UK, 1995; pp. 153–175.
6. Cartea, M.E.; Migdal, M.; Galle, A.M.; Pelletier, G.; Guerche, P. Comparison of sense and antisense methodologies for modifying the fatty acid composition of *Arabidopsis thaliana* oilseed. *Plant Sci.* **1998**, *136*, 181–194. [CrossRef]
7. Katavic, V.; Friesen, W.; Barton, L.D.; Gossen, K.K.; Giblin, E.M.; Luciw, T. Improving erucic acid content in rapeseed through biotechnology: What can the Arabidopsis FAE1 and Yeast SLC1-1 genes contribute? *Crop Sci.* **2001**, *41*, 739–747. [CrossRef]
8. McVetty, P.B.E.; Scarth, R. Breeding for improved oil quality in *Brassica* oilseed species. *J. Crop Prod.* **2002**, *5*, 345–369. [CrossRef]
9. Vles, R.O.; Gottenbos, J.J. Nutritional characteristics and food uses of vegetable oils. In *Oil Crops of the World*; Downey, R.K., Röbbelen, G., Ashri, A., Eds.; McGraw-Hill: New York, NY, USA, 1989; pp. 63–86.
10. Scarth, R.; McVetty, P.B.E. Designer oil canola. A Review of New Food-Grade Brassica Oils with Focus on High Oleic, Low Linolenic Types. In Proceedings of the 10th International Rapeseed Congress, Canberra, Australia, 26–29 September 1999.
11. Röbbelen, G.; Thies, W. Biosynthesis of seed oil and breeding for improved oil quality of rapessd. In *Brassica Crops and Wild Allies*; Tsunoda, S., Hinata, K., Gómez-Campo, C., Eds.; Japan Scientific Societies Press: Tokyo, Japan, 1980; pp. 121–132.
12. Sharafi, Y.; Majidi, M.M.; Goli, S.A.H.; Rashidi, F. Oil content and fatty acids composition in *Brassica* species. *Int. J. Food Prop.* **2015**, *18*, 2145–2154. [CrossRef]
13. Lühs, W.; Friedt, W. Present state and prospects of breeding rapeseed (*Brassica napus*) with a maximum erucic acid content for industrial applications. *Fat Sci. Technol.* **1994**, *96*, 137–146.
14. Uppström, B. Seed Chemistry. In *Brassica Oilseeds. Production and Utilization*; Kimber, D.S., McGregor, D.I., Eds.; CAB International: Wallingford, Oxon, UK, 1995; pp. 217–242.
15. Nieschlag, H.J.; Wolff, I.A. Industrial uses of high erucic oils. *J. Am. Oil Chem. Soc.* **1971**, *48*, 723–727. [CrossRef]
16. Kramer, J.K. *High and Low Erucic Acid in Rapeseed Oils*; Academic press: New York, NY, USA, 2012.
17. Gupta, S.K. Brassicas. In *Breeding Oilseed Crops for Sustainable Production: Opportunities and Constraints*; Academic press: New York, NY, USA, 2016.
18. Rizzo, W.B.; Leshner, R.T.; Odone, A.; Dammann, A.L.; Craft, D.A.; Jensen, M.E.; Jennings, S.S.; Davis, S.; Jaitly, R.; Sgro, J.A. Dietary erucic acid therapy for X-linked adrenoleukodystrophy. *Neurology* **1989**, *39*, 1415. [CrossRef]
19. Moser, H.W.; Moser, A.B.; Hollandsworth, K.; Brereton, N.H.; Raymond, G.V. "Lorenzo's oil" therapy for X-linked adrenoleukodystrophy: Rationale and current assessment of efficacy. *J. Mol. Neurosci.* **2007**, *33*, 105–113. [CrossRef]
20. Sasongko, N.D.; Möllers, C. Toward increasing erucic acid content in oil seed rape (*Brassica napus* L.) through the combination of genes for high oleic acid. *J. Am. Oil Chem.' Soc.* **2005**, *82*, 445–449. [CrossRef]
21. Downey, R.K. *Brassica* oilseed breeding-achievements and opportunities. *Plant Breed. Abstr.* **1990**, *60*, 1165–1170.
22. Cartea, M.E.; Soengas, P.; Picoaga, A.; Ordás, A. Relationships among *Brassica napus* (L.) germplasm from Spain and Great Britain as determined by RAPD markers. *Genet. Resour. Crop Evol.* **2005**, *52*, 655–662. [CrossRef]

23. Rodríguez, V.M.; Cartea, M.E.; Padilla, G.; Velasco, P.; Ordás, A. The nabicol: A horticultural crop in nortwestern Spain. *Euphytica* **2005**, *142*, 237–246. [CrossRef]

24. Cartea, M.E.; Rodríguez, V.M.; De Haro, A.; Velasco, P.; Ordás, A. Variation of glucosinolates and nutritional value in nabicol (*Brassica napus pabularia* group). *Euphytica* **2008**, *159*, 111–122. [CrossRef]

25. Padilla, G.; Cartea, M.E.; Rodríguez, V.M.; Ordás, A. Genetic diversity in a germplasm collection of *Brassica rapa* subsp *rapa* L. from northwestern Spain. *Euphytica* **2005**, *145*, 171–180. [CrossRef]

26. Padilla, G.; Cartea, M.E.; Velasco, P.; De Haro, A.; Ordás, A. Variation of glucosinolates in vegetable crops of *Brassica rapa*. *Phytochemistry* **2007**, *68*, 536–545. [CrossRef]

27. Obregón-Cano, S.; Cartea, M.E.; Moreno, R.; de Haro-Bailón, A. Variation in glucosinolate and mineral content in Galician germplasm of *Brassica rapa* L. cultivated under Mediterranean conditions. *Acta Hortic.* **2018**, *1202*, 157–164. [CrossRef]

28. De Haro, A.; Fernández, G.; Baladrón, J.J.; Ordás, A. Estudio de la variabilidad respecto a componentes nutritivos en brassicas gallegas. In Proceedings of the VI Congreso de la Sociedad Española de Ciencias Hortícolas, Barcelona, Spain, April 1995; pp. 123–128.

29. Getinet, A.; Rakow, G.; Downey, R.K. Agronomic performance and seed quality of Ethiopian mustard in Saskatchewan. *Can. J. Plant Sci.* **1996**, *76*, 387–392. [CrossRef]

30. De Haro, A.; Domínguez, J.; García-Ruiz, R.; Velasco, L.; Del Río, M.; Muñoz, J.; Fernández-Martínez, J. Registration of six Ethiopian mustard germplasm lines. *Crop Sci.* **1998**, *38*, 558. [CrossRef]

31. Velasco, L.; Nabloussi, A.; De Haro, A.; Fernández-Martínez, J.M. Development of high-oleic, low-linolenic acid Ethiopian-mustard (*Brassica carinata*) germplasm. *Theor. Appl. Genet.* **2003**, *107*, 823–830. [CrossRef] [PubMed]

32. Del Río-Celestino, M.; Font, R.; De Haro-Bailón, A. Inheritance of high oleic acid content in the seed oil of mutant Ethiopian mustard lines and its relationship with erucic acid content. *J. Agric. Sci.* **2007**, *145*, 353–365. [CrossRef]

33. Woods, D.L.; Capcara, J.J.; Downey, R.K. The potential of mustard (*Brassica juncea* (L.) Coss.) as an edible oil crop on the Canadian prairies. *Can. J. Plant Sci.* **1991**, *71*, 195–198. [CrossRef]

34. Sivaraman, I.; Arumugam, N.; Singh Sodhi, Y.; Gupta, V.; Mukhopadhyay, A.; Pradhan, A.K.; Burma, P.; Pental, D. Development of high oleic and low linoleic acid transgenics in a zero erucic acid *Brassica juncea* L. (Indian mustard) line by antisense suppression of the fad2 gene. *Mol. Breed.* **2004**, *13*, 365–375. [CrossRef]

35. Garcés, S.R.; Mancha, M. One step lipid extraction and fatty acid methyl esters preparation from fresh plant tissues. *Anal. Biochem.* **1993**, *211*, 139–143. [CrossRef]

36. SAS Institute. *SAS OnlineDoc*, version 8; SAS Institute, Inc.: Cary, CA, USA, 2000.

37. Steel, R.D.G.; Torrie, J.H.; Dickey, D.A. *Principles and Procedures in Statistics: A Biometrical Approach*, 3rd ed.; Mc Graw Hill: New York, NY, USA, 1997.

38. Mandal, S.; Yadav, S.; Singh, R.; Begum, G.; Suneja, P.; Singh, M. Correlation studies on oil content and fatty acid profile of some Cruciferous species. *Genet. Resour. Crop Evol.* **2002**, *49*, 551–556. [CrossRef]

39. Murphy, D.J. The use of conventional and molecular genetics to produce new diversity in seed oil composition for the use of plant breeders-progress, problems and future prospects. *Euphytica* **1995**, *85*, 433–440. [CrossRef]

40. Velasco, L.; Goffman, F.D.; Becker, H.C. Variability for the fatty acid composition of the seed oil in a germplasm collection of the genus. *Brassica. Genet. Resour. Crop Evol.* **1998**, *45*, 371–382. [CrossRef]

41. Knowles, P.F. Genetics and breeding of oil crops. In *Oil Crops of the World*; Röbbelen, G., Downey, R.K., Ashri, A., Eds.; McGraw-Hill: New York, NY, USA, 1989; pp. 361–374.

© 2019 by the authors. Licensee MDPI, Basel, Switzerland. This article is an open access article distributed under the terms and conditions of the Creative Commons Attribution (CC BY) license (http://creativecommons.org/licenses/by/4.0/).

Article

The Effect of Processing Methods on Phytochemical Composition in Bergamot Juice

Domenico Cautela [1], Filomena Monica Vella [2] and Bruna Laratta [2,*]

[1] Stazione Sperimentale per le Industrie delle Essenze e dei derivati dagli Agrumi (SSEA)—Azienda Speciale della Camera di Commercio di Reggio, via T. Campanella, 12-89125 Reggio Calabria, Italy; dcautela@ssea.it
[2] Consiglio Nazionale delle Ricerche (CNR), Istituto di Ricerca degli Ecosistemi Terrestri (IRET), via P. Castellino, 111-80131 Napoli, Italy; monica.vella@iret.cnr.it
* Correspondence: bruna.laratta@cnr.it; Tel.: +39-081-6132329

Received: 29 August 2019; Accepted: 9 October 2019; Published: 11 October 2019

Abstract: Experimental and epidemiological studies show a positive relation between consumption of citrus juices and reduction of risk for some chronic disorders, such as diabetes and cardiovascular diseases. In particular, the bergamot juice is characterized by noticeable amounts of phytochemicals such as flavanone glycosides, limonoids, and quaternary ammonium compounds, all health-beneficial biomolecules. In vitro and in vivo studies have shown anti-inflammatory, cholesterol-lowering, and anti-diabetic activities attributed to these compounds depending on their chemical structure. However, nutritional content of bergamot juice may vary as consequence of different processing techniques, thus needing to address this claim. For this reason, the objective of this research was to evaluate the effects of different processing systems on the proximate constituents, the composition, and the antioxidant activity of the correspondent juices. Overall, the results indicate that the process employed may influence the chemical composition and the functional properties of the ended juice. Screw press method produced a juice with greater content of flavanone glycosides (ranged from 37 to 402 mg/L) and limonoid aglycones (ranged from 65 to 67 mg/L) than the other processes ($p < 0.001$). However, the process used for extraction of bergamot juice did not affect significantly the N,N-dimethyl-L-proline content ($p < 0.5$). Moreover, the screw press juice showed the highest antioxidant activity with EC_{50} value of 9.35 µg/mL, thus suggesting that this method maintains for health the nutritional quality of a fresh-pressed juice.

Keywords: bergamot juice; juice processing; flavanone glycosides; limonoids; stachydrine; antioxidant

1. Introduction

Citrus fruits not only contain large amounts of ascorbic acid but are rich sources of bioactive compounds that exert wide varieties of biological functions to human health, including antioxidant, antiinflammation, antimutagenicity, anti-carcinogenicity, and anti-aging [1]. The functional components of citrus fruit include flavanones, limonoids, and quaternary ammonium compounds [2–5] Flavonoids and limonoids have shown many pharmacological properties: anticancer, cardiovascular, and anti-inflammatory and antioxidant activity [3,6–8]. Recent studies have demonstrated that N,N-dimethyl-L-proline (stachydrine), a quaternary ammonium compound presents in *Citrus* genus fruits in considerable quantities, protects endothelial against the injury induced by anoxiarcoxygenation [9] and, more, it ameliorates high-glucose induced endothelial cells senescence through the modulation of the Sirtuin 1 pathway [10].

The fruit of bergamot (*Citrus bergamia Risso et Poit.*, Rutaceae) is intensively cultivated in a limited coastal area of Reggio Calabria province (southern Italy), due to both climate and environmental conditions that are favorable for its cultivation.

The medicinal properties of bergamot were empirically used in the past for traditional uses that include improving cardiovascular function. Nowadays several clinical trials suggest that bergamot derived extract supplementation may produce beneficial effects in subjects with moderate hypercholesterolemia by reducing the plasma lipids and improving the lipoprotein profile [11,12].

Commonly, the bergamot fruit is used mostly for production of cold pressed essential oil (BEo), obtained from the peel by wash-scraping the fruit. BEo is widely employed in perfumery, cosmetic, pharmaceutical, and food industries [13]. Unlike essential oil, which is very appreciate, bergamot juice (BJ) is considered a by-product since it has not found its marketability due to scare information on its composition. The BJ is often used to adulterate other citrus juices that have a wider market like lemon juice, indeed the bitter acid taste, color, and aroma of BJ is very close to lemon juice [14].

More recently an increasing interest in BJ marketability arose from its content in bioactive compounds such as flavonoids, limonoids [15,16], and quaternary ammonium compounds [5].

Different industrial methods applied for other citrus fruits (oranges and grapefruits) could be employed also to extract the bergamot juice from the endocarp in addition to the hand squeezing method.

The technology used in citrus juice processing is similar throughout the world and operates on whole fruits (such as, FMC In Line system), which extracts the juice and the cold pressed peel oil at the same time or on rasped fruits as the screw press. An overview of citrus processing technologies was described by Cautela ed al [17]. Differences among processing systems and equipment employed in juice production arise both from local traditions and from morphological differences among citrus species.

The quality and phytochemical contents of the juice recovered vary according to juice processing methods [17]. One of the major quality variables of commercial juice is the mechanical pressure used to extract the juice from the fruit. Extraction conditions (hard/soft squeezing) will determine relative levels of juice and peel components. Some of them, such as flavanones, are present in much higher concentration in the peel than in the part considered edible by the consumers [18,19].

The processing techniques at the industrial scale like pasteurization, concentration, and freezing could also modify the initial nutritional and antioxidant content of citrus juices as observed for orange juice processing [20].

Some studies report the composition of bergamot juice obtained with manual juicing technologies, while there are limited reports on evaluation of the phytochemical content and compositional characteristics of juices obtained with industrial juicing processes [15,16].

The objectives of this research were to address changes in phytochemicals content and proximate constituents due to differences in the methods of juice extraction on the same batch of bergamot fruit.

Moreover, the antioxidant of the fresh BJ juices, obtained using different juice extraction methods, was to correlate with its phytochemicals content, in order to evaluate the effect of several readily available juicing techniques on the heathy properties of the obtained juices.

2. Materials and Methods

2.1. Study Design

The effect of different processing methods on fresh bergamot juice (BJ) was evaluated by determination of proximate constituents (titratable acidity, total soluble solids, pectins, and free sugars) and by the occurrence of specific phytochemicals (flavanone glycosides, limonoid aglycones, and stachydrine) playing several biological functions in humans. In vitro assay, free radical scavenging activity (RSA) was also determined (Figure 1).

Fresh bergamot fruits were harvested in January and February 2019 from the experimental field of SSEA agency (Stazione Sperimentale per le Industrie delle Essenze e dei derivati dagli Agrumi, Reggio Calabria, Italy) according to fruits ripening. Once harvested, a random sampling (~10 kg of bergamots) was carried out in order to obtain the batch of fruits juice for laboratory (Extraction process 1: Ep-1). The other harvested fruits (~300 kg) were sent to a local industry that kindly processed the fruits to produce industrial juices using two different extraction systems (FMC in-line extractor (Ep-2) and screw press extractor (Ep-3)).

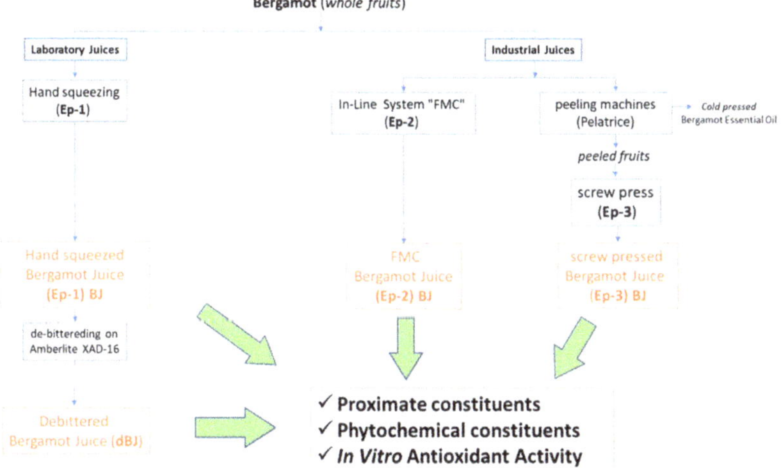

Figure 1. Schematic diagram showing the experimental design applied in this study.

Laboratory juice (n = 3) was prepared by hand squeezing using a manual squeezer (Extraction process 1: Ep-1), filtered through a stainless-steel filter with 1.18 mm (mesh diameter) and placed in 100 mL aliquots in plastic bags, and stored at −20 °C until used.

Three samples of laboratory juice were debittered by Polystyrene-DVB Resins (Amberlite XAD-16) batch process according to Calvarano et al [21]. The treatment was carried out in a 250 mL beaker containing 25 mL BJ and 5 g of exchange resin, with constant stirring for 20 min at 4 °C. Resin was removed by centrifugation for 10 min at 2000 rpm and at 4 °C.

Industrial juices were obtained using two different extraction systems (Ep-2 and Ep-3):

(a) The Ep-2 system consists of separating the essential oil from the juice in a single operation by using an FMC in-line extractor (FMC Corporation, Fairway Avenue Lakeland, FL-USA.). The FMC squeezer is designed to separate the fruit into three different parts: peel and seeds; peel oil and washing water; and juice. In this extractor, a whole fruit is positioned in one of opposed inter-fitting cup halves which were provided with means for cutting the peel or rind at diametrically opposed areas. Concurrent movement of one or both cups toward each other compresses the fruit and produces the diametrically opposed cuts. Cutting and compressing ruptures the juice sacks and the juice flows through one of the cut areas of the peel, through a screen, and to a collecting trough.

(b) In the Ep-3 system, oil extraction precedes juice extraction. In this case the essential oil is extracted from the whole fruit by rasping, using a machine called "pelatrice." The rasped fruits are then moved into a juice extractor (POLYCITRUS ZX2, Fratelli Indelicato S.r.l., Catania, Italy). This machine consists of a screw press extractor extracting juice by crushing and pressing whole fruit with a rolling screw. The screw press type operates by pressing the screw horizontally while conveying the pomelo flesh along the perforated cylindrical container. Juices flow out via the perforated cylinder and the wastes are ejected at the end plate.

Detailed descriptions of these processing were described by Cautela et al. [17]. In particular, the industrial juices were sampled during on-line processes, where 1 L of sample was firstly collected at regular intervals and then filtered through a stainless-steel filter (1.18 mm mesh diameter). The fruit juices were placed in 100 mL aliquots in plastic bags and stored at −20 °C until used.

2.2. Sample Preparation

Bergamot fruits were harvested and processed within 24 h. Before the preparation of laboratory juice, the fruits (3 batches of ~3 kg) were first washed with water (Milli-Q grade), to remove dust and pollutants from the exocarp, then drained and finally dried on paper.

For industrial juices, on receival to local industry, bergamot fruits were washed with plain water to remove leaves and dust then drained before processing. A batch of ~150 kg of fruits was employed from FMC In Line juicing extraction. Meanwhile the remaining fruits (1 batch ~150 kg) were rasped employing "pelatrice" machine according to processing line of the industry. The rasped fruits were further washed with water prior to screw press juice extraction in order to reduce the oil content.

2.3. Reagents and Standards

N,*N*-dimethylformamide (anhydrous, 99.8%), 2,2-diphenyl-1-picrylhydrazyl (DPPH), ascorbic acid (≥99.0%) , flavonoids analytical standard with purity ≥97.0% (neoeriocitrin, eriocitrin, narirutin, naringin, hesperidin, neohesperidin), limonoid aglycones analytical standard with purity ≥95.0% (limonin, nomilin), and the HPLC-grade solvents were purchased from Sigma Chemical Co. (St. Louis, MO, USA). *N*,*N*-dimethyl-*L*-proline (≥95.0%) was purchased from Extrasynthese (Genay, France). Analytical-grade water (resistivity ≥18 MΩ cm) and all other solvents and reagents of analytical grade were obtained from Carlo Erba Reagents (Milan, Italy).

2.4. Proximate Constituents

The proximate constituents' determinations were conducted in triplicate on all samples, according to the International Federation of Fruit Juice Producers (IFU) methods [22] of the European Fruit Juice Association (AIJN). The juices were thawed at 4 °C prior to analysis. The analytical determinations were carried out using the following methods: soluble solids, expressed as %*w*/*w* (°Brix), were determined according to IFU-8 method employing a RE20B refractometer (Mettler Toledo SpA, Novate Milanese Italy); total acidity, expressed as citric acid monohydrate (g/L), was obtained by titration against NaOH 0.25 N until pH of 8.1 according to IFU-3 method; ascorbic acid was assessed according to IFU-17 method by titration with 2,6-dichloroindophenol; total and water-soluble pectins were evaluated according to IFU-26 method; sugars (glucose, fructose, and sucrose) were determined spectrophotometrically by enzymatic assay kits (R-Biopharm AG, Darmstadt, Germany) according to IFU-55 and IFU-56 methods.

2.5. Analysis of Flavanone Glycosides

Flavanone glycosides content of the thawed juice were determined according to Cautela et al. [14]. Briefly, 10 mL of thawed juice was shaken with 20 mL of a 1:1 (*v*/*v*) mixture of 0.25 M *N*,*N*-dimethylformamide/ammonium oxalate and 20 mL of analytical-grade water and then filtered on 0.45 μm PTFE Pall filters.

The analyses were performed using a Surveyor autosampler LC system, interfaced with a PDA detector (Thermo Finnigan, Waltham, MA, USA) equipped with Xcalibur 3.1 software (Thermo Fisher Scientific, Waltham, USA).

A volume of 5 μL was employed for the analysis on a Supelco Spherisorb ODS2 HPLC Column (250 × 4.6 mm), and the column was thermostatically controlled at 35 °C. The elution was conducted, as already reported [23], by employing 0.3% acetic acid solution (solvent A) and acetonitrile (solvent B). A gradient elution was performed as the following: the initial solvent was 90% A and 10% B; the gradient elution was changed from 10% to 20% B in a linear mode for 15 min; this composition was maintained at isocratic flow for 10 min; the solvent B reached 50% in 10 min and from 50 to 90% B in 10 min.

PDA data were recorded in the 200–600 nm range, the identification of the flavanone glycosides was based on retention time and PDA spectra, and quantification was achieved by external standard calibration. Standard calibration curves were prepared in a concentration range 0.001–0.100 mg/mL. The calibration curves with the external standards were obtained using concentration (mg/mL) with respect to the area obtained from the integration of the PDA peaks at a wavelength of 284 nm. The quantification was achieved by comparison with the calibration curve obtained with standard solutions. The reproducibility of the detector response at each concentration level was evaluated by a triplicate injection of standard mix and expressed as percentage of relative standard deviations (RSD%). The RSDs were expected to be less than 2%. The limits of detection (LOD) were established at a signal to noise ratio (S/N) of 3. The limits of quantification (LOQ) were established at a signal to noise ratio (S/N) of 10. LOD and LOQ were experimentally verified by the nine injections of reference compounds in LOQ concentrations.

2.6. Determinations of N,N-dimethyl-L-proline (stachydrine) by HPLC-ESI-MS/MS

Stachydrine content was determined according to Servillo et al. [24] without any sample preparation except the centrifuged of the thawed juice diluted 1:25 (*v/v*) with 0.1% formic acid. Further HPLC-MS/MS analyses were performed on an Agilent 1100 Series liquid chromatograph equipped with an LC-MSD SL quadrupole ion trap. The separations were executed with a 150×3.0 mm i.d., 5 μm, Supelco Discovery-C8 column, at a flow rate of 100 μL/min. The chromatography was conducted isocratically with 0.1% formic acid in water. Samples of 20 μL of standard solutions or diluted juice sample were injected.

The conditions for ESI-MS/MS analyses, made in positive ion mode, utilizing nitrogen as the nebulizing and drying gas, were nebulizer pressure, 30 psi; drying temperature, 350 °C; and drying gas, 7 L/min. The ion charge control (ICC) was applied with the target set at 10,000 and maximum accumulation time at 20 ms. In order to obtain efficient collision induced fragmentations of the positively charged parent ion, the ion trap, molecular weight cutoff, and amplitude potential and other instrumental parameters were previously optimized for each analyte. The retention time (expressed in min) and peak areas of the monitored fragment ions were determined by the Agilent software Chemstation, version 4.2 (Agilent Technologies Inc., Santa Clara, CA USA).

The quantification was achieved by comparison with the calibration curve obtained with standard solutions.

2.7. Determinations of Limonoid Aglycones in BJ

The determinations of limonoid aglycones in BJ were conducted according to Esposito et al. [25]. Briefly, 10 mL of the juice sample was centrifuged at 2500 g for 10 min and loaded on Millipore C18 Sep Pak cartridge rinsed with 2 mL of methanol and then 5 mL of deionized water before the use. The cartridge was further washed with 20 mL of deionized water and limonoid aglycones were eluted with 50 mL of methanol. The extracts were dried using a rotary evaporator (IKA RV8, IKA®-Werke GmbH & Co. KG, Staufen, Germany) and resuspended in 1 mL of acetonitrile.

The limonoid aglycones were quantitated by RP chromatography using a Surveyor LC pump, a Surveyor autosampler, coupled with a photodiode array detector (PDA (Thermo Finnigan, Waltham, MA, USA), equipped with Xcalibur 3.1 software (Thermo Fisher Scientific, Waltham, MA USA).

Separations were achieved using a C-18 column (Phenomenex Luna 5 μ C18, 250×4.60 mm) and monitoring the wavelengths at 250 nm. The mobile phase consisted of a gradient program that began at methanol 10%, acetonitrile 0%, and water 90% and ended at methanol 41%, acetonitrile 10%, and water 49% in 45 min. The flow rate was 1 mL/min and the injection volume was 5 μL. Sample peak identifications were achieved by comparing retention times from the sample peak with those of standards run under identical conditions. Concentrations were determined using the external standard method.

2.8. In Vitro Antioxidant Activity

The free radical scavenging activity (RSA) of the bergamot juice was evaluated by 2,2′-diphenyl-1-picrylhydrazyl (DPPH) assay according to the procedure of Blois [26]. Prior of spectrophotometrically measurement, thawed juices were centrifuged at 10,000 rpm in an 5804R Eppendorf centrifuge (Eppendorf srl, Milano, Italy) for 5 min at room temperature and diluted with analytical-grade ultrapure H_2O with a concentration ranging from 5 to 20 µL/mL.

Briefly, 150 µL of bergamot diluted juices were mixed with 1.35 mL of 60 µM DPPH methanolic solution. The absorbance reduction at 517 nm of the DPPH was determined continuously for 40 min. The RSA was calculated as a percentage of DPPH discoloration, using the following equation:

$$\% \text{ RSA} = \left[\frac{(A_{DPPH} - A_s)}{A_{DPPH}} \right] \times 100 \tag{1}$$

where A_S is the absorbance of the solution when the extract was added and A_{DPPH} is the absorbance of the DPPH solution. The extract concentration (EC) necessary to achieve a 50% of radical DPPH inhibition (EC_{50}) was obtained by plotting the RSA percentage as a function of juice concentrations and was expressed as µL/mL.

Ascorbic acid was used as a reference standard (positive control) and dissolved in ultrapure H_2O to prepare a 100 µg/mL stock solution. The antioxidant activity of standard ascorbic was tested at various concentrations (10, 15, 25, 50, 60 µg/mL).

To standardize DPPH results, the antioxidant activity index (AAI), proposed by Scherer and Godoy [27], was calculated as follows:

$$AAI = DPPH \text{ concentration in reaction mixture}/EC_{50}. \tag{2}$$

Samples were classified as showing poor antioxidant activity (AAI < 0.5), moderate (0.5 < AAI < 1.0), strong (1.0 < AAI < 2.0), and very strong (AAI > 2.0) as reported by Scherer and Godoy [27].

2.9. Statistical Analysis

Statistical analysis was performed using the XLSTAT software, version 2016 (Addinsoft, Paris, France. All samples were analyzed in triplicates and the results were expressed as mean ± standard deviation (SD) after a normality distribution Kolmogorov–Smirnov test. Means, SD, calibration curves, and linear regression analyses (R^2) were determined using Microsoft Excel 2013 (Microsoft Corporation, Redmond, WA, USA).

Statistical comparisons were carried out by analysis of variance (ANOVA) and post hoc Tukey–Kramer tests. A p value less than 0.05 was considered statistically significant. All tests were two tailed.

The Pearson's correlation test was performed by using Microsoft Office Excel 2016 software. The correlation coefficients (r values, $p < 0.05$) were obtained to reveal the relationships between antioxidant activity index (AAI) and the content of flavanone glycosides, ascorbic acid, *N,N*-dimethyl-*L*-proline (ProBet), and limonoid aglycones.

3. Results

3.1. Differences in the Proximate Constituents Between the Juices Obtained with Different Processing Systems

The use of different juice extraction systems had influence on proximate constituents of bergamot juice and the outcomes of this study are summarized in Table 1. Bergamot juice was very acidic and mainly contained citric acid, which contributes significantly to the composition of this parameter. Acidity, as citric acid monohydrate, ranged from 41 g/L for Ep-3 juice to 43 g/L in Ep-1. The content of total soluble solids expressed as degree Brix (*%w/w*) did not exceed 10% in laboratory and industrial juices. Brix values of harsh (EP-3) and soft (EP-1) squeeze juices remained relatively consistent.

The average content of free sugars ranged from 9 g/L of glucose and fructose and 17 g/L of sucrose both in laboratory than in commercial squeezing juice.

Table 1. Proximate constituents (mean ± standard deviation, n = 3) of bergamot juice obtained using 3 different extraction methods: hand squeezing (Ep-1); FMC (Ep-2); screw press of peeled fruits (Ep-3).

	Ep-1	Ep-2	Ep-3
Total soluble solids (%)	9.5 ± 0.4 [a]	9.0 ± 0.2 [a]	9.1 ± 1.2 [a]
Acidity * (g/L)	43.0 ± 3.0 [a]	42.5 ± 3.5 [a]	41.0 ± 3.9 [a]
L-ascorbic acid (mg/L)	421 ± 25 [a]	415 ± 30 [a]	416 ± 33 [a]
Pectins (mg/L)	562 ± 34 [a]	640 ± 43 [a]	780 ± 75 [b]
Water-soluble pectins (mg/L)	278 ± 37 [a]	292 ± 65 [a]	473 ± 81 [b]
Sucrose (g/L)	18.0 ± 3.9 [a]	16.8 ± 4.5 [a]	16.7 ± 5.5 [a]
Glucose (g/L)	13.1 ± 1.5 [a]	12.6 ± 2.2 [a]	9.0 ± 2.5 [a]
Fructose (g/L)	12.6 ± 1.9 [a]	12.2 ± 1.9 [a]	9.7 ± 3.8 [a]

Means in a row without a common superscript letter differ (p < 0.05) as analyzed by two-way ANOVA and the TUKEY test. * as citric acid monohydrate.

It was clear that the different methods employed for the extraction of BJ did not affect significantly the quality traits of the juice, as no significant differences were detected in total soluble solids, titratable acidity, and sugar contents. Total soluble solids ranged from 9.0 ± 0.2% to 9.5 ± 0.4% in BJ obtained by FMC (Ep-2) and hand squeezed (Ep-1), respectively. The greatest contribution to the total soluble solids content was due to citric acid and this fact was also the evidence for the highest acidity value showed in Ep-1.

The average content of L-ascorbic acid in BJ was 415 mg/L and was similar in all juice. The effect of the system of extraction was also unrelated as regards L-ascorbic acid content.

Although it was interesting to note that harsh squeeze processed juice (EP-3) had considerable higher amount (p < 0.005) of pectic fractions (total and water-soluble pectins) than fresh hand squeezed juices (Ep-1) and soft squeeze processed juice (Ep-2) (Table 1).

3.2. Phytochemical Content in Bergamot Juices Obtained with Different Juice Extraction Systems

BJ was characterized by noticeable amounts of flavanone-7-O-neohesperidosides (naringin, neoeriocitrin, and neohesperidin) and lower amounts of flavanone-7-O-rutinosides (narirutin). The typical flavanone glycosides pattern of bergamot juice was shown in Figure 2 (panels b–d). The detection limits (LOD) found using RP-HPLC-DAD analysis were 0.5 mg/L, while the calculated limits of quantification (LOQ) were 1 mg/L for each flavanone glycosides. The validation process of HPLC analysis showed a good resolution of all components (Figure 2a), excellent linearity, as confirmed by the correlation coefficient R^2, ranging from 0.990 to 0.998, and a good precision, at the concentration level tested, since the coefficient of variation (CV) values were <5% for all the analytes. Considering the involvement of procedures as extraction, filtration and dilution of sample, the precision and the accuracy of the analytical method were acceptable with intra/inter assay coefficient of variation below 2.3% and 3.9%.

Results reported in Table 2 and Figure 2 suggested that BJ obtained by screw press (Ep-3) had significantly (p < 0.001) higher levels of flavanone glycosides (neoeriocitrin, naringin, neohesperidin) compared to hand squeezed juices (Ep-1) and BJ processed by and FMC (Ep-2). No significant variation in flavanone glycosides content was noticed in the hand squeezed juices and BJ obtained by Ep-2 (p < 0.1).

Naringin was the most abundant flavonoid present (394 ± 83; 175 ± 58; and 97 ± 13 mg/L in Ep-3, Ep-2, and Ep-1, respectively). The content of neoeriocitrin ranged from 114 ± 20 mg/L in Ep-1 juice to 402 ± 122 mg/L in Ep-3 juice, while the concentration of neohesperidin was found to be about 1.5- to 1.7-fold lower than naringin content.

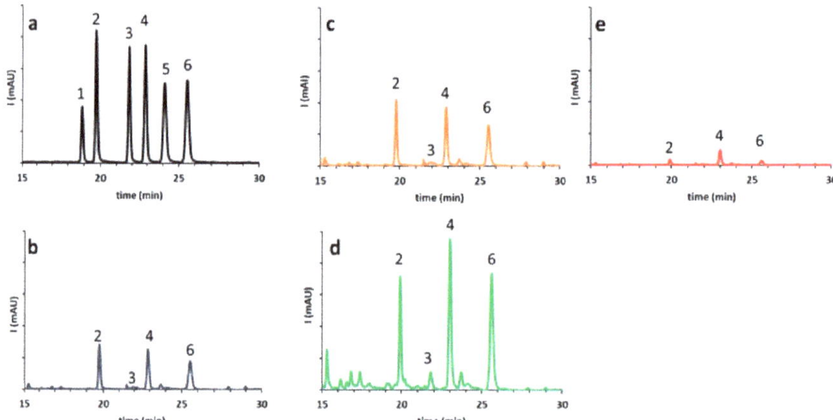

Figure 2. Comparison of RP-HPLC-DAD chromatograms at 284 nm of flavanone glycosides in bergamot juice (BJ) obtained using 3 processing methods: (**a**) Standard mixture of eriocitrin (1), neoeriocitrin (2), narirutin (3), naringin (4), hesperidin (5), neohesperidin (6). (**b**) Hand squeezed BJ (Ep-1); (**c**) FMC–BJ (Ep-2); (**d**) BJ obtained by screw press of peeled fruits (Ep-3); (**e**) debittered bergamot juice (dBJ).

Table 2. Concentration (mean ± standard deviation, n = 3) of flavanone glycosides, quaternary ammonium compounds, and limonoid aglycones in bergamot juice obtained using 3 different juice extraction methods: hand squeezing (Ep-1); FMC (Ep-2); screw press of peeled fruits (Ep-3), and in debittered bergamot juice (dBJ).

	Ep-1	Ep-2	Ep-3	dBJ
flavanone glycosides				
neoeriocitrin (mg/L)	114 ± 20 [a]	219 ± 64 [b]	402 ± 122 [c]	7 ± 2 [d]
narirutin (mg/L)	18 ± 3.0 [a]	18 ± 3.5 [a]	37 ± 3.9 [b]	<0.1
naringin (mg/L)	97 ± 13 [a]	175 ± 58 [b]	394 ± 83 [c]	5 ± 1 [d]
neohesperidin (mg/L)	66 ± 15 [a]	124 ± 69 [b]	279 ± 78 [c]	4 ± 2 [d]
Quaternary ammonium compounds				
N,N-dimethyl-L-proline (mg/L)	436 ± 49 [a]	569 ± 126 [a]	667 ± 147 [a]	395 ± 38 [a]
limonoid aglycones				
Limonin (mg/L)	20 ± 4 [a]	37 ± 13 [a]	65 ± 14 [b]	5 ± 1 [c]
Nomilin (mg/L)	17 ± 4 [a]	42 ± 11 [b]	67 ± 12 [b]	3 ± 1 [d]

Means in a row without a common superscript letter differ ($p < 0.05$) as analyzed by two-way ANOVA and the TUKEY test.

Data from Table 2 showed that N,N-dimethyl-L-proline was the main proline derived osmoprotectant compounds present in BJ. The process used for extraction of BJ did not affect significantly the N,N-dimethyl-L-proline content in juice ($p < 0.5$).

As reported in Table 2, the content of limonoid aglycones, limonin, and nomilin, increased significantly in the screw press (Ep-3) BJ with an average content of 65 mg/L. BJ obtained by FMC (Ep-2) did not significantly differ from hand squeezed (Ep-1) juice as regarding the limonin content ($p < 0.05$), while nomilin appeared to be lower in the Ep-1 juice.

Since bitterness in BJ was primarily related to two classes of compounds: neohesperidose O-flavanone glycosides (neoeriocitrin, naringin, and neohesperidin) and limonoids (limonin, nomilin), the content of these compound was evaluated in BJ treated with polystyrene-DVB resins (Amberlite XAD-16) batch process. In Table 2 the content of flavanone glycosides and limonoid aglycones of debittered BJ samples was also reported. The effect of resin treatment significantly reduced ($p < 0.001$) the amount of the flavanone glycosides, limonin, and nomilin (Figure 2, Table 2), but seemed to not affect L-ascorbic acid and *N,N*-dimethyl-L-proline, which did not significatively differ ($p < 0.5$) from fresh hand squeezed BJ (Table 2).

3.3. In Vitro Antioxidant Activity

The antioxidant activity of the BJ juices obtained using 3 different juice extraction methods were estimated by DPPH radical scavenging assay (Figure 3), ascorbic acid was used as the reference standard. The antioxidant activity of debittered bergamot juice (dBJ) was also evaluated in order to find a possible correlation between the free radical scavenging activity (RSA) of fresh processed BJ and its' content in phytochemicals.

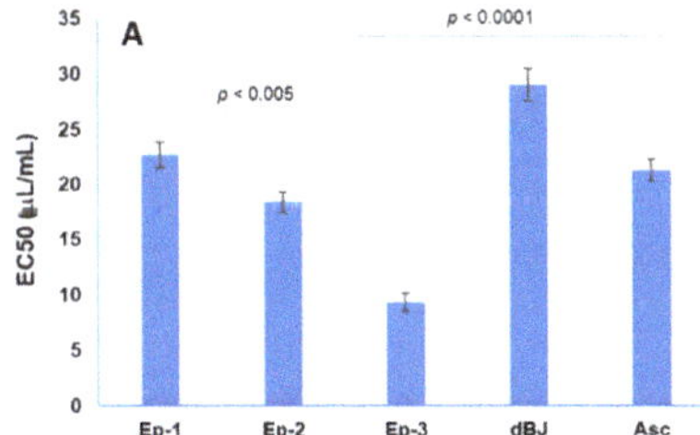
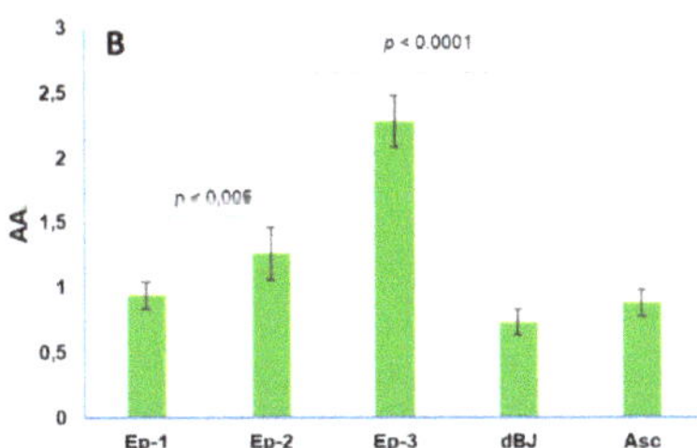

Figure 3. Antioxidant activity expressed as EC_{50} (**A**) and antioxidant activity index (AAI) (**B**) of bergamot juice obtained using 3 different juice extraction methods: hand squeezing (Ep-1); FMC (Ep-2); screw press of peeled fruits (Ep-3); and in debittered bergamot juice (dBJ). Asc: ascorbic acid (positive control). Error bar represent standard deviation ($n = 3$).

Ascorbic acid, at a concentration of 20 µg/mL, exhibited a percentage inhibition of 51.1% and a calculated EC_{50} value of 21.25 µg/mL.

After 40 min, screw press (Ep-3) BJ showed the highest DPPH radical scavenging activity, (EC_{50} was 9.35 µg/mL, $p < 0.0001$). The EC_{50} of hand squeezed (Ep-1) BJ was significatively lower than FMC juice (Ep-2) ($p < 0.005$). Debittered bergamot juice (dBJ) showed an EC_{50} value of 29 µg/mL, significantly higher than the other juice ($p < 0.0001$). It is mainly due to the L-ascorbic acid content, as previously reported in Table 2.

A very strong antioxidant activity (AAI > 2.0) was observed only in screw press (Ep-3) BJ, while the AAI of FMC juice (Ep-2) was significantly higher than Ep-1 (hand squeezed) and debittered bergamot juice (dBJ) ($p > 0.05$). The present results demonstrate stronger activity of EP-3 BJ than that of the standard antioxidant ascorbic acid ($p > 0.0001$).

Moreover, the BJ debittered with exchange resin showed significantly lower ($p < 0.0001$) total antioxidant activity due to minor content of flavanone glycosides and limonoid aglycones (Table 2, Figure 3). The exchange resin process led to a significant decrease ($p < 0.005$) in the total concentration of these compounds compared with samples of fresh juice, while no effects were observed on L-ascorbic acid, which ranged from 387 to 402 mg/L, and *N,N*-dimethyl-L-proline (Table 2).

Correlations between the antioxidant activity index (AAI) and the content of flavanone glycosides, ascorbic acid, *N,N*-dimethyl-L-proline (ProBet), and limonoid aglycones were calculated using the Pearson test in order to find out if parameters were statistically correlated (Figure 4). Strong positive correlations ($r > 0.95$) were determined for all compounds with the same exception: only a high negative correlation was observed for ascorbic acid content and AAI.

Results in Figures 3 and 4 indicated that the antioxidant activity index of BJ was mainly due to ascorbic acid content and increased together with the flavanone glycosides and limonoid aglycones content according to processing of juice extraction.

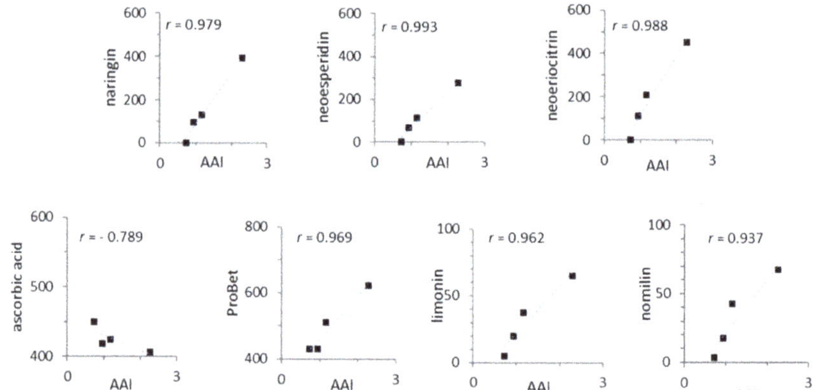

Figure 4. Pearson test correlation coefficients (*r*) among the antioxidant activity index (AAI) and the content (mg/kg) of flavanone glycosides, ascorbic acid, quaternary ammonium compounds, and limonoid aglycones in the BJ ($p < 0.05$).

4. Discussion

The objectives of this study were to evaluate the effects of different juice extraction systems on proximate constituents, bioactive components composition, and antioxidant activity of bergamot juice.

Beneficial activities of the components of citrus juices for human health were primarily ascribed to its antioxidant activity mainly due to L-ascorbic acids and polyphenols, like flavanone glucosides [16]. Despite these beneficial effects, the unprocessed fresh bergamot juice showed sour taste due to a high citric acid content (Table 1) and a bitter taste due to limonoids and flavanone-7-O-neohesperidosides which were also the most dominant bitter principles (Table 2).

Flavanones usually occurred as O-glycosyl derivatives, with the sugar moiety bound to the aglycone hydroxyl group at either C7 or C3. Among these compounds, the O-diglycosides were a dominant category and their structures were usually characterized by the linkage of either neohesperidose or rutinose to the flavonoid skeleton. The bitterness caused by flavanone-7-O-neohesperidosides was often referred to as 'primary' bitterness, while flavanone-7-O- rutinosides were tasteless [28].

The amount of the flavanone glycosides depends on the species and of the productive processes used (time and intensity of extraction, line production technology, and quality of the fruit used). It is known that a higher content of flavonoids and limonoids were found in albedo and membranes than in juice sacs [29]. Since there were several methods of extraction juice, each processing method may have its own characteristics in terms of the concentration of bioactive compounds as well as juice quality (proximate constituents).

Hand squeezing (Ep-1) was compared with two industrial squeezing methods: a "soft" juice extraction process (FMC single-strength extraction method, Ep-2) and a "hard" juice extraction process (Ep-3) operated by a screw press of peeled fruits, to evaluate their influences on health components of BJ.

4.1. Proximate and Phytochemical Content in Bergamot Juices Obtained with Different Extraction Systems

As regarding the proximate constituents of Ep-1 BJ (hand squeezed bergamot juice), results were in accordance with the previous report [14]. As showed in Table 1, juice extraction techniques used in this study did not significantly affect same proximate constituents related to juice quality parameters: soluble solids, acidity, L-ascorbic acid, and free sugars.

The range of variability of total acidity of the samples analyzed, expressed as g/L of monohydrated citric acid, was 37–41 g/L. These values were in good agreement with those previously reported by

Calvarano et al [21] and Cautela et al. [14]. As for total acidity content, in this study, significant differences were not observed due to different juice extraction techniques considering both soluble solid and free sugars content (Table 1). If compared to the other citrus juices, the amount of free sugars in bergamot was comparable with that of grapefruit and lemon [14,22].

In citrus juices, pectin was one of the major components of the suspended cloud material that imparts desirable appearance, texture, and flavor [30]. Since the amount of pectins was higher in the albedo and in the membranes, the content of pectins present in the juice depended on the juice extraction system used. Highest values were found in BJ obtained by the "hard extraction process" (Ep-3) compared to the juices obtained by the "soft extraction process". Results from Table 1 also showed that juice extractor pressure affected the water-soluble and total pectins content that increased about 1.5-folds in Ep-3 BJ.

Based on the contents of L-ascorbic acid found in all samples, BJ was a good source of vitamin C; in fact, a serving of BJ (240 mL) would cover the adult reference daily intake (RDI).

A glass of BJ also provided approximately 100 mg of stachydrine (*N,N*-dimethyl-*L*-proline) making this compound a valid biomarker of orange or other citrus juice consumption [31,32]. Furthermore, *N,N*-dimethyl-*L*-proline could contribute to the fine direct/indirect regulation, the endothelial function via the modulation of nitric oxide synthase and cellular senescence pathway [9,10].

A randomized, crossover, double-blind, controlled study performed in overweight and obese subjects showed that stachydrine was a useful biomarker to differentiate the intake of orange juice containing different amount of flavanones [32]. Furthermore, the study reported a positive correlation between the consumption of orange juice with a high content of flavanones and the improved relation of oxidative stress and inflammatory biomarkers [32].

The content of flavanone glycosides in hand squeeze (Ep-1) BJ (Table 2) were in good agreement with the data of flavonoid content in bergamot juice reported in literature [14–16]. The most abundant flavanone glycosides in bergamot fruits were naringin, neohesperidin, and neoeriocitrin. Among these, naringin was found to lower total cholesterol and low-density lipoprotein cholesterol levels in plasma [33] and has been also evaluated for its probable protective actions on pre-neoplastic lesions [34].

A clear technological effect on the flavanone glycosides (Table 2, Figure 2) and limonoid aglycones (Table 2) could be suggested, depending on the juice extraction system used.

Limonoid aglycones showed numerous pharmacological activities [7] and were the most abundant in seeds and peels [16,25]. This could explain the significatively high level ($p < 0.05$) of limonin and nomilin in EP-3 BJ compared to other juices.

The increase of flavanone glycosides with increasing extraction pressure were in agreement with Gil-Izquierdo et al. [20], who reported the effect of processing techniques at an industrial scale on orange juice antioxidant and beneficial health compounds.

4.2. In Vitro Antioxidant Activity

In our study the total antioxidant activity of BJ was evaluated using radical scavenging assays based on single electron transfer (SET) mechanisms (2,2-diphenyl-1-picrylhydrazyl, DPPH assays).

The DPPH assay measured the so-called radical-scavenging activity (RSA) which is the ability of extract constituents to scavenge reactive species to stop the initiation or propagation of oxidizing chain reactions [26].

The interaction or synergistic effect among the nutrients and/or bioactive compounds contained in citrus fruits could contribute to their antioxidant activity. Figure 3 illustrated the results of the antioxidant activity obtained for BJ tested in the present study.

Screw press (Ep-3) BJ exhibited the lowest EC_{50} value. Since the ascorbic acid and stachydrine content in all juice was comparable. Furthermore, the high flavanone glycosides, stachydrine, and limonoid aglycones content in screw press (Ep-3) BJ may also contribute to its potent DPPH radical activity. In fact, data reported in Figure 3, confirmed that the debittered bergamot juice (dBJ), had significant higher EC_{50} value. The contribution of antioxidant activity, in dBJ, was due to the ascorbic acid and stachydrine content since, in batch operations, cross-linked divinyl benzene-styrene resin reduced the flavanone glycosides and limonoid aglycones up to 90% (Table 2). The debittering processing with these resins had no effect on the minerals, acid, and amino acids content of the juice, as reported by Kimbal for navel orange juice [35].

In the present study, the contribution of ascorbic acid to the antioxidant activities of BJ was found to be moderate (Figure 4). Miller and Rice-Evans [36] had underlined the significant contributory role of polyphenols in the total antioxidant activity of long-life orange juice even if ascorbic acid was present at a higher concentration (Table 1).

5. Conclusions

BJ can be a good dietary source of nutrients like L-ascorbic acid and phytochemicals with antioxidant and health properties. However, nutritional content of bergamot juice varied as consequence of different processing techniques. Our results indicated that the juice extraction processes employed could influence the chemical composition and functional properties of BJ. The industrial processed juice, obtained by conditions (Ep-3), was markedly different from the fresh squeezed juice in the same proximate constituents as pectic substances and phytochemicals content.

Results from this study suggest that extracting juice under harsh conditions (Ep-3) increased the amount of phytochemical content and total antioxidant activity than FMC and hand squeezing juicing process.

Several limitations of the present study must be considered. This research looked at the effect of juice processing on the phytochemical content of the resulting juice. Further studies are necessary to understand the influence of the processing conditions (thermally or non-thermally) on increased shelf life of BJ.

Despite these beneficial effects, the unprocessed fresh bergamot juice showed a bitter taste due to limonoids and flavanone-7-O-neohesperidosides, hence a sensory evaluation should be addressed in order to evaluate consumer acceptance of processed BJ.

As the health benefits from the phytonutrients was better understood by addressing their bioavailability, future efforts will therefore aim to compare the different types of juices considering the bioavailability of phytonutrients in clinical trials.

Author Contributions: Author Contributions: Conceptualization, D.C and B.L.; formal analysis, D.C., F.M.V.; investigation, D.C. and F.M.V. writing-original draft, D.C. and B.L.

Funding: This research received no external funding.

Conflicts of Interest: The authors declare no conflict of interest.

References

1.   Codoñer-Franch, P.; Valls-Bellés, V. Citrus as functional foods. *Curr. Top. Nutraceutical Res.* **2010**, *8*, 173–184.

2.   Peterson, J.; Dwyer, J. Flavonoids: Dietary occurrence and biochemical activity. *Nutr. Res.* **1998**, *18*, 1995–2018. [CrossRef]

3.   Benavente-García, O.; Castillo, J. Update on uses and properties of citrus flavonoids: New findings in anticancer, cardiovascular, and anti-inflammatory activity. *J. Agric. Food Chem.* **2008**, *56*, 6185–6205. [CrossRef] [PubMed]

4.   Manners, G.D. Citrus limonoids: Analysis, bioactivity, and biomedical prospects. *J. Agric. Food Chem.* **2007**, *55*, 8285–8294. [CrossRef] [PubMed]

5.   Servillo, L.; Giovane, A.; Balestrieri, M.L.; Bata-Csere, A.; Cautela, D.; Castaldo, D. Betaines in fruits of citrus genus plants. *J. Agric. Food Chem.* **2011**, *59*, 9410–9416. [CrossRef] [PubMed]

6.   Li, C.; Schluesener, H. Health-promoting effects of the citrus flavanone hesperidin. *Crit. Rev. Food Sci. Nutr.* **2017**, *57*, 613–631. [CrossRef]

7.   Kim, J.; Jayaprakasha, G.K.; Patil, B.S. Limonoids and their anti-proliferative and anti-aromatase properties in human breast cancer cells. *Food Funct.* **2013**, *4*, 258–261. [CrossRef]

8. Yu, J.; Wang, L.; Walzem, R.L.; Miller, E.G.; Pike, L.M.; Patil, B.S. Antioxidant activity of citrus limonoids, flavonoids, and coumarins. *J. Agric. Food Chem.* **2005**, *53*, 2009–2014. [CrossRef]

9. Yin, J.; Zhang, Z.-W.; Yu, W.-J.; Liao, J.-Y.; Luo, X.-G.; Shen, Y.-J. Stachydrine, a Major Constituent of the Chinese Herb *Leonurus heterophyllus* Sweet, Ameliorates Human Umbilical Vein Endothelial Cells Injury Induced by Anoxia-Reoxygenation. *Am. J. Chin. Med.* **2010**, *38*, 157–171. [CrossRef]

10. Servillo, L.; D'Onofrio, N.; Longobardi, L.; Sirangelo, I.; Giovane, A.; Cautela, D.; Castaldo, D.; Giordano, A.; Balestrieri, M.L. Stachydrine ameliorates high-glucose induced endothelial cell senescence and SIRT1 downregulation. *J. Cell. Biochem.* **2013**, *114*, 2522–2530. [CrossRef]

11. Toth, P.P.; Patti, A.M.; Nikolic, D.; Giglio, R.V.; Castellino, G.; Biancucci, T.; Geraci, F.; David, S.; Montalto, G.; Rizvi, A.; et al. Bergamot reduces plasma lipids, atherogenic small dense LDL, and subclinical atherosclerosis in subjects with moderate hypercholesterolemia: A 6 months prospective study. *Front. Pharmacol.* **2016**, *6*, 299. [CrossRef] [PubMed]

12. Mannucci, C.; Navarra, M.; Calapai, F.; Squeri, R.; Gangemi, S.; Calapai, G. Clinical Pharmacology of *Citrus bergamia*: A Systematic Review. *Phyther. Res.* **2017**, *31*, 27–39. [CrossRef] [PubMed]

13. Khan, I.A.; Abourashed, E.A. *Leung's Encyclopedia of Common Natural Ingredients Used in Food, Drugs, and Cosmetics*, 3rd ed.; John Wiley & Sons Inc.: Hoboken, NJ, USA, 2009; pp. 91–93.

14. Cautela, D.; Laratta, B.; Santelli, F.; Trifirò, A.; Servillo, L.; Castaldo, D. Estimating bergamot juice adulteration of lemon juice by High-Performance Liquid Chromatography (HPLC) analysis of flavanone glycosides. *J. Agric. Food Chem.* **2008**, *56*, 5407–5414. [CrossRef] [PubMed]

15. Gattuso, G.; Caristi, C.; Gargiulli, C.; Bellocco, E.; Toscano, G.; Leuzzi, U. Flavonoid glycosides in bergamot juice (*Citrus bergamia Risso*). *J. Agric. Food Chem.* **2006**, *54*, 3929–3935. [CrossRef] [PubMed]

16. Russo, M.; Arigò, A.; Calabrò, M.L.; Farnetti, S.; Mondello, L.; Dugo, P. Bergamot (*Citrus bergamia Risso*) as a source of nutraceuticals: Limonoids and flavonoids. *J. Funct. Foods* **2016**, *20*, 10–19. [CrossRef]

17. Cautela, D.; Castaldo, D.; Servillo, L.; Giovane, A. Enzymes in citrus juice processing. In *Enzymes in Fruit and Vegetable Processing: Chemistry and Engineering Applications*; Bayindirli, A., Ed.; CRC Press/Taylor and Francis: Boca Raton, FL, USA, 2010; pp. 194–214.

18. Tomas-Barberan, F.A.; Clifford, M.N. Flavanones, chalcones and dihydrochalcones. Nature, occurrence and dietary burden. *J. Sci. Food Agric.* **2000**, *80*, 1073–1080. [CrossRef]

19. Gouws, C.A.; Georgouopoulou, E.; Mellor, D.D.; Naumovski, N. The effect of juicing methods on the phytochemical and antioxidant characteristics of the purple prickly pear (*Opuntia ficus indica*)—Preliminary Findings on Juice and Pomace. *Beverages* **2019**, *5*, 28. [CrossRef]

20. Gil-Izquierdo, A.; Gil, M.I.; Ferreres, F. Effect of processing techniques at industrial scale on orange juice antioxidant and beneficial health compounds. *J. Agric. Food Chem.* **2002**, *50*, 5107–5114. [CrossRef]

21. Calvarano, M.; Postorino, E.; Gionfriddo, F.; Calvarano, I. Sulla deamarizzazione del succo di bergamotto. *Essenz. Deriv. Agrum.* **1995**, *45*, 384–398.

22. Methods of Analysis. *International Federation of Fruit Juice Producers (IFU)*; Methods of Analysis: Paris, France, 2001.

23. Vella, F.M.; Cautela, D.; Laratta, B. Characterization of Polyphenolic Compounds in Cantaloupe Melon By-Products. *Foods* **2019**, *8*, 196. [CrossRef]

24. Servillo, L.; Giovane, A.; Balestrieri, M.L.; Cautela, D.; Castaldo, D. Proline derivatives in fruits of bergamot (*Citrus bergamia Risso* et Poit): Presence of *N*-methyl-L-proline and 4-hydroxy-L-proline betaine. *J. Agric. Food Chem.* **2011**, *59*, 274–281. [CrossRef] [PubMed]

25. Esposito, C.; Tridente, R.; Balestrieri, M.L.; Laratta, B.; Servillo, L.; Castaldo, D. Distribuzione dei limonoidi nelle diverse parti del frutto di bergamotto. *Ess. Deriv. Agric.* **2006**, *76*, 3–6.

26. Blois, M.S. Antioxidant determinations by the use of a stable free radical. *Nature* **1958**, *181*, 1199–1200. [CrossRef]

27. Scherer, R.; Godoy, H.T. Antioxidant activity index (AAI) by the 2,2-diphenyl-1-picrylhydrazyl method. *Food Chem.* **2009**, *112*, 654–658. [CrossRef]

28. Horowitz, R.M. Taste effects of flavonoids. In *Plant Flavonoids in Biology and Medicine, Biochemical, Pharmacological, and Structure-Activity*; Cody, V., Middleton, E., Jr., Harborne, J., Alan, R., Eds.; Liss: New York, NY, USA, 1986; pp. 163–175.

29. McIntosh, C.A.; Mansell, R.L. Three-dimensional distribution of limonin, limonoate A-ring monolactone, and naringin in the fruit tissues of three varieties of *Citrus paradisi*. *J. Agric. Food Chem.* **1997**, *45*, 2876–2883. [CrossRef]

30. Croak, S.; Corredig, M. The role of pectin in orange juice stabilization: Effect of pectin methylesterase and pectinase activity on the size of cloud particles. *Food Hydrocoll.* **2006**, *20*, 961–965. [CrossRef]

31. Lloyd, A.J.; Beckmann, M.; Favé, G.; Mathers, J.C.; Draper, J. Proline betaine and its biotransformation products in fasting urine samples are potential biomarkers of habitual citrus fruit consumption. *Br. J. Nutr.* **2011**, *106*, 812–824. [CrossRef] [PubMed]

32. Rangel-Huerta, O.D.; Aguilera, C.M.; Perez-de-la-Cruz, A.; Vallejo, F.; Tomas-Barberan, F.; Gil, A.; Mesa, M.D. A serum metabolomics-driven approach predicts orange juice consumption and its impact on oxidative stress and inflammation in subjects from the BIONAOS study. *Mol. Nutr. Food Res.* **2017**, *61*, 1600120. [CrossRef]

33. Jung, U.J.; Kim, H.J.; Lee, J.S.; Lee, M.K.; Kim, H.O.; Park, E.J.; Kim, H.K.; Jeong, T.S.; Choi, M.S. Naringin supplementation lowers plasma lipids and enhances erythrocyte antioxidant enzyme activities in hypercholesterolemic subjects. *Clin. Nutr.* **2003**, *22*, 561–568. [CrossRef]

34. Kaur, J.; Kaur, G. An insight into the role of citrus bioactives in modulation of colon cancer. *J. Funct. Foods* **2015**, *13*, 239–261. [CrossRef]

35. Kimball, D.A.; Norman, S.I. Processing Effects during Commercial Debittering of California Navel Orange Juice. *J. Agric. Food Chem.* **1990**, *38*, 1396–1400. [CrossRef]

36. Miller, N.J.; Rice-Evans, C.A. The relative contributions of ascorbic acid and phenolic antioxidants to the total antioxidant activity of orange and apple fruit juices and blackcurrant drink. *Food Chem.* **1997**, *60*, 331–337. [CrossRef]

© 2019 by the authors. Licensee MDPI, Basel, Switzerland. This article is an open access article distributed under the terms and conditions of the Creative Commons Attribution (CC BY) license (http://creativecommons.org/licenses/by/4.0/).

Article

Influence of Variety and Storage Time of Fresh Garlic on the Physicochemical and Antioxidant Properties of Black Garlic

M. Ángeles Toledano Medina [1], Jesús Pérez-Aparicio [1], Alicia Moreno-Ortega [1,2,*] and Rafael Moreno-Rojas [2]

1   Instituto de Investigación y Formación Agraria y Pesquera, Avda. Félix Rodríguez de la Fuente sn, Palma del Río, Córdoba, 14700 Andalusia, Spain
2   Department of Food Science and Technology. Campus Rabanales, Darwin Building, University of Córdoba, 14014 Andalusia, Spain
*   Correspondence: aliciamorenoortega@hotmail.com

Received: 28 June 2019; Accepted: 31 July 2019; Published: 3 August 2019

Abstract: Black garlic is made from the fresh kind, submitting it to a controlled temperature (~65 °C) and humidity (>85 °C) for a prolonged period of time. The aim of this study was to assess the differences in the process and in the final product as a result of employing three garlic varieties (*Spanish Roja, Chinese Spring* and *California White*), and to check the influence of the storage time on fresh garlic in the quality of the final product by using garlic obtained in two different agricultural seasons, that of the current year (2014) and of the previous one (2013). The results revealed some differences in the parameters analysed during the manufacturing of the black garlic from the three varieties used, and even according to the harvest in question. However, when comparing initial and final values of the samples, a very similar evolution in their acidity, reducing sugars, °Brix, pH, polyphenol content, and antioxidant capacity was noted.

Keywords: black garlic; variety garlic; storage; acidity; reducing sugars; °Brix; polyphenol content; antioxidant capacity

---

1. Introduction

The genus Allium belongs to the Alliacae family. Garlic (*Allium sativum*), onion (*Allium cepa*), leek (*Allium porrum*), and chives (*Allium schoenoprasum*) were some more famous herbs from this genus [1].

Black garlic is obtained by a multi-step heating process at a controlled temperature and humidity during a variable period of time from raw garlic [2–7]. During the production process, a series of physico-chemical changes of the fresh garlic are produced. The garlic bulbs turn into a darker product, its acidity values increase, it loses its characteristic pungency, and it develops an intense sweet taste. In addition, the total polyphenol content and antioxidant capacity increase during the black garlic process [8]. The final product is characterized by its black colour, sweet-sour taste with a balsamic touch that reminds us of raisins, and for not having the typical garlicky taste that is rejected by many consumers.

This product derived from garlic has been aim for many studies. Some recent publications have described the beneficial effects of black garlic in the prevention or improvement of cardiovascular diseases, diabetes, obesity, or cancerigenous processes, among others [2–4]. However, there are few publications providing results on the manufacturing process of this product and especially on the influence of factors like the garlic variety used or the time elapsing after its harvest on the quality of the final product obtained. In keeping with recent studies [9], 70 °C is the temperature of choice for black garlic production process with high relative humidity conditions (80%–90% RH).

The loss of characteristic smell and flavor of fresh garlic is one of the most important quality aspects in black garlic. This loss of fresh garlic flavor largely conditions the duration of the black garlic process, however, if the duration of the process is longer, the acidity values and production costs are higher [8]. The characteristic flavor of fresh garlic is produced when garlic cloves are handled, as sulfur compounds are produced by enzyme alliinase action on S-alk(en)yl substituted cysteine sulfoxides, with S-allylcysteinesulfoxide (alliin) being the most abundant sulfur compound in fresh garlic [10].

Previous studies demonstrated than S-alk(en)yllcysteine sulfoxides content depends mainly on genetic factors and post-harvest storage conditions of garlic bulbs, while the edaphoclimatic conditions during the plant growth have a lower influence [11]. Hornickova et al. (2010) [11] showed a significant increase of S-alk(en)ylcysteine sulfoxides content after a prolonged storage of fresh garlic. However, these results were obtained after eight weeks of storage at 5 °C. Some companies that market fresh garlic carry out storage at −2 °C, in order to keep the garlic in dormant state (dormancy) and to avoid germination of the garlic clove (sprouting).

In this direction, the aim proposed in this study was to evaluate the employment of three garlic varieties (*Spanish Roja.*, *Chinese Spring* and *California White*) harvested at the same time in two seasons, that of the current year and that of the previous one, and stored at −2 °C after harvest, in the process and in the quality of obtaining black garlic.

2. Materials and Methods

To manufacture the black garlic, three fresh garlic varieties were used (*Spanish Roja*, *California White* and *Chinese Spring*) from two different harvests (fresh garlic from the current year and that coming from the previous one). The garlic from the previous year's harvest was kept under refrigeration at −2 °C up to its processing.

During the black garlic manufacturing process, all the samples were submitted to the same temperature (72 ± 2 °C) and relative humidity (close to 90%) conditions.

Samples of the three varieties were analysed at the beginning and during the manufacturing of the black garlic (at 14, 21, and 34 days of manufacture). In each control, three samples were taken. In each sampling, the following physicochemical properties were determined: soluble solids (°Brix), pH, acidity (% citric acid), content in reducing sugars, total polyphenol concentration, and antioxidant capacity.

2.1. Analysis of pH, Soluble Solids (°Brix), Acidity, and Reducing Sugars

The pH analysis was made directly on the ground, homogenized sample using a potentiometer Crison Basic 20 model. To determine the content in soluble solids (°Brix), it was necessary to filter approximately 5 g of the ground sample and the liquid from it in an Abbe refractometer.

Acidity was expressed as % of citric acid. Approximately 15 g of the ground sample was weighed, and sufficient distilled water added to facilitate its homogenization, and next it was filtered for its evaluation with NaOH 0.25 N.

The reducing sugars were determined in accordance with the Rebelein method. Then, 2 mL of the sample homogenized in a beaker of precipitate was taken. To prevent interference from other substances with reducing properties, 16 mL of distilled water, 1 mL of 15% trihydrated potassium hexacyanoferrate (II), and 1 mL of 30% zinc sulphate were added to the 2 mL of sample. This was left in repose for 10 min and 5 mL of this solution was taken, to which 10 mL of 30% potassium iodide, 10 mL of 16% sulphuric acid, and 10 mL of starch paste were next added, after which it was evaluated with sodium thiosulphate 0.55 N, turning a creamy-white colour. The volume employed of thiosulphate in a blank measurement was subtracted from the volume of thiosulphate used and was compared with a calibration line made by titrating with thiosulphate different solutions of glucose in distilled water, thus obtaining the content (g/Kg) of reducing sugars in the sample.

2.2. Analysis of Polyphenol Content and Antioxidant Capacity

The samples of each control were lyophilized and 0.3 g of lyophilized sample in 10 mL of a 50% *v/v* of ethanol and distilled water was used. It was shaken for one hour in a rotating carousel. It was subsequently filtered using a buchner funnel with a whatman filter over a kitasato flask connected to a vacuum pump. The filtered extract was made up to 25 mL with a 50% *v/v* hydroalcoholic solution. Two extractions per each sample were made.

The polyphenol concentration was determined with the Folin–Ciocalteu method [12]. To a 25 mL volumetric flask, 0.5 mL of the extract, 10 mL of distilled water, 1 mL of Folin–Ciocalteu reagent, and 3 mL of 20% *p/v* sodium carbonate were added and topped up with distilled water. The mixture was heated at 50 °C for 5 min to accelerate the colouring reaction. Next, it was cooled down with water and a reading at 765 nm in a spectrophotometer was taken. The result was compared with a calibrated curve made by taking solutions of gallic acid of 75, 100, 200, 250, and 300 ppm. The results were expressed by considering the dilution of the sample (0.3 g in 25 mL) in grams of gallic acid equivalent per kg of lyophilized sample.

The antioxidant capacity was determined following the ABTS (2,2′-azino-bis(3-ethylbenzo thiazoline-6-sulphonic acid)) radical method [13] and was expressed in mmol equivalent to the Trolox. A total of 2.557 mL was prepared of a solution of the reagent 7 mM ABTS in distilled water, to which 0.333 mL of a solution of 2.5 mM of potassium persulphate in distilled water was added. The solution prepared was stored in darkness for 16 h—the time needed for the formation of the radical (ABTS+). Next, 0.15 mL of the ABTS+ solution was diluted in 15 mL of ethanol and its absorbance value (A_0) adjusted to 734 nm at 7 min. Into a 1 cm light path cuvette, 0.980 mL of the ABTS + solution and 0.02 mL of the sample extract were placed. This was shaken and the absorbance reading taken at 734 nm and at 7 min (A_1). The inhibition percentage was calculated by the following expression:

$$\% \text{ inhibition} = (A_0 - A_1) * 100/A_0.$$

Subsequently, a calibration curve was drawn with different Trolox concentrations and the inhibition percentage was obtained expressed in mmol equivalent to Trolox per kg of lyophilized sample.

2.3. Statistical Analysis

The results obtained were analysed by analysis of variance. A different model for each year was used. Two variation factors were established: the garlic variety with three fixed levels (each of the varieties used in the study), and the control time with four fixed levels (the initial one and each of the controls made at 14, 21, and 34 days of the process). The interaction between both variation factors was included in the model. The analysis of variance was made with a $p < 0.001$ probability error. To determine which levels of each factor presented significant differences, the mean values obtained in the different controls were compared in accordance with Tukey's test.

3. Results and Discussion

Table 1 shows the values calculated of the mean and standard of the mean in the reducing sugars, acidity, °Brix, pH, antioxidant capacity, and polyphenol content. The mean values were grouped according to the control during the black garlic manufacturing process (Initial, 14, 21, and 34 days), depending on the garlic variety used (*Spanish Roja*, *Chinese Spring*, and *California White*) for each year in which the garlic employed was harvested (2013 and 2014).

Table 1. Acidity, pH, soluble solids (°Brix), reducing sugars, antioxidant capacity, and polyphenols (grams of gallic acid equivalent per kg of lyophilized sample) of the garlic varieties *Spanish Roja*, *Chinese Spring*, and *California White* in two consecutive harvests.

Day	Acidity			pH			Brix		
	Var. Spanish Roja	*Var. Chinese Spring*	*Var. California White*	*Var. Spanish Roja*	*Var. Chinese Spring*	*Var California White*	*Var. Spanish Roja*	*Var. Chinese Spring*	*Var. California White*
2013									
0	0.48 ± 0.014 a	0.64 ± 0.02 b	0.52 ± 0.02 a	5.88 ± 0.01 a	5.96 ± 0.02 b	5.78 ± 0.01 c	43.5 ± 0.55 a	30.5 ± 0.15 b	40.67 ± 0.18 c
14	1.09 ± 0.02 c	1.13 ± 0.01 c	1.89 ± 0.1 d	4.68 ± 0.01 d	4.66 ± 0 e	3.71 ± 0.01 f	40.5 ± 0.56 c	33.1 ± 0.47 d	40.3 ± 0.1 c
21	1.94 ± 0.01 d	2.19 ± 0.03 e	1.92 ± 0.01 d	3.96 ± 0.05 g	4.03 ± 0.02 g	3.84 ± 0.01 h	52.27 ± 0.07 e	35.77 ± 0.09 f	47.1 ± 0.12 g
34	3.35 ± 0.13 f	4.38 ± 0.06 g	3.08 ± 0.05 h	3.3 ± 0.35 i	3.34 ± 0.2 i	3.4 ± 0.01 j	55.75 ± 0.25 h	42.5 ± 0.5 i	48.6 ± 0.36 j
2014									
0	0.42 ± 0.03 a	0.79 ± 0.02 b	0.33 ± 0.01 c	5.87 ± 0.05 a	6.25 ± 0.03 b	5.91 ± 0.02 a	39.03 ± 0.03 a	33.83 ± 0.2 b	36.23 ± 0.19 c
14	1.42 ± 0.11 d	1.11 ± 0.04 e	1.23 ± 0.01 f	4.39 ± 0.05 c	4.65 ± 0 d	4.29 ± 0 e	42.73 ± 0.3 d	32.67 ± 0.35 e	40.33 ± 0.43 f
21	1.88 ± 0.05 gh	1.78 ± 0.06 g	1.94 ± 0.03 h	4.21 ± 0.05 f	4.19 ± 0.01 f	3.85 ± 0.02 g	42.6 ± 0.15 g	33.87 ± 0.3 h	44.17 ± 0.09 i
34	3.33 ± 0.1 i	4.46 ± 0.19 j	2.79 ± 0.02 k	3.47 ± 0.02 h	3.34 ± 0.02 i	3.54 ± 0.01 j	48.17 ± 0.73 j	38.28 ± 0.17 k	47.33 ± 0.08 j

Day	Sugars			Antioxidant capacity			Polyphenols		
2013									
0	1.57 ± 0.18 a	1.91 ± 0.2 a	2.72 ± 0.14 b	23.9 ± 2.79 a	62.61 ± 3.37 b	32.62 ± 2.76 c	3.26 ± 0.29 a	4.86 ± 0.24 b	3.56 ± 0.2 a
14	8.97 ± 0.45 c	9.07 ± 0.43 c	31.9 ± 0.37*d	52 ± 2.5 d	81.85 ± 3.52 e	116.14 ± 4.49 f	4.42 ± 0.2 c	6.75 ± 0.3 d	9.97 ± 0.91 e
21	28.77 ± 1.23 e	18.46 ± 0.99 f	28.66 ± 0.76 e	103.11 ± 4.0 g	135.74 ± 5.37 h	114.31 ± 4.98 f	10 ± 0.4 e	13.64 ± 0.52 f	9.36 ± 0.25 e
34	31.56 ± 0.69 d	21.61 ± 0.25 g	28.01 ± 0.51 e	97.21 ± 1.6 i	113.77 ± 5.49 f	81.99 ± 3.32 e	12.63 ± 0.26 g	15.79 ± 0.41 h	12.65 ± 0.64 g
2014									
0	1.17 ± 0.33 a	1.34 ± 0.17 a	0.81 ± 0.20 b	22.75 ± 2.25 a	85.22 ± 5.6 b	10.97 ± 1.42 c	3.48 ± 0.22 a	5.28 ± 0.13 b	4 ± 0.08 c
14	17.8 ± 0.1 c	10.77 ± 1.73 d	19.72 ± 0.26 e	107.68 ± 3.39 d	98.71 ± 4.2 b	94.07 ± 5.17 b	9.84 ± 0.57 d	7.86 ± 0.2 e	7.15 ± 0.23 f
21	24.37 ± 1.48 f	14.85 ± 0.83 g	25.05 ± 0.71 f	110.01 ± 3.68 d	125 ± 5.7 e	127.17 ± 2.89 e	11.14 ± 0.5 g	11.17 ± 0.4 g	11.17 ± 0.68 g
34	23.98 ± 0.71 f	15.7 ± 0.28 g	27.96 ± 0.62 h	98.21 ± 7.82 d	116.7 ± 4.4 e	90.99 ± 6.85 d	14.62 ± 0.29 h	14.16 ± 0.58 h	14.34 ± 0.36 h

0: initial control; 14: control at 14 days; 21: control at 21 days; 34: control at 34 days. Different letters (two-ways ANOVA) denote significant differences among the garlic variety with three fixed levels and the control time with four fixed levels for the same parameter.

The result of the analysis of variance was significant for both variation factors in each of the analyses made (acidity, pH, Brix, reducing sugars, antioxidant capacity, and polyphenol content) and in the two harvests in the study.

The crossing factor was only significant in the content of polyphenols in garlic of 2014 owing to, in the two last controls, the different garlic varieties not showing any significant differences.

To explain the results obtained with garlic stored during one-year (2013) in comparison with garlic obtained in the year of the study (2014), one should not discount the edaphoclimatic influences; for this reason, statistical comparison between them was avoided. Although, it is possible and achievable produce black garlic with fresh garlic stored for long periods without finding great differences in the physico-chemical characteristic in the final product. However, there were clear differences of reducing sugars content, mainly for the *California White* variety (0.81 g/kg lyophilized sample for 2014 samples and 2.72 g/kg lyophilized sample for 2013 samples), possibly because this variety could dehydrate more during storage. This also explains the acidity values shown for the *California White* variety at day 0 (0.33% citric acid for 2014 samples and 0.52% citric acid for 2013 samples). Finally, the differences between garlic from both harvests were reduced during the black garlic production process.

3.1. Acidity

During the manufacturing process, the acidity value was increasing progressively in all the varieties. The increase could be partly related to derived compounds from browning and heat treatment [14]. These results are in line with previous studies [8,14]. Zhang et al., (2016) [14] showed the acidity values during black garlic production process with different temperatures (60 °C, 70 °C, 80 °C, 90 °C). They reported increasing acidity values for the different process temperatures; if the process temperature was higher, the acidity increase was reached faster. The acidity value reported by Zhang et al. (2016) [14] with 70 °C and 33 to 36 days of the black garlic process was about 4%, similar to the value obtained in our study.

In garlic harvested in 2013, the mean acidity values were significantly higher in the initial control in the var. *Chinese Spring*. The acidity value was also significantly higher in the variety *Chinese Spring* in the controls made at 21 and 34 days. The var. *California White*, in the final control, had a significantly lower acidity than the rest of the varieties, although, in the remaining controls, the differences were not significant compared with those of the var. *Spanish Roja*. In garlic harvested and analysed in 2014, the results were similar, with the var. *Chinese Spring* being the one displaying significantly higher values in the initial and final controls. The var. *California White* gave a significantly lower value in the initial control and at 14 and 34 days (2.79).

The results confirm that, at the temperature established (72 °C), the ideal duration of the process should be under 34 days, as excessive acidity values are reached in all the varieties studied, or that acidity generated in the product should be neutralised. Zhang et al. (2016) [14] reported that an acidity higher than 4% produced an unpleasant acid taste in black garlic. According to our experience, the acidity limit should be around the values of 2% and 2.5%.

In the Figure 1, the results of the acidity are shown in a graph:

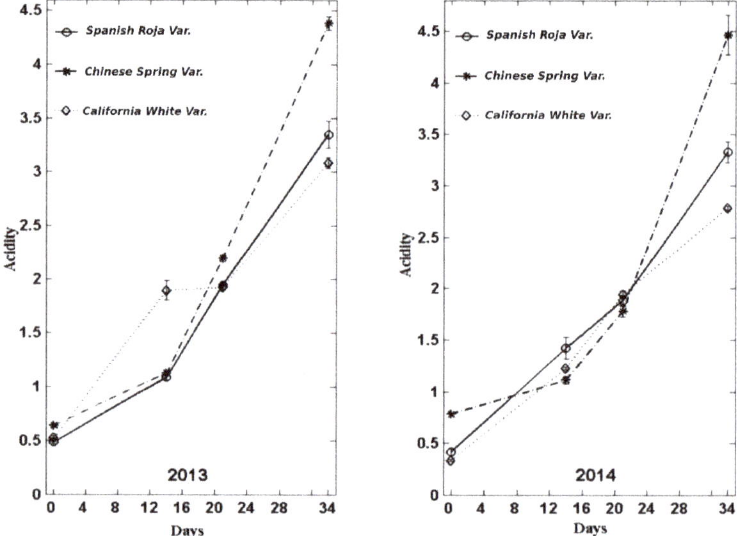

Figure 1. Evolution of acidity (% citric acid) during the manufacture of black garlic with three varieties and two types of garlic: one-year old garlic (2013) and garlic recently harvested (right).

3.2. Reducing Sugars

The evolution, the same as in the acidity results, showed an increasing trend from values close to 1 g/Kg up to values close to 30 g/Kg at 34 days. Similar results were reported by other studies with a substantial increase of reducing sugars during the production of black garlic. Choi et al. (2014) [15] showed a 10-fold increase in reducing sugar content during 35 days of maturity stage. Yuan et al. (2016) [16] observed an increase from 187.79% to 790.96%, explained mainly by changes in fructose and glucose content during the process, as black garlic showed similar sucrose concentration to fresh garlic. These considerable increases of fructose and glucose contents during black garlic production could be the result of the hydrolysis of fructans during the process, Yuan et al. (2016) [16] observed a decrease of 84.6% of total fructans content in black garlic in relation to fresh garlic. The increase of reducing sugar content explains the characteristic sweet taste of black garlic.

The reducing sugar content was significantly higher in the var. *California White* with respect to the other varieties in the initial control and at 14 days of the process in garlic of the harvest of 2013. However, in the last control, the *Spanish Roja* variety had a significantly somewhat higher value (31.96 g/Kg). In garlic of the 2014 harvest, the results were lower especially in the final control at 34 days, with the var. *California White* being the one showing a higher value compared with the other varieties. The var. *Chinese Spring* was the one showing the lowest content of reducing sugars out of the three varieties studied in garlic kept from the harvest of 2013 and in garlic of 2014.

In the Figure 2, these results are depicted:

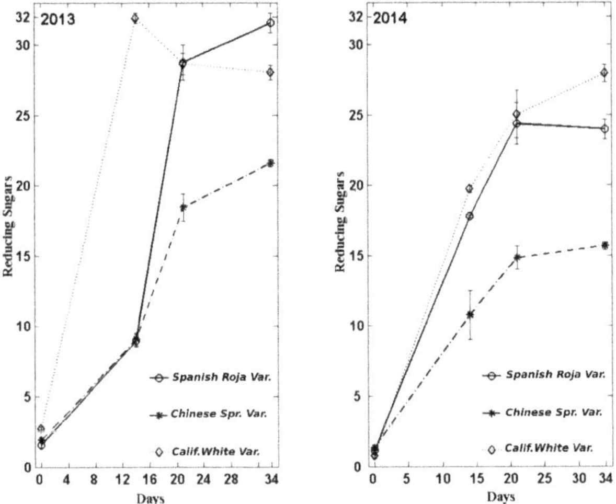

Figure 2. Evolution of the reducing sugar content (g/kg lyophilized sample) during the manufacture of black garlic with three varieties and two types of garlic: garlic aged one year and garlic recently harvested.

3.3. °Brix and pH

The results of °Brix were coherent with the results obtained in reducing sugars, both in garlic kept from the 2013 harvest and analysed in 2014 and in garlic harvested in 2014. Thus, the garlic variety with the lowest value of °Brix was the variety *Chinese Spring* in the initial control and during the whole manufacture process. The results showed higher values in samples from the 2013 harvest in practically all the controls made. Regarding their evolution, they tended to increase during the process.

The pH results, although they exhibited significant differences between varieties, had a greater similarity between varieties and between both harvests, 2013 and 2014. The Figure 3 shows the results obtained of the °Brix and pH.

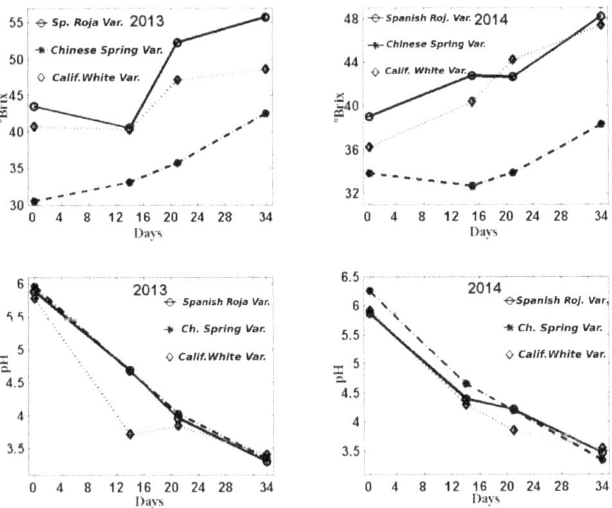

Figure 3. Evolution of °Brix (**top**) and pH (**bottom**) during black garlic manufacture with three varieties and two garlic types: one-year-old garlic (2013) and recently harvested garlic (2014).

Our results about the evolution of °Brix and pH during garlic storage are in line with those reported by Toledano-Medina et al., (2016) [8], Choi et al., (2014) [15], and Bae et al., (2014) [17].

3.4. Antioxidant Capacity

Antioxidant capacity evolution showed strong growth up until the maximum reached at 21 days. These values are in line with the results provided by different research works that described a 4.5-fold increase in the initial values [2,8,15]. Choi et al., (2014) [15] also reported lower antioxidant activity during black garlic production, mainly at 21 days of the process. The increase of antioxidant capacity could be related to the increase of total polyphenols, flavonoids, and intermediate compounds of Maillard reaction produced during heat treatment [15,18,19].

The antioxidant capacity results were significantly higher in the var. *Chinese Spring* in all the controls made with garlic from the harvest of 2013 and in the final control of those from the 2014 one. There were no notable differences between either type of garlic (2013 and 2014), and in both cases, a correction in the value obtained in the last control (34 days) with respect to the one made at 21 days was observed.

In the Figure 4, the results obtained are depicted.

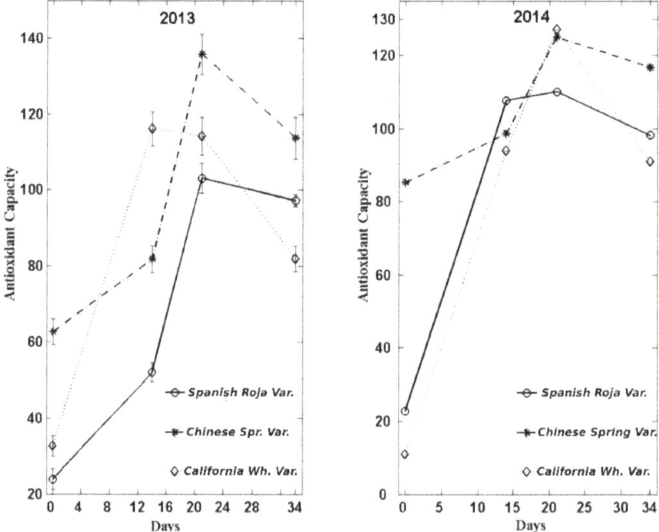

Figure 4. Evolution of antioxidant capacity (TROLOX mmol/kg lyophilized sample) during the manufacture of black garlic with three varieties and two types of garlic: one-year-old and recently harvested.

3.5. Polyphenol Content

In neither of the garlic types (2013 and 2014) were any notable differences seen, and, in this case, no correction in the value obtained in the last control (34 days) was observed. These results are in line with those obtained in previous studies [8,14,15], which showed increased levels of polyphenol content during the black garlic production. In addition, the important increase of the polyphenol content could be related to compounds derived from Maillard reaction, such as 5-hydroxymethylfurfural (5-HMF). Zhang et al. (2016) [14] evaluated the effect of different temperatures during black garlic production on 5-HMF content; they observed a high increase of this compound during the different heat treatments, especially when the temperature was up to 70 °C. Furthermore, Choi et al. (2014) [15] showed an increase of total content of flavonoids during black garlic production, which could explain the increase of total polyphenol content during the process. Lu et al. (2017) [19] reported uridine, adenosine, carbonile alcaloids, and 5-hydroxymethylfurfural as the compounds with higher polyphenolic content and antioxidant capacity in different ethyl acetate extracts of black garlic.

The results were significantly higher in the variety *Chinese Spring* in all the controls made with garlic from the 2013 harvest, although, in the last two controls of the 2014 harvest, the results did not give any significant differences between varieties.

The Figure 5 shows the results obtained.

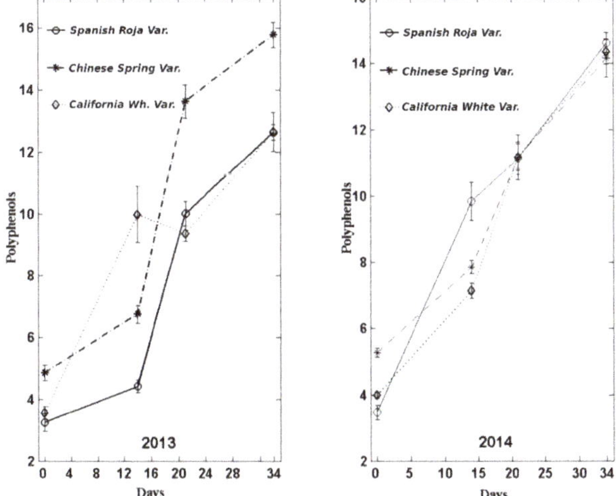

Figure 5. Evolution of the polyphenol content (g Gallic/Kg lyophilized sample) during the manufacture of black garlic with three varieties and two types of garlic: one-year-old and recently harvested.

4. Conclusions

With regard to the manufacturing process, regardless of the variety used and only taking into account the physicochemical parameter evolution, the values obtained were similar to those published by different authors [2–6,8].

In this work, it was demonstrated that black garlic can be made with garlic proceeding from a previous harvest that is kept dormant under refrigeration for one year. This possibility should additionally be reinforced with some procedure in order to differentiate black garlic made with fresh garlic from a recent season from that made with garlic kept from the previous one.

The analyses conducted show some differences between the previous year's garlic (2013) and that of the study (2014); for example, the 2013 garlic contained a higher content in reducing sugars, °Brix, and acidity, especially for the *California White* variety. The most probable reason for those differences could be a loss of humidity during storage and a resulting concentration.

With regard to the varieties employed, the garlic of the variety *Chinese Spring* exhibited a smaller amount of reducing sugars and °Brix, and a greater acidity than the rest of the varieties analysed in garlic of 2013 and 2014. Its antioxidant capacity and polyphenol content were higher than those of the rest of the varieties, especially in garlic of 2013.

As for the process, it is known that the temperature determines its duration. It should be pointed out that an excessive duration in the process is detrimental to the final product. First, its acidity becomes excessive as the process becomes protracted. Second, the product's antioxidant capacity diminishes after reaching a prior maximum value when the process is extended, although the polyphenol content goes on increasing.

Author Contributions: M.Á.T.M. and J.P.-A. performed the experimental stage. R.M.-R. and J.P.-A. designed this study. R.M.-R., J.P.-A., and A.M.-O. revised the manuscript. M.Á.T.M., J.P.-A., and A.M.-O. wrote this manuscript. All the listed authors have read and approved the submitted manuscript.

Funding: This research received no external funding.

Conflicts of Interest: The authors declare no conflict of interest.

References

1.   Rahman, K. Garlic and aging: New insights into an old remedy. *Ageing Res. Rev.* **2003**, *2*, 39–56. [CrossRef]
2.   Lee, Y.M.; Gweon, O.C.; Seo, Y.J.; Im, J.; Kang, M.J.; Kim, M.J.; Kim, J.I. Antioxidant effect of garlic and aged black garlic in animal model of type 2 diabetes mellitus. *Nutr. Res. Pract.* **2009**, *3*, 156. [CrossRef] [PubMed]
3.   Wang, D.; Feng, Y.; Liu, J.; Yan, J.; Wang, M.; Sasaki, J.; Lu, C. Black Garlic (Allium sativum) Extracts Enhance the Immune System. *Med. Aromat. Plant Sci. Biotechnol.* **2010**, *4*, 37–40.
4.   Jin-ichi, S.; Chao, L.; Einosuke, M.; Mami, T.; Katsunori, H. Processed Black Garlic (Allium sativum) Extracts Enhance Anti-Tumor Potency against Mouse Tumors. *Med. Aromat. J. Plant Sci. Biotechnol.* **2007**, *1*, 278–281.
5.   Jang, E.K.; Seo, J.H.; Lee, S.P. Physiological activity and antioxidative effects of aged black garlic (*Allium sativum* L.) extract. *Korean J. Food Sci. Technol.* **2008**, *40*, 443–448.
6.   Seo, Y.J.; Gweon, O.C.; Im, J.E.; Lee, Y.M.; Kang, M.J.; Kim, J.I. Effect of Garlic and Aged Black Garlic on Hyperglycemia and Dyslipidemia in Animal Model of Type 2 Diabetes Mellitus. *Prev. Nutr. Food Sci.* **2009**, *14*, 1–7. [CrossRef]
7.   Ryu, J.H.; Kang, D. Physicochemical properties, biological activity, health benefits, and general limitations of aged black garlic: A review. *Molecules* **2017**, *22*, 919. [CrossRef] [PubMed]
8.   Medina, T.M.A.; Prez-Aparicio, J.; Moreno-Rojas, R.; Merinas-Amo, T. Evolution of some physicochemical and antioxidant properties of black garlic whole bulbs and peeled cloves. *Food Chem.* **2016**, *199*, 135–139. [CrossRef] [PubMed]
9.   Kimura, S.; Tung, Y.C.; Pan, M.H.; Su, N.W.; Lai, Y.J.; Cheng, K.C. Black garlic: A critical review of its production, bioactivity, and application. *J. Food Drug Anal.* **2017**, *25*, 62–70. [CrossRef] [PubMed]
10. Block, E. The Organosulfur Chemistry of the GenusAllium Implications for the Organic Chemistry of Sulfur. *Angew. Chem. Int. Ed. Engl.* **1992**, *31*, 1135–1178. [CrossRef]
11. Horníčková, J.; Kubec, R.; Cejpek, K.; Velíšek, J.; Ovesná, J.; Stavělíková, H. Profiles of S-Alk(en)ylcysteine Sulfoxides in Various Garlic Genotypes. *Czech J. Food Sci.* **2010**, *28*, 298–308. [CrossRef]
12. Singleton, V.L.; Orthofer, R.; Lamuela-Raventós, R.M. Analysis of total phenols and other oxidation substrates and antioxidants by means of folin-ciocalteu reagent. *Methods Enzym.* **1998**, *299*, 152–178.
13. Re, R.; Pellegrini, N.; Proteggente, A.; Pannala, A.; Yang, M.; Rice-Evans, C. Antioxidant activity applying an improved ABTS radical cation decolorization assay. *Free Radic. Biol. Med.* **1999**, *26*, 1231–1237. [CrossRef]
14. Zhang, X.; Li, N.; Lu, X.; Liu, P.; Qiao, X. Effects of temperature on the quality of black garlic. *J. Sci. Food Agric.* **2016**, *96*, 2366–2372. [CrossRef] [PubMed]
15. Choi, I.; Cha, H.; Lee, Y. Physicochemical and Antioxidant Properties of Black Garlic. *Molecules* **2014**, *19*, 16811–16823. [CrossRef] [PubMed]
16. Yuan, H.; Sun, L.; Chen, M.; Wang, J. The Comparison of the Contents of Sugar, Amadori, and Heyns Compounds in Fresh and Black Garlic. *J. Food Sci.* **2016**, *81*, 1662–1668. [CrossRef] [PubMed]
17. Bae, S.E.; Cho, S.Y.; Won, Y.D.; Lee, S.H.; Park, H.J. Changes in S-allyl cysteine contents and physicochemical properties of black garlic during heat treatment. *LWT Food Sci. Technol.* **2014**, *55*, 397–402. [CrossRef]
18. Hwang, I.G.; Kim, H.Y.; Woo, K.S.; Lee, J.; Jeong, H.S. Biological activities of Maillard reaction products (MRPs) in a sugar–amino acid model system. *Food Chem.* **2011**, *126*, 221–227. [CrossRef]
19. Lu, X.; Li, N.; Qiao, X.; Qiu, Z.; Liu, P. Composition analysis and antioxidant properties of black garlic extract. *J. Food Drug Anal.* **2017**, *25*, 340–349. [CrossRef] [PubMed]

© 2019 by the authors. Licensee MDPI, Basel, Switzerland. This article is an open access article distributed under the terms and conditions of the Creative Commons Attribution (CC BY) license (http://creativecommons.org/licenses/by/4.0/).

Article

Variation of Adolescent Snack Food Choices and Preferences along a Continuum of Processing Levels: The Case of Apples

Elizabeth Svisco [1], Carmen Byker Shanks [1,*], Selena Ahmed [1] and Katie Bark [2]

[1]   Food and Health Lab, Montana State University, Bozeman, MT 59718, USA; sviscoe@gmail.com (E.S.); selena.ahmed@montana.edu (S.A.)

[2]   Montana Team Nutrition; Montana Team Nutrition, Montana State University, Bozeman, MT 59718, USA; kbark@mt.gov

*   Correspondence: cbykershanks@montana.edu; Tel.: +1-406-994-1952

Received: 21 December 2018; Accepted: 22 January 2019; Published: 1 February 2019

Abstract: Food processing is used for transforming whole food ingredients into food commodities or edible products. The level of food processing occurs along a continuum from unprocessed to minimally processed, processed, and ultra-processed. Unprocessed foods use little to no processing and have zero additives. Minimally processed foods use finite processing techniques, including drying, freezing, etc., to make whole food ingredients more edible. Processed foods combine culinary ingredients with whole foods using processing and preservation techniques. Ultra-processed foods are manufactured using limited whole food ingredients and a large number of additives. Ultra-processed snack foods are increasing in food environments globally with detrimental implications for human health. This research characterizes the choices, consumption, and taste preferences of adolescents who were offered apple snack food items that varied along a processing level continuum (unprocessed, minimally processed, processed, and ultra-processed). A cross-sectional study was implemented in four elementary school classrooms utilizing a buffet of apple snack food items from the aforementioned four food processing categories. A survey was administered to measure students' taste acceptance of the snacks. The study found that the students selected significantly ($p < 0.0001$) greater quantities of ultra-processed snack foods (M = 2.20 servings, SD = 1.23) compared to minimally processed (M = 0.56 servings, SD = 0.43) and unprocessed (M = 0.70 servings, SD = 0.37) snack foods. The students enjoyed the taste of ultra-processed snack foods (M = 2.72, SD = 0.66) significantly more ($p < 0.0001$) than minimally processed (M = 1.92, SD = 1.0) and unprocessed (M = 2.32, SD = 0.9) snack foods. A linear relationship was found between the selection and consumption quantities for each snack food item (R2 = 0.88). In conclusion, it was found that as processing levels increase in apple snack foods, they become more appealing and more heavily consumed by elementary school students. If applied broadly to snack foods, this conclusion presents one possible explanation regarding the high level of diet-related diseases and nutrient deficiencies across adolescents in America. Food and nutrition education, food product development, and marketing efforts are called upon to improve adolescent food choices and make less-processed snack food options more appealing and accessible to diverse consumers.

Keywords: nutrition; processed; snack food; food; ingredients; preferences; eating behaviors; healthy snacks; adolescent snacking

1. Introduction

Processed foods have increasingly dominated the United States' food supply, food environments, and diets over the past five decades [1] with detrimental implications for nutrition and human health [2–5]. The trend of increased processed food consumption characterizes the nutrition transition as a shift towards nutrient-poor diets filled with ultra-processed food items [2]. The overconsumption of ultra-processed foods that are often formulated using saturated or trans fats, excess sugar, excess salt, and artificial ingredients has had negative implications for diet-related health outcomes, including increased incidence of obesity, Type 2 Diabetes, cardiovascular disease, and cancer [2–5]. One in five children ages 6 to 19 and living in the United States are classified as obese with a body mass index at or above the ninety-fifth percentile [6]. A recent article demonstrates that a 10% increase in consumption of ultra-processed foods among study participants was associated with a significant increase (greater than 10%) in overall cancer and breast cancer [5]. Along with the United States, the nutrition transition and associated nutrition and health outcomes are occurring in communities globally [7].

Ultra-processed foods comprise 58% of the consumed calories and 90% of the added sugar intake in the American diet [1]. It was found that meals prepared with ultra-processed and processed food ingredients had one third more added sugar, one fourth more saturated fat and sodium, less than half the fiber content, and two thirds more calories when compared to meals prepared from the processed and minimally processed groups [4]. For adolescents, regardless of country of residence, obesity was positively correlated with the taste of fat and sugar-enriched food products [8]. Eliminating ultra-processed products from a diet can improve consumer health with studies reporting decreases in attention deficit hyperactivity disorder (ADHD) symptoms [9], a reduction in insulin resistance in children with Type 2 Diabetes [10], and decreased mineral deficiencies in children with autism [11].

Food processing (Table 1) can be characterized at different levels along a continuum from unprocessed, minimally processed, and processed culinary ingredients up to processed foods and ultra-processed foods. The different levels of food processing vary according to the degree of manipulation including the utilization of new technologies and the input of artificial ingredients that take foods further from their natural state [1]. At the extreme end, ultra-processed food items often have little physical resemblance to the fresh and wholesome food items that they were originally derived from [12]. The addition of extra processing steps and artificial ingredients transforms food products such as meats, fruits, and vegetables into new foods that can be both visually and nutritionally different compared to their original whole food state [4]. Unprocessed foods include fresh fruits, vegetables, and meats that have undergone limited processing including cleaning, portioning, chilling, grating, etc. [13]. Minimally processed foods have undergone a small amount of processing such as drying, freezing, or fermentation [13]. Processed culinary ingredients include sugar, oil, fat, salt, and other ingredients that are originally taken from plants and then milled, pressed, or pulverized to be used in cooking or baking [13]. Processed products are whole food ingredients that have been combined with processed culinary ingredients using processing methods along with preservation techniques such as salting, pickling, smoking, curing, etc. [13]. Lastly, ultra-processed foods are created using ingredients that are not used in culinary work [1], including artificial flavors, colors, sweeteners, emulsifiers, and preservatives. Ultra-processed products can be fortified with micronutrients and are formulated to be "ready-to-eat" versions of whole food items [13].

Table 1. Classification of the food processing levels adapted from Juul and Hemmingsson [13].

Classification	Description	Examples
Unprocessed	Undergone limited processing including chilling, slicing, grating, and packaging.	Fresh fruit and vegetables, unsalted nuts and seeds, grains, milk, and pulses.
Minimally Processed	A small amount of processing including drying, freezing, pasteurizing, gas and vacuum packing, fat reduction, and fermentation.	Frozen produce, dried beans, dried fruits, unsweetened fruit juices, pasteurized milk, coffee, and plain yogurt.
Processed Culinary Ingredient	Include ingredients that are originally taken from plants or nature and are then milled, pressed, pulverized, stabilized, or purified to be used in cooking or baking.	Vegetable oils, fats, butter, cream, sugar, sweeteners, salts, and flour.
Processed	Include the combination of culinary ingredients with whole foods to increase the taste or durability using processing and preservation techniques such as canning with oils, salting, pickling, smoking, and curing.	Fruits preserved in syrups, canned meats in brine, reconstituted meat, and cheese.
Ultra-Processed	Created using little whole food ingredients and a large number of additives, including artificial flavors, colors, sweeteners, emulsifiers, and preservatives. Processing techniques include baking, frying, moulding, and hydrogenation. Ultra-processed products can be fortified with micronutrients and are formulated to be "ready-to-eat" versions of whole food items.	Cereals, grain bars, hot dogs, cookies, candies, chips, crackers, soft drinks, and sauces.

Given that snack foods are a large part of the adolescent diet and impact health outcomes, it is important to understand adolescent preferences for snack foods along a continuum of processing. Currently, there is limited research that investigates adolescent preferences for snack food items from various processing levels. This study seeks to address the aforementioned research gap by examining adolescent food choices, consumption, and taste preferences for snack food items from the continuum of processing from unprocessed to minimally processed, processed, and ultra-processed. The overall research question of this study is: *What are the food choices, consumption, and taste preferences of adolescent students who are offered snack food items that vary along a continuum of processing levels (unprocessed, minimally processed, processed, and ultra-processed)?*

It is hypothesized that the study participants will select and consume greater quantities of the processed and ultra-processed apple snack food items when compared to the unprocessed and minimally processed snack food items on the basis of their food preferences. The findings from this study have the potential to increase the awareness of the impact of food processing on the incidence of obesity and chronic illness and decrease the consumption of ultra-processed foods in America. Through a clear understanding of what ultra-processed snack foods are, how they impact consumption and taste preferences, and potential health implications from overconsumption, consumers will hopefully be encouraged to choose less-processed foods for their diet. The findings from this study also have the potential to enhance and support health, nutrition education, and school foodservice programs by providing them with concrete results regarding ultra-processed snack food preferences and consumption.

Background

Processed and ultra-processed foods are often high in calories while being low in nutritional value [14]. Ultra-processed foods are void of naturally occurring ingredients and are not fresh food items [4]. Ultra-processed foods are characterized as rich in solid fats, sugar, sodium, artificial ingredients, additives, and chemicals instead of protein, fiber, vitamins, and minerals [15,16]. The ingredients in ultra-processed food products are added for enhanced taste, visual appeal, or preservation [14]. Examples of ultra-processed ingredients include high fructose corn syrup, trans fats, monosodium glutamate, artificial colors, brominated vegetable oil, artificial sweeteners, and nitrates/nitrites.

The top five food companies in the United States responsible for the sales of packaged foods (Kraft, PepsiCo, Nestle, Mars, and Kellogg) control approximately 25% of US food sales [15]. These food companies are a part of the processed food industry that manufacture ultra-processed foods including lunch meats, sweetened beverages, candy bars, cereals, and snack bars. For example, Kraft Foods is responsible for 6.8% of all packaged food sales in the United States [15]. The prevalence of processed and ultra-processed foods in diets in the United States has rapidly increased due to key food environment characteristics including enhanced desirability (such as flavor and appeal), affordability, convenience (such as a decreased preparation time, portability, and an extended shelf-life), and efficiency [17,18].

It is well recognized that food processing has the potential to enhance the sensory desirability of foods by altering their physical structure, taste, aroma, and texture [17]. With the development of new food processing technologies coupled with consumer demand, the number of processed products available has notably increased as chemists continue to develop and add flavor enhancers and other artificial ingredients to foods [19,20]. For example, in the 1960s, processed potatoes, including potato chips, made up 35% of potato use and today this number has increased to 64% [21]. Research supports that ultra-processed foods have addictive qualities, which stems from hyper-palatable taste and unnaturally high levels of pleasure associated with consumption that ultimately impact behavioral characteristics and can lead to overeating [22]. The addictive qualities of ultra-processed foods are magnified at a young age and are heavily influenced by an early introduction to them along with a high frequency of consumption [22]. Child-focused marketing has increased this issue. For example, in 2012 4.6 billion dollars was spent on fast food advertising and 6–11-year-old children viewed an average of 3.2 fast food advertisements per day [23].

The most frequently consumed ultra-processed foods by Americans include packaged snack foods such as cookies, salty snacks, and breakfast cereals [12]. Research demonstrates that snacking is on the rise and predicts future increases in the consumption of ultra-processed snack food items [24]. Specifically, between 2016 and 2017, the number of Americans snacking five or more times per day increased from 11.5% to 14.2% [24].

Today, adolescents have greater exposure to highly processed snack foods than previous generations. A recent study demonstrated that adolescents consume an average of 4.3 processed snacks per day [25]. From 1977 to 1978, adolescents consumed about 300 calories per day from snacks; this number increased to 526 calories from 2005 to 2006 [26]. These 526 calories from snacks make up between 22 and 38% of the recommended daily calories for an adolescent female and between 16 and 33% for an adolescent male, depending on their activity level and exact age [27]. The key drivers of adolescent food choices for snacks include convenience, accessibility, and desirability (i.e., pleasurable to consume during social or leisurely times) [28].

The production of ultra-processed foods presents challenges to the sustainability of food systems and sustainable diets because of the energy and materials required for their processing and packaging. Overall, the production of food requires 10% of the total U.S. energy budget, 80% of freshwater, and 50% of all U.S. land [29]. At the same time, food production represents food waste challenges due to food loss on farms, in retail stores, in food operations, and in households as a result of inefficiency throughout all stages in the food system [29]. In 2008, it was found that food processing specifically led to 16% of food loss during manufacturing [29]. Snack food packaging is designed to preserve, contain, trace, and increase the appeal of snack food items for Americans. Packaging innovations are leading to heightened quantities of waste ending up in landfills [30]. In 2010, 21% of the total waste generation was made up of food while 16% was from paper, 5% was from glass, and 17% was from plastics, which are all used in food packaging [31]. This solid waste leads to chemical leaching into the soils (hence poisoning soils for food to be grown in) or groundwater and increased greenhouse gases being released into the atmosphere [30].

The increased consumption of ultra-processed foods and associated health outcomes is driving a need for more nutrition education regarding ultra-processed foods as well as the development and marketing of less-processed snack foods. The five stages of Popkins' [2] nutrition transition highlight the need for a shift from dietary patterns characterized by processed foods and high sugar and fat consumption to desired dietary patterns that include the reduced consumption of processed foods coupled with increased consumption of fruits and vegetables. This change in dietary patterns away from high fat and sugar processed foods is associated with reduced body weight along with a reduced risk for diet-related chronic disease (obesity, Type 2 Diabetes, cardiovascular disease, and cancer) that ultimately extends aging and improves overall health [2–5]. In order to foster the desired societal and behavior change away from the consumption of ultra-processed foods, nutrition education efforts are increasing regarding the impacts of ultra-processed foods and benefits of healthy snack foods, particularly fruits and vegetables [31]. The United States Department of Agriculture and MyPlate promote sliced vegetables, fresh fruits, dried fruits, and unsweetened applesauce as healthy snacks for children along with lean proteins, nuts, and other whole food items [32]. Fruit and vegetable consumption provides people with dietary fiber, which can lower obesity rates and the incidence of cardiovascular disease [31]. Vitamins, minerals, and phytochemicals, essential for human health, are also supplied through fruits and vegetables and can produce anti-inflammatory, antioxidant, and phytoestrogen properties to support bodily functions [31]. Companies are learning and responding to consumer demand and nutrition research by developing and marketing less-processed food snacks that better support desired nutrition and health outcomes. Examples include fruit and nut trail mix with zero added sweeteners or preservatives, sweet potato tortilla chips made from only three whole ingredients, fruit snacks with only fruit as an ingredient, and dried chickpeas [24].

2. Materials and Methods

The cross-sectional research examined adolescent food choices for unprocessed, minimally processed, processed, and ultra-processed versions of apple snack food items. Adolescent students were presented with a one-time, unlimited buffet of four apple snack food options. Each buffet included four of eight apple snack foods (two varieties of raw apples, two brands of dried apple slices, two brands of cinnamon applesauce, and two apple-based fruit snack brands) to reduce bias within and between the processing types. Adolescent snack food choices were measured by student selection, consumption, and taste preferences. The selection and consumption data were measured through food weight while food preference data was measured through a survey.

2.1. Subjects

The data were collected at three schools in Southwest Montana and included participation from four fourth grade classrooms made up of 25, 28, 24, and 20 students, respectively, for a total of 97 research subjects. The data were collected during class periods lasting 45 minutes each. Middle schools were selected based upon a convenience sample of middle schools within a one-hour driving distance for the research team. The middle school principals and teachers in Southwest Montana were contacted by the research team with the study details. Several schools declined to participate because research initiatives did not align with the schools' individual research initiatives. The willing principals and teachers were asked about the number of nine to ten-year-old students within the middle school and the number of fourth grade classrooms that could potentially participate in the study. The ideal class size was approximately 25 students per class. Three middle schools were selected and they consented to participate. The participating ages and grade level were selected to fill a gap in sensory research that targets participants in the beginning stages of adolescence.

Consent was given by the principal of each school and each individual student. The principals were able to review the methodology and content of the study. According to the school's protocol for research, verbal consent was given by each student after a document was read aloud to the participating class. The consenting information outlined the study purpose and methods and that the participation was voluntary. Students were excluded if they were not the correct age, were allergic to any of the food products, or were not willing to participate.

The table below outlines the descriptive characteristics of the schools where students involved in the study were enrolled (Table 2). The demographics between schools varied, with School 1 having

217 total students with 92.7% Caucasian, 51% males, 49% females, and 17.51% of students participating in the National School Lunch Program [33,34]. Two fourth grade classes were used at School 1. School 2 included 199 students with 87.6% Caucasian, 58% females, 42% males, and 10.6% of students participating in the National School Lunch Program [33,34]. School 3 included 448 students with 89.1% Caucasian, 52% males, 48% females, and 37.1% participating in the National School Lunch Program [33,34].

Table 2. Demographics from the three schools involved in the research study [33,34].

School	Number of Total Students Enrolled in School	Number of Classrooms and Students Participating in Study	Race (Entire school)	Gender (Entire school)	National School Lunch Program Participation (Entire school)
1	164	2 (25 and 28 students)	92.7% Caucasian	51% Male and 49% Female	16.50%
2	161	1 (28 students)	87.6% Caucasian	42% Male and 58% Female	19.3%
3	448	1 (24 students)	89.1% Caucasian	52% Male and 48% Female	37.1%

Adolescence is defined from 10 to 19 years of age [35]. The test subjects were between nine and ten years old. This age group encompasses the beginning stages of adolescence, an important age group to target when examining the formation of snack food choices [36]. It is critical for adolescents to establish healthy snacking habits going into early adulthood.

2.2. Snack Food Product Samples

The research study included eight different apple-based snack food items with two types selected for each processing level (Table 3). Apples were selected as the main food item because they are within the top two most-consumed fruits annually for youth in America [37]. Specifically, out of the top ten consumed fruits, raw apples are the second most consumed and applesauce is the seventh most consumed.

Table 3. Eight apple snack food items used for the research study were randomly assigned to the classrooms so that each classroom received one product from each processing level in a randomized buffet order. Foods were classified into processing levels [13].

Food Product	Processing Level	Processing Level Criteria
Gala Apple Slices Red Delicious Apple Slices	Unprocessed	• Whole food • Sliced and washed
Bare Brand Dried Apple Chips UNFI Dried Apples	Minimally Processed	• Dried or slowly baked • Zero added ingredients
Musselman's Cinnamon Applesauce Seneca Cinnamon Applesauce	Processed	• Mashed • Combined with processed culinary ingredients (sugar, water, cinnamon) • Jarred for preservation
Welch's Apple Medley Fruit Snacks (WAMFS) Western Family Fruit Snax (WFFS)	Ultra-processed	• WAMFS: 18 ingredients such as artificial food coloring, corn syrup, lactic acid, ascorbic acid, and alpha tocopherol acetate [38] • WFFS: 15 ingredients including corn syrup and artificial coloring [39] • Artificial ingredients such as colors, preservatives, sweeteners, and stabilizers • Fortified with micronutrients

Red Delicious and Gala apples were selected because they are the top two most-produced apple varieties in America, with more than 54 million 42-pound units of Red Delicious apples produced in 2011 and nearly 33 million 42-pound units of Gala apples produced in 2011 [40]. Bare Dried Apple

Chips and UNFI Dried Fruit Apples were selected as minimally processed snack food items because Bare Apple Chips are widely accessible to consumers while being sold at some of the nation's largest retail stores including Walmart [41]. UNFI Dried Apples are sold by a large distributing company called United Natural Foods that distributes to grocery stores across the United States [42]. Musselman and Seneca brand cinnamon apple sauces were chosen because they are widely distributed throughout America as snack food items and can be purchased at some of the nation's largest retail chains including Walmart [41]. Western Family Fruit Snax and Welch's Apple Medley Fruit Snacks were chosen because Welch's is among the top five fruit snack brands in America [43] and both brands are well-known in the United States while having numerous channels to connect their snack foods to many Americans.

Only one brand item from each processing group was available to the students in each classroom for the research study, which equaled four products being tested at once. The eight apple snack food items and their ordering were randomized so each classroom received a different buffet spread.

The grocery stores from which the food items were purchased were randomly selected from a list of local food outlets that sold the desired product. At the selected grocery stores, whole apples were selected to represent similar shape, size, and coloring, forming a uniform sample for the research subjects. At the selected grocery stores, apple chips, applesauce, and fruit snacks were selected with the furthest expiration dates so that the products were fresh for the study participants.

2.3. Basic Protocol

The data collection times were scheduled apart from meals or snacks so that hunger levels did not influence the amount of food that was taken or consumed by study participants. The consenting process included a signed letter of approval from all school principals involved that stated parents were informed of the research study, approval of an expedited Institutional Review Board from Montana State University (approval ES102517), and a verbal message relayed to student participants. This message was read to all classrooms before the study began and informed students that their information would be anonymous and that they did not have to participate in the study if they did not wish to. The teachers, principals, and students were informed of what foods were being sampled in order to avoid issues with food allergies and protect the students. To further ensure the safety of the students, all raw items were washed, disposable gloves were worn by team members handling food, surfaces were wiped down, and serving utensils were available to avoid contamination of food items. All four snack food items were placed in bowls and put on the buffet line in front of the classroom. Apples were sliced instead of whole due to the greater appeal for sliced fruit versus whole fruit when working with children [44]. The whole apples were sliced using an apple slicer to create 6 equal-sized wedges per apple, which increased uniformity and efficiency in preparation. The dried apples, applesauce, and fruit snacks were placed in similar bowls. Fruit snack packages were individually opened and poured in a bowl so that packaging did not influence the decisions. The bowls were filled with enough snack food item for two serving sizes for each student, as specified by the Nutrition Facts panel. More was added if a classroom ran out and extra weight of new product was accounted for. The snack food labels were used to determine the serving sizes for each product; one serving size was equivalent to one fruit snack bag (27 grams for Western Family and 24 grams for Welch's), 1/2 cup of dried apple chips (18 grams for Bare Brand and 30 grams for UNFI), 1/2 cup of applesauce (118 grams), and 1/2 medium apple per student (4 slices or 84 grams).

Four snack food bowls were placed on scales to weigh portion sizes selected by the students. Signs were placed in front of each bowl to label them with either a "plain" or "catchy" name in order to follow the behavioral economics techniques of "using catchy names" and "using signage" to increase the selection and consumption of food items [45]. Both the signage types were randomly assigned to 2 classrooms each. The snack food "catchy names" were created by a fourth-grade student. All the names are shown below in Table 4.

Table 4. Names used for labeling snack food options on the buffet line.

Product Classification	Plain Name	Catchy Name
Unprocessed	Apples	All-star Apples
Minimally Processed	Dried Apples	Capitan Apple Crisps
Processed	Cinnamon Applesauce	Apple Super Sauce
Ultra-Processed	Fruit Snacks	Fruity Tooty Fruit Snacks

The four snack foods were compared with each classroom of students to capture all the levels of food processing and compare preferences when all the options were available to the research subjects [46–48]. Each student was assigned a number, which was written on their plate and bowl and then used to record and match the student selection, consumption, and survey results. The researchers handed materials to students and recorded the weight in grams on the scale for each snack food item as each student selected their portion. The students were told not to share their food and that they did not have to finish everything on their plates and in their bowls. Figure 1 summarizes the overall design and outline of the research setup.

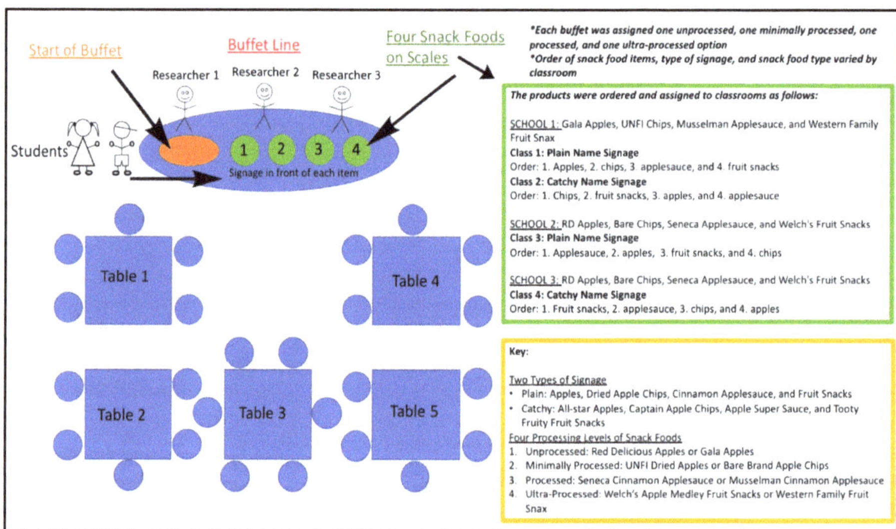

Figure 1. The figure above displays the classroom setup of the research and explains how snack food items were assigned, ordered, and presented to the students based on the classroom number.

After the students finished snacking, they were asked to bring their plates and bowls (with leftover food) to the weighing station where the researchers measured the waste in grams for each product individually. To weigh the applesauce, an empty bowl was tared on the scale and then each student's bowl was placed on the scale with the remaining applesauce in it to get the accurate weight of the applesauce alone. To weigh the three snack food items on the plate, a plate was tared on the scale and then each item from each student's plate was scraped onto the plate individually to weigh it.

Next, the subjects were asked to record their plate number on and complete a survey to rate their satisfaction for all four food products (Table 5). The "Tried It, Liked It, Loved It" survey was based upon a survey initially developed by Montana Food Corps [49] and then modified by researchers at the Montana State University Food and Health Lab and Montana Team Nutrition and validated in various school nutrition studies [50].

Table 5. The Tried It, Liked It, Loved It Survey.

Snack Food	Tried It	Liked It	Loved It	Did Not Try [2]
Example entry [1]		X		
Raw apple slices				
Dried Apples				
Cinnamon Applesauce				
Fruit Snacks				

[1] Students were provided with the chart above and were asked to place an X in the box that explained their satisfaction with each snack food. [2] The categories "Tried It," "Liked It," "Loved It," "Did Not Try" were pilot tested in a previous study [49] and directions were thoroughly explained to the students prior to the study. The students understood that "tried it" would be the correct answer if they did not like the food.

2.4. Data Analysis

All data was entered into Excel (Microsoft Excel, Redmond, WA, United States of America, 2017). This included the survey results, which were coded using numbers to represent "did not try," "tried," "liked," or "loved" in order to analyze the results. Data were also categorized in Excel based on the use of either "plain" or "catchy" names. A one-way ANOVA test (analysis of variance used to analyze differences among group means in a sample) was used to compare taste preferences to the selection and consumption data (percentages of a serving) and to the two labeling types (plain or catchy names) after confirmation of normal distribution was complete. All calculations in the table below (Table 6) were made and recorded in an Excel spreadsheet.

Table 6. Calculations made to compare the selection and consumption. All weight measurements were in grams and calculated for each student and food product.

Measurement	Calculation
Food Selection	(Weight of snack-filled bowl on scale) − (Scale weight after food selection is made per student)
Food Consumption	(Snack food selection weight) − (Food waste weight)
Percent Consumed	(Weight consumed)/(Weight selected)
Percent of Serving Consumed	(Weight consumed)/(Serving size weight)
Percent of Serving Selected	(Weight selected)/(Serving size weight)

Data from schools were aggregated. JMP (JMP® SAS Institute Inc., Cary, IL, USA) was used to look for descriptive statistics and correlation within the results. Descriptive statistics were used to calculate the means, ranges, and standard deviations for the selection data, consumption data, and survey responses. A one-way ANOVA test was used to compare taste preferences to the selection and consumption data (percentages of a serving) and to the two labeling types (plain or catchy names) used. Significance was set at $p \leq 0.05$.

3. Results

3.1. Snack Food Selection

Students selected significantly more fruit snacks (ultra-processed food) compared to the other three snack food options (processed, minimally processed, and unprocessed) based upon comparing the percentage of a serving that was selected by each student for all snack food items ($p \leq 0.0001$). A mean of 70.5% of a suggested serving size was selected for the unprocessed snack food (standard deviation of 0.37), 56.3% of a suggested serving size was selected for the minimally processed snack food (standard deviation of 0.43), 87.6% of a suggested serving size was selected for the processed snack food (standard deviation of 0.49), and 202% of a suggested serving was selected for the ultra-processed snack food (standard deviation of 1.23). This means that the subjects selected the greatest quantities of ultra-processed fruit snacks and the smallest quantities of minimally processed dried apple chips.

The signage type of "plain" versus "catchy" names did not create a significant difference in food selection ($p = 0.9590$).

3.2. Snack Food Consumption

The students consumed significantly more fruit snacks (ultra-processed food) compared to the other three snack food options (processed, minimally processed, and unprocessed) based upon comparing the percentage of a serving that was consumed by each student for all snack food items ($p \leq 0.0001$). A mean of 53% of a suggested serving size was consumed for the unprocessed snack food (standard deviation of 0.38), 40% of a suggested serving size was consumed for the minimally processed snack food (standard deviation of 0.36), 72% of a suggested serving size was consumed for the processed snack food (standard deviation of 0.52), and 186.7% of a suggested serving was consumed for the ultra-processed snack food (standard deviation of 1.15). Therefore, the students consumed the greatest quantities of ultra-processed fruits snacks and the smallest quantities of minimally processed dried apples. Note that the students selected and consumed less minimally processed snack foods (dried apple chips) than unprocessed snack foods (apple slices). Besides this comparison between unprocessed and minimally processed snack foods, the selection and consumption increased as the processing continuum increased from unprocessed to ultra-processed. The signage type of "plain" versus "catchy" names did not create a significant difference in food consumption ($p = 0.7712$). Figure 2 below shows how processing levels affected snack food consumption levels.

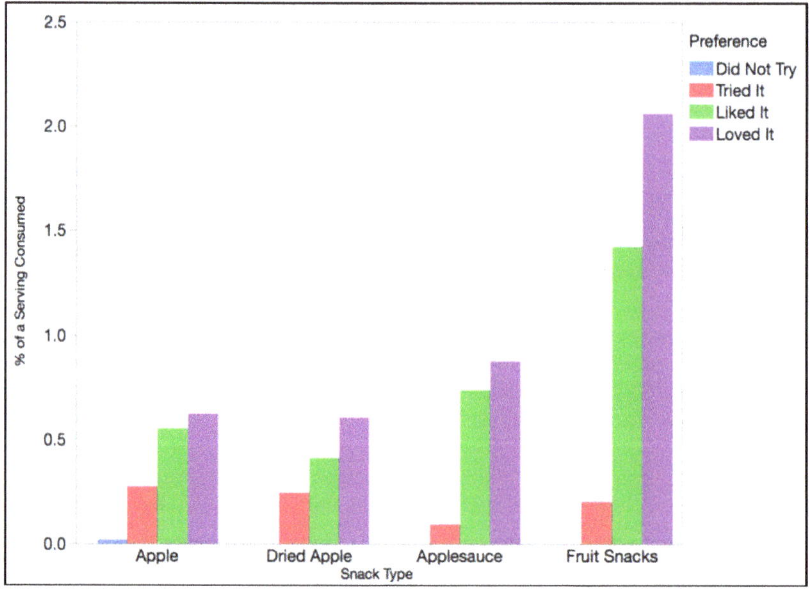

Figure 2. The percent of a serving consumed is detailed based on the four food processing levels and snack food options available to the students. It is important to note that since the selection had a linear relationship with consumption, the figure comparing selection to snack type was nearly the same as the one for consumption.

A linear relationship was found between the selection and consumption quantities for each snack food item ($R2 = 0.88$, $p < 0.0001$). For all four processing categories, the amount of food selected directly influenced the amount of food that was consumed (Figure 3). Table 7 below outlines the selection and consumption data results for each snack food type and also lists the serving sizes used as references when calculating percentages.

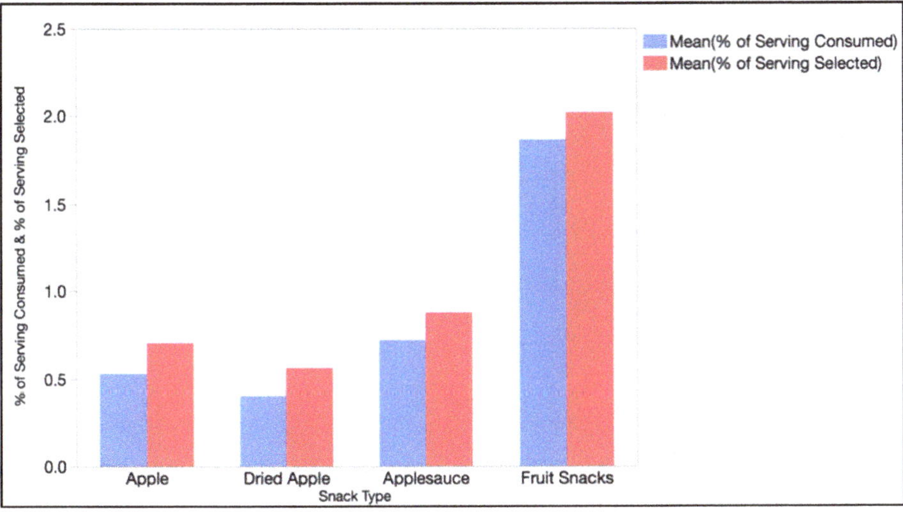

Figure 3. Linear relationship between the apple snack food selection and consumption.

Table 7. Results from the selection and consumption data originally recorded in grams and calculated into a percentage of a serving as specified on the product packaging.

Snack Food Item	Processing Category	Results	Serving Size
Apple	Unprocessed	• Mean of 70.5% of a serving selected • Mean of 53% of a serving consumed	Both: 4 slices or $\frac{1}{2}$ an apple or 84 grams
Dried Apple	Minimally Processed	• Mean of 56.3% of a serving selected • Mean of 40.1% of a serving consumed	Bare: $\frac{1}{2}$ cup or 18 grams UNFI: $\frac{1}{2}$ cup or 30 grams
Apple Sauce	Processed	• Means of 87.6% of a serving selected • Mean of 72.2% of a serving consumed	Both: $\frac{1}{2}$ cup or 118 grams
Fruit Snacks	Ultra-Processed	• Mean of 202.1% of a serving selected • Mean of 186.7% of a serving consumed	Western Family: one bag: 27 grams Welch's: one bag: 24 grams

3.3. Snack Food Preferences

Student preference was calculated using the survey results and a coding scale where 0 = did not try, 1 = tried it, 2 = liked it, and 3 = loved it. Therefore, when all the survey results were combined on JMP for each snack food item or processing level, a mean closer to three exemplified higher subject taste satisfaction while a mean closer to 0 exemplified lower subject taste satisfaction. The students enjoyed the taste of ultra-processed snack foods (mean of 2.72 with a standard deviation of 0.66) significantly more ($p < 0.0001$) than other processing levels including the processed snack foods (mean of 2.48 with a standard deviation of 0.89), minimally processed foods (mean of 1.92 with a standard deviation of 1.0), and unprocessed foods (mean of 2.32 with a standard deviation of 0.9). In accordance with the selection and consumption results, satisfaction increased moving upward on the processing continuum, except when comparing the unprocessed and minimally processed categories (Figure 4). More specifically, for the unprocessed snack food option, 6.2% of the students did not try it, 10.3% tried it, 27.8% liked it, and 55.7% loved it. For the minimally processed snack food option, 9.3% of the students did not try it, 26.8% tried it, 26.8% liked it, and 37.1% loved it. For the processed snack food option, 5.2% of the students did not try it, 11.3% tried it, 13.4% liked it, and 70.1% loved it. For the ultra-processed snack food option, 3.1% of the students did not try it, 2.1% tried it, 14.4% liked it, and 80.4% loved it. The highest percentage was found for subjects who loved fruit snacks (80.4%) and the lowest "loved it" result was found for the dried apple chips (37.1%). These numbers are listed below in Table 8.

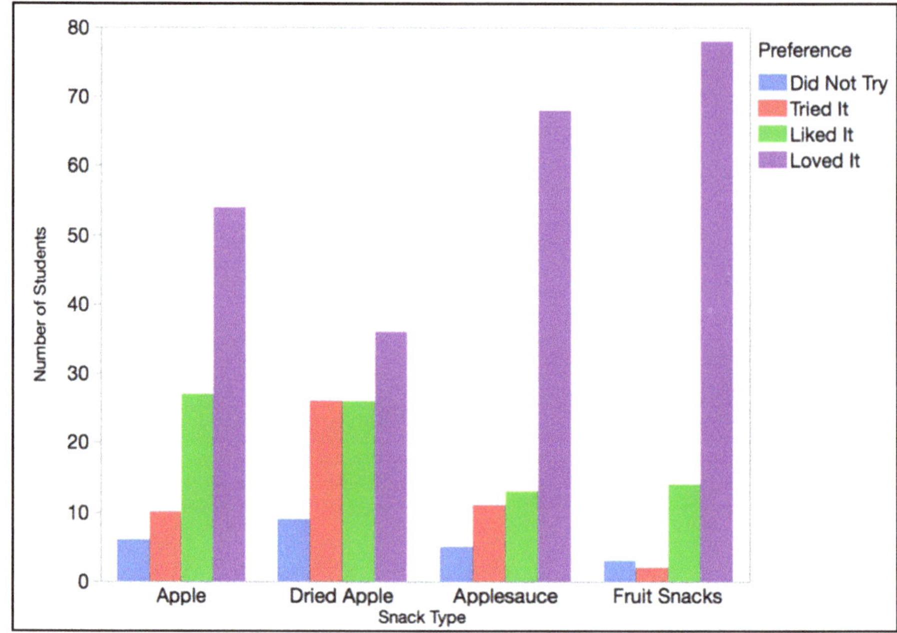

Figure 4. Tried It, Liked It, Loved It Survey results that portray the number of students who did not try, tried, liked, or loved each apple snack food for the four processing levels.

Table 8. Results from the "Tried It, Liked It, Loved It" Survey taken by the students to display preferences based on the snack food item and processing category.

Snack Food Item	Processing Category	Results
Apple	Unprocessed	• 6.2% did not try • 10.3% tried it • 27.8% liked it • 55.7% loved it
Dried Apple	Minimally Processed	• 9.2% did not try • 26.8% tried it • 26.8% liked it • 37.1% loved it
Apple Sauce	Processed	• 5.2% did not try • 11.3% tried it • 13.4% liked it • 70.1% loved it
Fruit Snacks	Ultra-Processed	• 3.1% did not try • 2.1% tried it • 14.4% liked it • 80.4% loved it

4. Discussion

Prior to this study, several research gaps existed regarding youth taste preferences which the current study has contributed to addressing. Previous sensory studies often focus on a limited age range with most involving either babies and toddlers or adults. Though there are published studies on sensory analysis conducted with toddlers and preschoolers [51–53], fewer sensory studies involve youth that are in the elementary, middle, and high school age range. For example, a study involved children between the ages of five and ten along with their mothers as study participants to compare preferences for creaminess and perception of fat in foods [54]. As flavor preferences may differ depending on exposure to different food items throughout the lifespan or nutritional goals based on growth and development, it is critical to evaluate taste preferences across all ages.

There are many research studies that focus on preferences for fruits and vegetables. For example, a study compared the taste of various vegetables after preparing them using different cooking styles or methods in order to determine how this influenced preference [55]. Another study provided children and adults with a familiar fruit and a novel fruit to compare appetitive and familiarity ratings by sensory stages [56]. The focus on fruit and vegetable research is a result of current US Dietary Guidelines to increase intake [57]. Though there are studies focusing on fruit and vegetable snacks, it was determined that further research needed to be conducted with grade school youth and snack food items to understand adolescent food preferences for these snacking options.

Two research studies were found that compare the taste of ultra-processed food products to less-processed food products. One taste test study involving a processed food product aimed to determine whether children perceived food with nutrition claims on their labels as healthier or tasting differently [58]. This study focused on how marketing and packaging affect taste rather than how ingredients and processing affect taste. The participants were asked which product was healthier and which tasted better when two identical products were placed in front of them (one involving a health claim and one without) [58]. Another study compared the taste and acceptance of whole-grain versus refined pancakes and tortillas to see if whole-grain options could replace refined products for grade school children [59]. The study was voluntary at lunch and compared these options by asking participants about overall liking, taste, color, softness, and ranking using a hedonic facial scale [59]. To examine consumption and connect consumption to preference, plate waste was collected [59]. In this study, no differences were noted in consumption of whole-wheat pancakes when compared to refined wheat pancakes, while consumption of whole-wheat tortillas was lower than refined products [59]. Though the second study did compare an ultra-processed product to a less-processed one, it involved lunch products and did not involve snack food products.

Research has also been conducted to compare various components associated with ultra-processed foods including intensity of sugary or salty flavors or preference for high-fat products. A study using 4- to 6-year-old children failed to confirm that the children who are sensitive to bitter tastes would report a higher intake of sweets and a lower intake of savory fats [60]. This study included four one-hour taste tests at dinnertime along with a final taste test including measurement of body composition [60]. Another study examined children and their mothers' preferences for creaminess and perception of fat in pudding along with concentrations of sucrose in water [54]. It was found that children preferred higher sucrose amounts in water and lower fat content in pudding compared to their mothers [54]. The methodology of this study included having participants taste different concentrations of sucrose and fat in water and pudding and then rank the samples based on intensity of sweetness and creaminess [54]. Overall, it is evident that more research needed to be conducted involving children throughout adolescent years and taste preferences associated with ultra-processed snack foods versus less-processed snack foods.

The *Variation of Adolescent Snack Food Choices and Preferences along a Continuum of Processing Levels: The Case of Apples* study was structured in such a way that it would address the gaps in research and provide new insight. As with the studies mentioned above, this study incorporates a taste test, plate waste data, hedonic rating for the taste of food items, and a variety of processed snack food products to make comparisons. Like the tortilla study mentioned above, it was found that the participants were drawn to highly processed food items and consumed more of them [59]. It was also found that the participants were drawn to snack food items that were more highly processed, which could have had to do with taste qualities associated with sugar, salt, and fats [54,60]. Therefore, this research builds upon prior research and expands the realm to include the following results.

This research study provides evidence about the snack food preferences of adolescents along a continuum of unprocessed, minimally processed, processed, and ultra-processed snack foods. The key findings from this study highlight the following behaviors and preferences of the study participants regarding snack food consumption: the participants selected and consumed more processed and ultra-processed snack foods when given an array of snacking options, the participants enjoyed the taste of ultra-processed and processed snack foods when compared to less-processed options of similar flavor, and the participants consumed greater quantities of snack foods when they selected greater portion sizes. Strategies are necessary to increase the desirability of less-processed snack food options that have higher nutrient density, fiber, and water content.

This study demonstrates that over two suggested servings of fruit snacks were consumed on average by students while half a serving or less of apples and dried apples were consumed. The correlation of snack food choice with taste preference suggests that the research subjects consumed high amounts of ultra-processed foods in order to please their palettes and hunger levels. The survey results highlight that over 80% of the students loved the taste of ultra-processed fruit snacks while only 55% loved the apple slices and 37% loved the dried apples. Since taste has an influence on food choices [22], these results show that adolescents are drawn towards the flavors and textures of more-processed products. The hyper palatable flavors associated with ultra-processed foods come from salts, sugars, fats, and additional additives. The artificial additives that can enhance flavor include high fructose corn syrup, mono sodium glutamate (MSG), and other chemically-derived ingredients [61,62].

When looking at the selection and noticing that higher quantities of ultra-processed and processed snack foods were chosen by subjects before tasting them, the participants likely associated more-processed products with better taste before trying them. Previous exposure to these snack food items or the physical appearance of the items may have influenced the participants' food choices. For example, the bright colors, shapes, and gooey texture associated with fruit snacks or the smooth, mashed texture of applesauce may have encouraged students to select more of these items. Also, the noticeable cinnamon flavor (smell, cinnamon speckles, and physical sign) of the applesauce could have swayed students to select less or more of it due to previous exposure to cinnamon and how much they enjoy the spice. Being able to eat the applesauce out of a bowl with a spoon also gives the snack food item a unique appeal. Another component that could have influenced selection is peer pressure. It was noticed that students became particularly excited about fruit snacks and relayed this excitement to their peers. Noticing how much was selected for each snack food product by peers and hearing their comments about the items could have encouraged them to select more or less of particular options.

When physically comparing fresh apple slices to fruit snacks or dried apples, a serving size of fresh apples takes up a much larger visual space due to water and fiber content in whole apples. Additionally, the water and fiber content have the ability to nourish the body and satiate hunger. It was also noted that in every class involved in the study, at least one student asked for a second serving of the fruit snacks after all the students had served themselves at the buffet line. Out of the four snack food options, the fruit snacks were the only snack food item that the students continued to ask for more of. It was also the only snack food item that needed to be refilled on the buffet line by the research team due to high levels of student selection.

Since the students selected, consumed, and enjoyed minimally processed dried apple chips less than unprocessed apples, it is evident that they did not like the physical appearance or taste of dried apples when compared to unprocessed apples. Other factors that could have influenced the results include previous exposure to apples, fruit snacks, and applesauce. Dried apple chips are not as mainstream at grocery stores compared to the other options, therefore they might have been a new snack food to a lot of the students. Also, dried apples are relatively more expensive than the other apple snack foods, which may have influenced the amount of families who were able to purchase them for their children and expose them to the snacks on an earlier or regular basis.

It is important to note that selection correlated with consumption for all of the snack food options, regardless of processing level. It was originally hypothesized that the students would select equal amounts of all products but consume greater amounts of more-processed options. When students had the ability to choose the portion size, they selected and consumed larger quantities of food than the recommended serving size. These results support individually packaged snack foods and smaller servings for adolescents in an effort to avoid overeating. This is particularly true for ultra-processed products that are higher in calories, fat, sugar, and sodium while being lower in fiber, protein, vitamins, and nutrients.

4.1. Recommendations for the Snack Food Environment

Based on this research, it is recommended that changes be made to improve the food environment and adolescent snack food trends in America. First, since the subjects were drawn to and selected greater quantities of ultra-processed foods, it is necessary to improve the appeal of nutritious snack food items that are less processed. In order to efficiently do so, government food policies, labeling laws, food safety standards, and food company values all need to be shifted to encourage health and sustainability. With a collaborative effort, snack food product development, marketing, advertising, packaging, and distribution can all move to support positive growth in the American food system. An example of this effort includes increasing advertising for whole foods such as vegetables, fruits, nuts, and seeds for snack foods. Also, food labels can be monitored and redesigned to highlight the artificial ingredients and provide consumers with simplified nutrition information and realistic portion sizes.

Since it was also identified that the subjects enjoyed the taste of ultra-processed snack foods compared to less-processed options, it is evident that the taste of less-processed snack foods should be enhanced. Food companies can do this by working with their research and development teams to create flavorful and equally tasty products without chemicals, preservatives, and other artificial ingredients. This can be achieved by experimenting with natural spices and flavors or cooking techniques. Food science and research goals should always prioritize consumer safety and health. Also, flashier packaging, unique labels, and naturally derived coloring could be used to excite adolescents about less-processed and more nutrient-dense snacking options.

It is also recommended to improve snacking in America by teaching citizens how to produce and create healthier snack foods on their own. This would improve access to healthy snack foods and reduce prices for families living in food deserts, which are defined as locations lacking access healthful and affordable foods [53]. If Americans were encouraged to grow their own food and create snacks on their own, they would be able to provide healthy snacks for their families. Examples include drying local meat and making spiced jerky, growing fruits for fresh, dried, mashed, or frozen fruit snacks, growing vegetables to be snacked on with hummus or baked into chips, and roasting nuts and seeds with cultural spices. Snack foods can be prepared at home and packaged using reusable containers to be brought on-the-go.

Lastly, educating Americans about nutrient density, how to evaluate snack food products based on nutrition, and the dangers of processing can help prepare them to handle unhealthy food environments. Peer pressure, uncertainty about what foods to consume, and misinformation about health claims can all lead to unhealthy snacking and, therefore, chronic illness and obesity. When consumers purchase snack foods, they should feel comfortable reading labels, understanding ingredients, and analyzing nutrition facts. For this to be accomplished, information about labels, food processing, and healthy snacking should be incorporated into school health classes. Government tools such as videos, pamphlets, or booklets should also be developed and widespread for families to understand snack foods. The desire to eat healthy snacks and sway consumer choice starts with education and having the knowledge to understand why ultra-processed snacks can be dangerous or unhealthy in a food environment is key to changing behavior. As Americans continue to demand healthier snacks, the food system will respond by creating, marketing, and selling less ultra-processed foods that support more sustainable diets.

4.2. Limitations

The limitations to this study included having a small sample of students and not having random selection or assignment of testing subjects. A sample of fourth-grade students from three schools is not representative of all fourth-grade children and it is difficult to draw causal inference. Another limitation is that every child had different levels of previous exposure to processed foods and different taste preferences. The food environment both at school and at home can have a large impact on adolescent food choices due to culture, peer pressure, food prices, and availability. Lastly, it is a limitation that only one whole food item (an apple) was used because certain subjects may have enjoyed apple-based products more than others. If a student did not enjoy apples, the results of the study would not have accurately portrayed their snack food preferences across the processed snack food continuum.

5. Conclusions

Overall, it was found that the processing levels of snack food items have the ability to influence adolescent taste preferences along with the selection and consumption quantities. More specifically, ultra-processed and processed foods have a large appeal for adolescents, potentially leading to overconsumption and unhealthy snacking decisions. Unprocessed and minimally processed food options are not chosen as frequently as processed and ultra-processed foods when all four processing options are made available to an audience of adolescent children.

Author Contributions: Individual author contributions are as follows: conceptualization, E.S., C.B.S., S.A., and K.B.; methodology, E.S., C.B.S., S.A., and K.B.; software, E.S., C.B.S., and S.A.; validation, C.B.S., S.A., and K.B.; formal analysis, E.S., C.B.S., and S.A.; investigation, E.S., C.B.S., and S.A.; resources, C.B.S., and S.A.; data curation, E.S., C.B.S., S.A.; writing—original draft preparation, E.S., C.B.S., S.A., K.B.; writing—review and editing, E.S., C.B.S., S.A., and K.B.; visualization, E.S., C.B.S., and S.A.; supervision, C.B.S.; project administration, E.S., C.B.S., and S.A.; funding acquisition, C.B.S., and S.A.

Funding: This research was funded by the National Institute of General Medical Sciences of the National Institutes of Health under Award Number P20GM103474 and Award Number 5P20GM104417. The content presented here is solely the responsibility of the authors and does not represent the official views of the National Institutes of Health.

Conflicts of Interest: The authors declare no conflict of interest. The funders had no role in the design of the study; in the collection, analyses, or interpretation of data; in the writing of the manuscript, or in the decision to publish the results.

References

1. Steele, E.M.; Popkin, B.M.; Swinburn, B.; Monteiro, C.A. The share of ultra-processed foods and the overall nutritional quality of diets in the US: Evidence from a nationally representative cross-sectional study. *Popul. Health Metr.* **2017**, *15*, 6. [CrossRef] [PubMed]
2. Popkin, B.M. The shift in stages of the nutrition transition in the developing world differs from past experiences! *Public Health Nutr.* **2002**, *5*, 205–214. [CrossRef] [PubMed]
3. Monge, A.; Lajous, M. Ultra-processed foods and cancer. *BMJ (Clin. Res. Ed.)* **2018**, *360*, k599. [CrossRef] [PubMed]
4. Monteiro, C.A.; Levy, R.B.; Claro, R.M.; de Castro, I.R.; Cannon, G. Increasing Consumption of Ultra-Processed Foods and Likely Impact on Human Health: Evidence from Brazil. *Public Health Nutr.* **2011**, *14*, 5–13. [CrossRef] [PubMed]
5. Fiolet, T.; Srour, B.; Sellem, L.; Kesse-Guyot, E.; Allès, B.; Méjean, C.; Deschasaux, M.; Fassier, P.; Latino-Martel, P.; Beslay, M.; et al. Consumption of ultra-processed foods and cancer risk: Results from NutriNet-Santé prospective cohort. *BMJ* **2018**, *360*, k322. [CrossRef] [PubMed]
6. Centers for Disease Control and Prevention. Available online: https://www.cdc.gov/healthyschools/obesity/facts.htm (accessed on 29 January 2018).
7. Popkin, B.M.; Adair, L.S.; Ng, S.W. Global Nutrition Transition and the Pandemic of Obesity in Developing Countries. *Nutr Rev.* **2012**, *70*, 3–21. [CrossRef] [PubMed]

8.   Ahrens, W. Sensory taste preferences and taste sensitivity and the association of unhealthy food patterns with overweight and obesity in primary school children in Europe—A synthesis of data from the IDEFICS study. *Flavour* **2015**, *4*, 8. [CrossRef]

9.   Pelsser, L.M.; Frankena, K.; Toorman, J.; Savelkoul, H.F.; Dubois, A.E.; Pereira, R.R.; Haagen, T.A.; Rommelse, N.N.; Buitelaar, J.K. Effects of a restricted elimination diet on the behaviour of children with attention-deficit hyperactivity disorder (INCA study): A randomised controlled trial. *Lancet* **2011**, *377*, 494–503. [CrossRef]

10. Cai, W.; Ramdas, M.; Zhu, L.; Chen, X.; Striker, G.E.; Vlassara, H. Oral advanced glycation endproducts (AGEs) promote insulin resistance and diabetes by depleting the antioxidant defenses AGE receptor-1 and sirtuin 1. *Proc. Natl. Acad. Sci. USA* **2012**, *109*, 15888–15893. [CrossRef]

11. Dufault, R.; Lukiw, W.J.; Crider, R.; Schnoll, R.; Wallinga, D.; Deth, R. A macroepigenetic approach to identify factors responsible for the autism epidemic in the United States. *Clin. Epigenet.* **2012**, *4*, 6. [CrossRef]

12. Steele, E.M.; Baraldi, L.G.; da Costa Louzada, M.L.; Moubarac, J.C.; Mozaffarian, D.; Monteiro, C.A. Ultra-processed foods and added sugars in the US diet: Evidence from a nationally representative cross-sectional study. *BMJ Open* **2016**, *6*, e009892. [CrossRef] [PubMed]

13. Juul, F.; Hemmingsson, E. Trends in consumption of ultra-processed foods and obesity in Sweden between 1960 and 2010. *Public Health Nutr.* **2015**, *18*, 3096–3107. [CrossRef] [PubMed]

14. Poti, J.M.; Slining, M.M.; Popkin, B.M.; Kenan, W.R. Where are kids getting their empty calories? Stores, schools, and fast food restaurants each play an important role in empty calorie intake among US children in 2009–2010. *J. Acad. Nutr. Diet.* **2014**, *114*, 908–917. [CrossRef] [PubMed]

15. Moodie, R.; Stuckler, D.; Monteiro, C.; Sheron, N.; Neal, B.; Thamarangsi, T.; Lincoln, P.; Casswell, S.; Lancet NCD Action Group. Profits and pandemics: Prevention of harmful effects of tobacco, alcohol, and ultra-processed food and drink industries. *Lancet* **2013**, *381*, 670–679. [CrossRef]

16. Drewnowski, A.; Fulgoni, V.L. Nutrient density: Principles and evaluation tools. *Am. J. Clin. Nutr.* **2014**, *99*, 1223S–1228S. [CrossRef] [PubMed]

17. Augustin, M.A.; Riley, M.; Stockmann, R.; Bennett, L.; Kahl, A.; Lockett, T.; Osmond, M.; Sanguansri, P.; Stonehouse, W.; Zajac, I.; et al. Role of food processing in food and nutrition security. *Trends Food Sci. Technol.* **2016**, *56*, 115–125. [CrossRef]

18. Seto, K.C.; Ramankutty, N. Hidden linkages between urbanization and food systems. *Science* **2016**, *352*, 943–945. [CrossRef] [PubMed]

19. ACS Chemistry for Life, American Chemical Society National Historic Chemical Landmarks. Available online: https://www.acs.org/content/acs/en/education/whatischemistry/landmarks/usda-flavor-chemistry.html (accessed on 5 January 2018).

20. Monteiro, C.A.; Moubarac, J.C.; Cannon, G.; Ng, S.W.; Popkin, B. Ultra-processed products are becoming dominant in the global food system. *Obes. Rev.* **2013**, *14*, 21–28. [CrossRef]

21. USDA Economic Research Center. Available online: https://www.ers.usda.gov/topics/crops/vegetables-pulses/potatoes.aspx (accessed on 18 April 2017).

22. Gearhardt, A.N.; Davis, C.; Kuschner, R.; Brownell, K.D. The Addiction Potential of Hyperpalatable Foods. *Curr. Drug Abuse Rev.* **2011**, *4*, 140–145. [CrossRef]

23. Harris, J.L.; Schwartz, M.B.; Musell, C.R.; Dembek, C.; Liu, S.; LoDolce, M.; Heard, A.; Fleming-Milici, F.; Kidd, B. Fast Food Facts 2013: Measuring Progress in Nutrition and Marketing to Children and Teens. Available online: www.fastfoodmarketing.org/media/fastfoodfacts_report.pdf (accessed on 29 January 2019).

24. Information Resources, Inc. Available online: https://www.iriworldwide.com/IRI/media/video/2017%20State%20of%20the%20Snack%20Food%20Industry.pdf (accessed on 11 July 2017).

25. Larson, N.I.; Miller, J.M.; Watts, A.W.; Story, M.T.; Neumark-Sztainer, D.R. Adolescent snacking behaviors are associated with dietary intake and weight status. *J. Nutr.* **2016**, *146*, 1348–1355. [CrossRef]

26. Sebastian, R.S.; Goldman, J.D.; Wilkinson Enns, C. *Snacking Patterns of US Adolescents: What We Eat in America, NHANES 2005–2006*; Food Surveys Research Group Dietary Data Brief No. 2; Food Surveys Research Group: Washington, DC, USA, 2010.

27. USDA Center for Nutrition Policy and Promotion. Available online: https://www.cnpp.usda.gov/dietary-guidelines (accessed on 16 April 2017).

28. Savige, G.; MacFarlane, A.; Ball, K.; Worsley, A.; Crawford, D. Snacking behaviours of adolescents and their association with skipping meals. *Int. J. Behav. Nutr. Phys. Act.* **2007**, *4*, 36. [CrossRef] [PubMed]
29. Gunders, D. *Wasted: How America Is Losing up to 40 Percent of Its Food from Farm to Fork to Landfill*; Natural Resources Defense Council: New York, NY, USA, 2012; Volume 26.
30. Marsh, K.; Bugusu, B. Food packaging—Roles, materials, and environmental issues. *J. Food Sci.* **2007**, *72*, R39–R55. [CrossRef] [PubMed]
31. Slavin, J.L.; Lloyd, B. Health Benefits of Fruits and Vegetables. *Adv. Nutr.* **2012**, *3*, 506–516. [CrossRef] [PubMed]
32. USDA. Available online: https://www.choosemyplate.gov/ten-tips-snack-tips-for-parents (accessed on 8 April 2018).
33. U.S. Department of Commerce. Available online: https://www.census.gov/ (accessed on 3 April 2017).
34. Office of Public Instruction. Available online: https://apps.opi.mt.gov/opireportingcenter/frmPublicReports.aspx?ProcName=procPublicReportingFreeReducedSelect&ScreenTitle=Free/Reduced%20Eligibility%20Data&SelectStateFy=1 (accessed on 25 September 2017).
35. Sacks, D. Age limits and adolescents. *Pediatr. Child Health* **2003**, *8*, 577. [CrossRef]
36. Story, M.; Neumark-Sztainer, D.; French, S. Individual and environmental influences on adolescent eating behaviors. *J. Am. Diet. Assoc.* **2002**, *102*, S40–S51. [CrossRef]
37. Produce for Better Health Foundation. Available online: http://www.PBHFoundation.org (accessed on 17 September 2017).
38. Welch's. Available online: https://www.welchsfruitsnacks.com/products/ (accessed on 4 October 2017).
39. Western Family. Available online: http://www.westernfamily.com/our-products/ (accessed on 5 October 2017).
40. USDA Economics, Statistics, and Market Information System. Available online: usda.mannlib.cornell.edu/MannUsda/viewDocumentInfo.do?documentID=1825 (accessed on 5 June 2017).
41. The NPD Group, Inc. The Checkout Penetration Index. Available online: https://www.npd.com/wps/portal/npd/us/news/press-releases/2017/only-three-businesses-can-say-at-least-84-percent-of-us-consumers-spent-with-them-in-2016/ (accessed on 24 September 2017).
42. United Natural Foods Inc. Available online: https://www.unfi.com/ (accessed on 4 October 2017).
43. Research and Markets. Available online: https://www.researchandmarkets.com/research/txpd37/global_sweet_and?w=4 (accessed on 15 May 2017).
44. Wansink, B.; Just, D.R.; Hanks, A.S.; Smith, L.E. Pre-sliced fruit in school cafeterias: Children's selection and intake. *Am. J. Prev. Med.* **2013**, *44*, 477–480. [CrossRef] [PubMed]
45. Smarter Lunchrooms Movement. Available online: https://www.smarterlunchrooms.org/about/research (accessed on 10 April 2017).
46. Guinard, J.X. Sensory and consumer testing with children. *Trends Food Sci. Technol.* **2000**, *11*, 273–283. [CrossRef]
47. Nederkoorn, C.; Jansen, A.; Havermans, R.C. Feel your food. The influence of tactile sensitivity on picky eating in children. *Appetite* **2015**, *84*, 7–10. [CrossRef]
48. Werthmann, J.; Jansen, A.; Havermans, R.; Nederkoorn, C.; Kremers, S.; Roefs, A. Bits and pieces. Food texture influences food acceptance in young children. *Appetite* **2015**, *84*, 181–187. [CrossRef]
49. Americorps Montana. Montana Healthy Food and Communities Initiative. Available online: https://mtfoodcorps.ncat.org/ (accessed on 3 October 2017).
50. Byker Shanks, C.; Bark, K.; Stenberg, M.; Roth, A. *Validation of Tried It, Liked It, Loved It Survey Among K-12 Students*; Montana State University: Bozeman, MT, USA, Unpublished work; 2017.
51. Houston-Price, C.; Butler, L.; Shiba, P. Visual exposure impacts on toddlers' willingness to taste fruits and vegetables. *Appetite* **2009**, *53*, 450–453. [CrossRef]
52. Noradilah, M.J.; Zahara, A.M. Acceptance of a test vegetable after repeated exposures among preschoolers. *Malays. J. Nutr.* **2012**, *18*, 67–75.
53. Popper, R.; Kroll, J.J. Conducting sensory research with children. *J. Sens. Stud.* **2005**, *20*, 75–87. [CrossRef]
54. Mennella, J.A.; Finkbeiner, S.; Reed, D.R. The proof is in the pudding: Children prefer lower fat but higher sugar than do mothers. *Int. J. Obes. (Lond.)* **2012**, *36*, 1285–1291. [CrossRef] [PubMed]

55. Poelman, A.A.M.; Delahunty, C.M.; de Graaf, C. Cooking time but not cooking method affects children's acceptance of Brassica vegetables. *Food Qual. Prefer.* **2013**, *28*, 441–448. [CrossRef]
56. Dovey, T.M.; Aldridge, V.K.; Dignan, W.; Staples, P.A.; Gibson, E.L.; Halford, J.C. Developmental differences in sensory decision making involved in deciding to try a novel fruit. *Br. J. Health Psychol.* **2012**, *17*, 258–272. [CrossRef]
57. United States Department of Agriculture. Available online: https://www.choosemyplate.gov/ (accessed on 11 April 2017).
58. Soldavini, J.; Crawford, P.; Ritchie, L.D. Nutrition Claims Influence Health Perceptions and Taste Preferences in Fourth- and Fifth-Grade Children. *J. Nutr. Educ. Behav.* **2012**, *44*, 624–627. [CrossRef]
59. Chu, Y.L.; Warren, C.A.; Sceets, C.E.; Murano, P.; Marquart, L.; Reicks, M. Acceptance of Two US Department of Agriculture Commodity Whole-Grain Products: A School-Based Study in Texas and Minnesota. *J. Am. Diet. Assoc.* **2011**, *111*, 1380–1384. [CrossRef] [PubMed]
60. Lease, H.; Hendrie, G.A.; Poelman, A.A.M.; Delahunty, C.; Cox, D.N. A Sensory-Diet database: A tool to characterise the sensory qualities of diets. *Food Qual. Prefer.* **2016**, *49*, 20–32. [CrossRef]
61. White, J.S. Straight talk about high fructose corn syrup: What it is and what it ain't. *Am. J. Clin. Nutr.* **2008**, *88*, 1716S–1721S. [CrossRef]
62. Rogers, P.J.; Blundell, J.E. Umami and appetite: Effects of monosodium glutamate on hunger and food intake in human subjects. *Physiol. Behav.* **1990**, *48*, 801–804. [CrossRef]

 © 2019 by the authors. Licensee MDPI, Basel, Switzerland. This article is an open access article distributed under the terms and conditions of the Creative Commons Attribution (CC BY) license (http://creativecommons.org/licenses/by/4.0/).

 *foods* 

Article

Carotenoid Biosynthesis in Oriental Melon (*Cucumis melo* L. var. *makuwa*)

Pham Anh Tuan [1,†], **Jeongyeo Lee** [2,†], **Chang Ha Park** [1], **Jae Kwang Kim** [3], **Young-Hee Noh** [2], **Yeon Bok Kim** [4], **HyeRan Kim** [2,5,*] and **Sang Un Park** [1,*]

[1] Department of Crop Science, Chungnam National University, 99 Daehak-ro, Yuseong-gu, Daejeon 34134, Korea; tuan_pham_6885@yahoo.com (P.A.T.); parkch804@gmail.com (C.H.P.)

[2] Plant Systems Engineering Research Center, Korea Research Institute of Bioscience and Biotechnology (KRIBB), 125 Gwahangno, Yuseong-gu, Daejeon 305-806, Korea; leejy@kribb.re.kr (J.L.); yhnoh87@kribb.re.kr (Y.-H.N.)

[3] Division of Life Sciences, College of Life Sciences and Bioengineering, Incheon National University, Yeonsu-gu, Incheon 406-772, Korea; kjkpj@inu.ac.kr

[4] Department of Medicinal and Industrial Crops, Korea National College of Agriculture & Fisheries, 1515, Kongjwipatjwi-Ro, Jeonju, Jeonbuk 54874, Korea; yeondarabok@korea.kr

[5] Systems and Bioengineering, University of Science and Technology, 217 Gajung-ro, Daejeon 34113, Korea

[*] Correspondence: kimhr@kribb.re.kr (H.K.); supark@cnu.ac.kr (S.U.P.); Tel.: +82-42-860-4345 (H.K.); +82-42-821-6730 (S.U.P.); Fax: +82-42-860-4149 (H.K.); +82-42-822-2631 (S.U.P.)

[†] These authors contributed equally to this work.

Received: 23 January 2019; Accepted: 13 February 2019; Published: 19 February 2019

Abstract: Full-length cDNAs encoding ξ-carotene desaturase (CmZDS), lycopene ε-cyclase (CmLCYE), β-ring carotene hydroxylase (CmCHXB), and zeaxanthin epoxidase (CmZEP), and partial-length cDNA encoding ε-ring carotene hydroxylase (CmCHXE) were isolated in Chamoe (*Cucumis melo* L. var. *makuwa*), an important commercial fruit. Sequence analyses revealed that these proteins share high identity and common features with other orthologous genes. Expression levels of entire genes involved in the carotenoid biosynthetic pathway were investigated in the peel, pulp, and stalk of chamoe cultivars Ohbokggul and Gotgam. Most of the carotenoid biosynthetic genes were expressed at their highest levels in the stalk, whereas carotenoids were highly distributed in the peel. The expression levels of all carotenoid biosynthetic genes in fruits of the native cultivar Gotgam chamoe were higher than those in the cultivar Ohbokggul chamoe, consistent with the abundant carotenoid accumulation in Gotgam chamoe fruits and trace carotenoid content of Ohbokggul chamoe fruit. Lutein and β-carotene were the dominant carotenoids; high levels (278.05 µg g^{-1} and 112.02 µg g^{-1} dry weight, respectively) were found in the peel of Gotgam chamoe. Our findings may provide a foundation for elucidating the carotenoid biosynthetic mechanism in *C. melo* and inform strategies for developing new chamoe cultivars with improved characteristics.

Keywords: β-carotene; carotenoids; *Cucumis melo* L. var. *makuwa*; chamoe; gene characterization; lutein

---

1. Introduction

Melon (*Cucumis melo* L.), which belongs to the Cucurbitaceae family, is one of the most highly consumed fruit crops worldwide because of its pleasant flavor and nutritional value. Melons provide a rich source of protein, minerals, vitamins, and a wide range of antioxidant compounds [1–3]. The fruit can be consumed as a salad or as juice and is used by the food industry in products such as jam, ice cream, and yogurt. Melon fruits are diverse in shape, size, color, and flavor. In Korea, oriental melon (*Cucumis melo* L. var. *makuwa*), commonly known as chamoe, is an important commercial fruit due to its vigorous growth, good quality, and unique flavor, and consumer demand for the fruit is high. Chamoe has also been used in traditional medicine as a liver tonic and for its cardio-protective, antidiabetic, anti-obesity, and anticancer properties [4,5]. The cultivar Ohbokggul chamoe, which has a golden-colored skin with silver lines and sweet white flesh, is one of the most popular fruits on the market. The native Korean cultivar chamoe, Gotgam, has green skin with distinctive green stripes running from end to end, and very thick, light-green flesh. Gotgam chamoe has greater flavor, nutrient content, and disease resistance than other chamoe cultivars [6]. Given these favorable characteristics, Gotgam chamoe has recently acquired agronomic relevance for melon breeding programs in Korea.

Carotenoids, which contain 40 carbon molecules and are formed through the condensation of isoprenoids, represent a diverse group of pigments in nature [7]. In plants, carotenoids contribute to yellow, orange, and red coloration, and play a major role in the quality of flowers and fruit. The biosynthesis of biomolecules as carotenoids could be related to agri-environmental factors [8]. Several studies have demonstrated a positive correlation between phytochemical biosynthesis and light intensity and the spectral quality of vegetables and microgreens produced in controlled environments [9]. Carotenoids are accessory pigments that harvest light for photosynthesis, protect the photosystem from photooxidation, and attract pollinators and agents of seed dispersal [10–12]. In addition, the oxidative cleavage of carotenoids produces apocarotenoids, which serve as development signals and antifungal agents and contribute to the flavor and aroma of flowers and fruit [13]. In terms of human health, carotenoids play an important protective role as antioxidants, and a diet containing carotenoid-rich vegetables and fruit can reduce the risk of cancer, cardiovascular disease, macular degeneration, cataracts, and ultraviolet-induced skin damage [14–17]. More than 50 carotenoids with β-ring end groups (e.g., β-carotene and β-cryptoxanthin) are precursors of vitamin A, which is one of the most important micronutrients affecting human health [18,19]. Vitamin A deficiency increases the risk of infectious disease, especially measles, diarrhea, and malaria, and is considered the most common public health problem among preschool-aged children [20,21]. The importance of carotenoids to human health has led to an increase in studies of vegetables and fruit that contain these compounds.

In higher plants, carotenoids are synthesized and localized in the plastids, while the corresponding genes are located in the nucleus. To date, genes involved in carotenoid biosynthetic pathways in higher plants have been described in detail [22]. The first step in the formation of carotenoids is the condensation of two geranylgeranyl diphosphate (GGDP) molecules to form phytoene, which is catalyzed by phytoene synthase (PSY) (Figure 1). Phytoene undergoes a series of four desaturations to form lycopene via ξ-carotene, which is catalyzed by two enzymes, phytoene desaturase (PDS) and ξ-carotene desaturase (ZDS). Lycopene is a branching point in the pathway and is cyclized to form α-carotene by lycopene β-cyclase (LCYB) together with lycopene ε-cyclase (LCYE) or to produce β-carotene by LCYB alone through two reactions. Thereafter, α-carotene and β-carotene are hydroxylated to produce lutein and zeaxanthin, respectively. These reactions are catalyzed by β-ring carotene hydroxylase (CHXB) and ε-ring carotene hydroxylase (CHXE). Further epoxidation of zeaxanthin by zeaxanthin epoxidase (ZEP) produces violaxanthin, which is used to synthesize plant hormone abscisic acid (ABA) through oxidative cleavage catalyzed by 9-*cis* epoxycarotenoid dioxygenase (NCED) [23]. Along the pathway, carotenoids can be cleaved by carotenoid cleavage dioxygenases (CCD) to produce a diverse set of apocarotenoids [13].

Here, full-length cDNAs encoding ZDS, LCYE, CHXB, and ZEP, and partial-length cDNA encoding CHXE were isolated in *C. melo*. The expression levels of genes involved in carotenoid biosynthesis and carotenoid accumulation were investigated in fruits of the cultivar Ohbokggul and the native cultivar Gotgam chamoe using quantitative real-time PCR and high-performance liquid chromatography (HPLC), respectively. Therefore, this study will help elucidate the carotenoid biosynthetic mechanism in *C. melo* and will provide valuable information for breeding chamoe cultivars with improved characteristics.

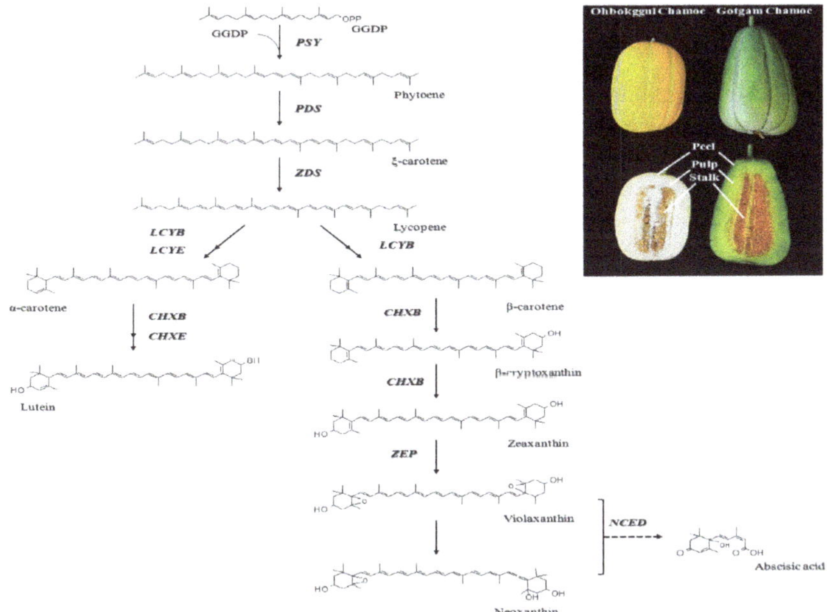

Figure 1. Carotenoid biosynthetic pathway in plants and photographs of Ohbokggul and Gotgam chamoes. GGDP (geranylgeranyl diphosphate); PSY (phytoene synthase); PDS (phytoene desaturase); ZDS (ξ-carotene desaturase); LCYB (lycopene β-cyclase); LCYE (lycopene ε-cyclase); CHXB (β-ring carotene hydroxylase); CHXE (ε-ring carotene hydroxylase); ZEP (zeaxanthin epoxidase); NCED (9-*cis* epoxycarotenoid dioxygenase).

2. Materials and Methods

2.1. Plant Materials

Two chamoe cultivars, *Cucumis melo* L. var. *makuwa* 'Ohbokggul' and *C. melo* L. var. *makuwa* 'Gotgam', were grown in a greenhouse at an experimental farm, and obtained from Nongwoo Bio (Korea) during the fruiting season in October 2012. Ohbokggul and Gotgam chamoes are differentiated by shape, size, and color (Figure 1). Three fruits of each cultivar were collected, and their peels, pulps, and stalks were separated. The samples were frozen in liquid nitrogen and stored at −80 °C until analysis.

2.2. Isolation of cDNAs Encoding Carotenoid Biosynthetic Genes

GenBank accession numbers U38550, NM_125085, NM_001125948, NM_180954, and U58919 were used as queries to search for homologous sequences in our internal chamoe transcriptome database (unpublished data). Full-length cDNAs encoding ZDS, LCYE, CHXB, and ZEP, and partial-length cDNA encoding CHXE were isolated in *C. melo* and designated as CmZDS, CmLCYE, CmCHXB, CmCHXE, and CmZEP (GenBank accession numbers: KF668331, KF668332, KF668333, KF668334, and KF668335, respectively).

2.3. Quantitative Real-Time PCR Analysis

Quantitative RT-PCR was performed for the precise analysis of transcript levels. Primers targeted to CmPSY, CmPDS, CmLCYB, CmCCD1, CmNCED, and CmACT2 (Accession Nos. GU361622, KC507802, GU457407, XM_004170465, JF838293, and AB033599, respectively) and five genes isolated in this study were designed using the Primer Quest computer program (http://eu.idtdna.com/Scitools/

Applications/Primerquest/), producing fragments of 80 to 90 bp (Table 1). Total RNA (5 μg) from each sample was combined with random hexamer primers in a SuperScript first-strand cDNA synthesis system according to the manufacturer's instructions (Invitrogen Life Technologies, Carlsbad, CA, USA). After cDNA synthesis, quantitative real-time PCR was performed using SYBR® Green SuperMix RT-PCR kit (IQ Sybr SYBR Green Super Mix, Bio-Rad, Hercules, CA, USA) on a MiniOption detection system (Bio-Rad, Hercules, CA, USA). Results were analyzed using Bio-Rad software (GeneXpression Macro Chromo4) and the comparative threshold cycle (Ct) method using CmACT2 as the reference according to the manufacturer's instructions for data normalization.

Table 1. Primers used in this study for quantitative real-time PCR analysis.

Gene Name	Primer Sequence (5′ to 3′)	
	Forward Primer	**Reverse Primer**
CmPSY	TGTGCAGAGTATGCCAAGAC	GTCCGCCTACACCATACATAAA
CmPDS	GGCTGGAGAAGTGGAGTTATTG	CCTCAGCTTAAAGCCAGAATACA
CmZDS	ACACTCCAGACGCAGATTTC	GCAATGATCCCTGTCCTTCA
CmLCYB	GTTTCTTCCCGAGCTGTTACT	GAGTTCCCTTTGCCATGATTTC
CmLCYE	TGGTCCAGATCTGCCATTTAC	CCGGCCATACATGCTCTATAC
CmCHXB	GCTGTCATGGCGGTTTATTAC	GGCACCAACAGAGAGAGAAA
CmCHXE	AATCGTTGCACTTGCCATATTC	GCTCCAGTAGTCATCCCAATG
CmZEP	GTAGAAGAATACGGGTTGCTGTA	CCGAGTCCAACTCCCAAATAA
CmCCD1	CATGATGAGACTCCTCCGATTAC	GATTTGGTCCCACCCTAACA
CmNCED	CAATCCTCTCTTCCAACCAACT	CTAGCGGAACCGTGATTGATAG
CmACT2	CTACGAACTTCCTGATGGACAAG	CCAATGAGAGATGGCTGGAATAG

2.4. Sequence Analysis

The deduced amino acid sequences of carotenoid biosynthetic genes from *C. melo* were analyzed for homology using the BLAST program and the NCBI GenBank database (http://www.ncbi.nlm.nih.gov/BLAST). Sequence alignments were carried out using BioEdit Sequence Alignment Editor, version 5.0.9 (Department of Microbiology, North Carolina State University, Raleigh, NC, USA). The predicted molecular mass of protein was calculated using an online website (http://www.sciencegateway.org/tools/proteinmw.htm).

2.5. Carotenoid Extraction and HPLC Analysis

Extraction and measurement of carotenoids by HPLC were performed as previously described by our group [24]. Briefly, carotenoids were released from the chamoe samples (0.02 g) by adding 3 mL of ethanol containing 0.1% ascorbic acid (*w/v*), vortex mixing for 20 s, and placing in a water bath at 85 °C for 5 min. The carotenoid extract was saponified with potassium hydroxide (120 μL, 80% *w/v*) in the 85 °C water bath for 10 min. After saponification, samples were placed immediately on ice, and cold deionized water (1.5 mL) was added. β-Apo-8′-carotenal (0.2 mL, 25 g/mL) was added as an internal standard. Carotenoids were extracted twice with hexane (1.5 mL) by centrifugation at $1200\times g$ to separate the layers. Aliquots of the extracts were dried under a stream of nitrogen and redissolved in 50:50 (*v/v*) dichloromethane/methanol before analysis by HPLC. The carotenoids were separated on a C30 YMC column (250 × 4.6 mm, 3 μm; Waters Corporation, Milford, MA, USA) by Agilent 1100 HPLC (Massy, France) equipped with a photodiode array (PDA) detector. Chromatograms were generated at 450 nm. Solvent A consisted of methanol/water (92:8 *v/v*) with 10 mM ammonium acetate. Solvent B consisted of 100% methyl *tert*-butyl ether. Gradient elution was performed at 1 mL/min under the following conditions: 0 min, 90% A/10% B; 20 min, 83% A/17% B; 29 min, 75% A/25% B; 35 min, 30% A/70% B; 40 min, 30% A/70% B; 42 min, 25% A/75% B; 45 min, 90% A/10% B; and 55 min, 90% A/10% B. Carotenoid standards were purchased from CaroteNature (Lupsingen, Switzerland). For quantification, calibration curves were created by plotting four different concentrations of carotenoid standards according to the peak area ratios with β-apo-8′-carotenal. Quantification was performed using calibration curves ranging from 0.3 to 5 µg/mL. The linear equations were y = 0.1178x − 0.027 for zeaxanthin, y = 0.1194x − 0.0063 for lutein, y = 0.0822x − 0.0003 for β-carotene, y = 0.0822x − 0.0003 for 9-cis β-carotene, y = 0.0822x − 0.0003 for 13-cis β-carotene, y = 0.0822x − 0.0003 for α-carotene, and y = 0.0884x − 0.0251 for β-cryptoxanthin.

2.6. Statistical Analysis

The data on expression levels of carotenoid biosynthetic genes were analyzed using the computer software Statistical Analysis System (SAS version 9.2). Treatment means were compared by Duncan's multiple range test.

3. Results

3.1. Sequence Analyses of Carotenoid Biosynthetic Genes from C. melo

CmZDS was composed of 1976 bp, with a 1731-bp open reading frame (ORF) encoding a protein of 576 amino acids (predicted molecular mass of 63.90 kDa; Figure S1). The closest homolog of CmZDS was ZDS from *Cucumis sativus* (98% identity and 99% similarity), followed by ZDS from *Cucurbita moschata* (93% identity and 96% similarity), ZDS from *Vitis vinifera* (89% identity and 95% similarity), and ZDS from *Citrus unshiu* (84% identity and 90% similarity). As shown in Figure S1, CmZDS contained a conserved dinucleotide-binding motif (GXGX$_2$GX$_3$AX$_2$LX$_3$GX$_6$EX$_5$GG) and a carotenoid-binding domain also found in other orthologous genes [25,26].

CmLCYE was 1958 bp long and had a 1602-bp ORF, encoding a protein of 533 amino acids with a predicted molecular mass of 58.81 kDa (Figure S2). CmLCYE shared 97% identity and 97% similarity with *Cucumis sativus* LCYE, 82% identity and 91% similarity with *Camellia sinensis* var. *assamica* LCYE, 83% identity and 90% similarity with *Glycine max* LCYE, and 79% identity and 88% similarity with *Vitis vinifera* LCYE. The deduced amino acid sequence of CmLCYE comprised a dinucleotide binding motif and two cyclase motifs, which are the common features of carotenoid cyclases [27,28].

CmCHXB consisted of 1292 bp with a 933-bp ORF and encoded a protein of 310 amino acids (predicted molecular mass of 34.78 kDa; Figure S3). CmCHXB exhibited 96% identity and 97% similarity with *Cucumis sativus* CHXB, 89% identity and 94% similarity with *Cucurbita moschata* CHXB, 77% identity and 86% similarity with *Vitis vinifera* CHXB, and 74% identity and 85% similarity with *Ipomoea nil* CHXB. Four conservatively spaced histidine motifs proposed to be involved in iron binding during hydroxylation reactions are marked in Figure S3 [29].

CmCHXE was composed of 933 bp encoding a partial 3′-end ORF of 148 amino acids. A BLAST search at the amino acid level showed that CmCHXE exhibited high homology to other CHXEs (Figure S4). Specifically, CmCHXE shared 97% identity and 97% similarity with *Cucumis sativus* CHXE, 84% identity and 92% similarity with *Vitis vinifera* CHXE, 88% identity and 94% similarity with *Fragaria vesca* subsp. *vesca* CHXE, and 83% identity and 93% similarity with *Daucus carota* subsp. *sativus* CHXE.

CmZEP was composed of 2514 bp with a 1998-bp ORF and encoded a protein of 665 amino acids with a predicted molecular mass of 73.20 kDa (Figure S5). CmZEP shared 98% identity and 98% similarity with *Cucumis sativus* ZEP, 95% identity and 96% similarity with *Citrullus lunatus* ZEP, 88% identity and 94% similarity with *Cucurbita moschata* ZEP, and 75% identity and 85% similarity with *Prunus armeniaca* ZEP. CmZEP displayed two short motifs typical of the lipocalin family of proteins and a phosphopeptide-binding domain (The forkhead-associated (FHA) domain), which are present in all known ZEP genes [30,31].

3.2. Expression Levels of Carotenoid Biosynthetic Genes in Ohbokggul and Gotgam Chamoes

Expression levels of carotenoid biosynthetic genes in the peel, pulp, and stalk of Ohbokggul were compared to those in Gotgam (Figure 2). In Ohbokggul chamoe, the highest expression levels of CmPSY were found in the stalk, with lower levels in the pulp and peel. This same pattern of expression was observed for *CmPDS*, *CmLCYB*, *CmLCYE*, *CmCCD1*, and *CmNCED*. Transcript levels of *CmCHXE* were highest in the peel and lowest in the stalk of Ohbokggul; *CmZDS*, *CmCHXB*, and *CmZEP* were expressed at similar levels in the peel, pulp, and stalk of Ohbokggul. In general, mRNA levels of carotenoid biosynthetic genes in all fruit parts were higher in Gotgam than in Ohbokggul. Transcription of most carotenoid biosynthetic genes (*CmPSY*, *CmPDS*, *CmCHXB*, *CmCCD1*, and *CmNCED*) was highest in the stalk and lowest in the peel of Gotgam chamoe. *CmLCYE* and *CmZEP* showed the highest expression levels in the pulp and peel, respectively. No differences in transcript levels of *CmZDS* and *CmCHXE* were found in the peel, pulp, or stalk of Gotgam chamoe.

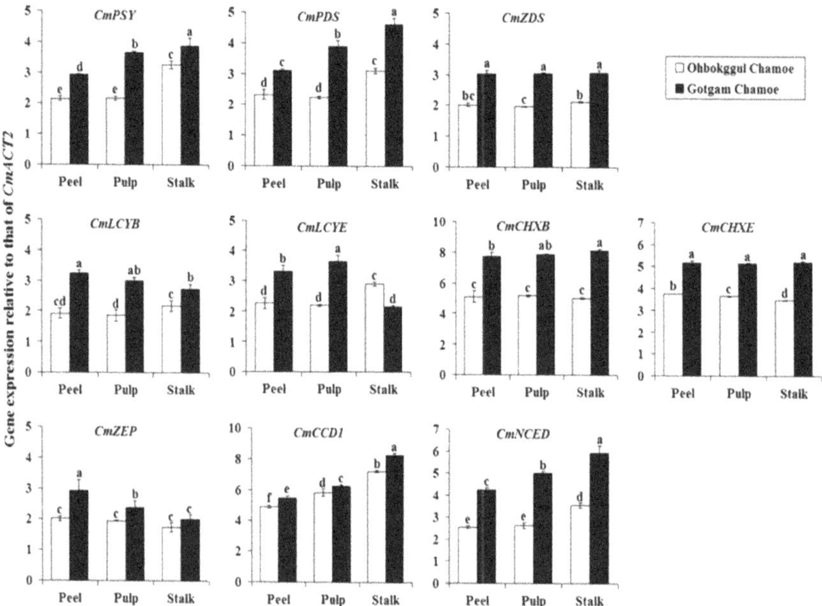

Figure 2. Expression levels of carotenoid biosynthetic genes in different parts of Ohbokggul and Gotgam chamoe fruit. Values are means; bars represent standard error from three independent measurements. The letters a, b, c, d, e, and f indicate significant differences at the 5% level by Duncan's multiple range test.

3.3. Analysis of Carotenoid Content in Ohbokggul and Gotgam Chamoes

The same Ohbokggul and Gotgam peel, pulp, and stalk materials used for quantitative real-time PCR were used to analyze carotenoid composition and content by HPLC (Table 2). Surprisingly, carotenoids were very poorly synthesized in Ohbokggul fruit, with only trace amounts of total carotenoids measured in the peel (0.89 µg g^{-1}), pulp (0.02 µg g^{-1}), and stalk (0.51 µg g^{-1}). In contrast, total carotenoid content was high in the peel of Gotgam chamoe (428.81 µg g^{-1}). Lutein and β-carotene were the dominant compounds in Gotgam peel (278.05 µg g^{-1} and 112.02 µg g^{-1}, respectively); lower concentrations of 9-*cis* β-carotene (10.27 µg g^{-1}), 13-*cis* β-carotene (11.82 µg g^{-1}), and β-cryptoxanthin (13.44 µg g^{-1}) were also detected in the peel. Similar to the peel, lutein and β-carotene were the major carotenoids synthesized in the pulp, while only β-carotene showed an appreciable concentration in the stalk of Gotgam chamoe.

Table 2. Carotenoid composition and content in different parts of Ohbokggul and Gotgam chamoe fruit (μg g^{-1} dry weight). The results are expressed as means $\pm$ standard error from three independent measurements. N.D., not detected.

Carotenoids	Ohbokggul Chamoe			Gotgam Chamoe		
	Peel	Pulp	Stalk	Peel	Pulp	Stalk
α-carotene	N.D.	N.D.	N.D.	2.54 ± 0.33	0.38 ± 0.01	N.D.
Lutein	0.45 ± 0.05	0.02 ± 0.00	0.07 ± 0.01	278.05 ± 23.51	14.16 ± 0.37	0.52 ± 0.11
β-carotene	0.27 ± 0.04	N.D.	0.33 ± 0.04	112.02 ± 10.69	6.45 ± 1.06	17.64 ± 3.94
9-*cis* β-carotene	0.02 ± 0.00	N.D.	0.02 ± 0.00	10.27 ± 0.69	0.50 ± 0.05	0.66 ± 0.15
13-*cis* β-carotene	0.07 ± 0.02	N.D.	0.04 ± 0.01	11.82 ± 1.56	0.96 ± 0.32	2.29 ± 0.53
β-cryptoxanthin	0.03 ± 0.01	N.D.	0.01 ± 0.00	13.44 ± 1.12	1.88 ± 0.19	2.23 ± 0.55
Zeaxanthin	0.05 ± 0.01	N.D.	0.05 ± 0.00	0.67 ± 0.07	0.11 ± 0.02	0.02 ± 0.00
Total	0.89 ± 0.14	0.02 ± 0.00	0.51 ± 0.07	428.81 ± 37.99	24.44 ± 2.01	23.35 ± 5.28

4. Discussion

In the present study, five carotenoid biosynthetic genes, *CmZDS*, *CmLCYE*, *CmCHXB*, *CmCHXE*, and *CmZEP*, were isolated from *C. melo*. Sequence analyses revealed that they shared high identity and common features with other orthologous genes. In addition, expression levels of entire genes involved in carotenoid biosynthetic pathways were investigated in different fruit parts of the Ohbokggul and Gotgam cultivars, the latter of which is a native Korean variety. *CmPSY*, which catalyzes the first committed and rate-limiting step in carotenoid biosynthesis [32,33], and most of the other carotenoid biosynthetic genes were expressed at their highest levels in the stalk. However, carotenoids were highly distributed in the peel, where tissue has direct exposure to light, suggesting the essential role of light in carotenoid accumulation in chamoe. On the other hand, PSY is often encoded by multiple genes which exhibit distinct expression and regulation in plants. There are two isoforms of PSY in tomato, where PSY1 is a chromoplast-specific isoform and PSY2 is a chloroplast-specific isoform [34]. It has been suggested that there is another isoform of *CmPSY* which directly regulates the carotenoid accumulation in the peel of chamoe. In addition, *CmCCD1*, which can cleave multiple carotenoid substrates at various positions, showed the highest expression level in the stalk [13]. Therefore, we hypothesize that the low content of carotenoids in the stalk of chamoe was because of the high activity of *CmCCD1* found in this part.

The expression levels of all carotenoid biosynthetic genes in Gotgam fruits were higher than those in Ohbokggul fruits, which probably led to the abundant carotenoid accumulation in Gotgam melons and the low carotenoid content in fruits of Ohbokggul. However, these higher expression levels of carotenoid biosynthetic genes cannot entirely account for the substantially higher total carotenoid content (up to 480-fold) in Gotgam peel compared to Ohbokggul peel. In addition, differences in carotenoid biosynthesis between the Ohbokggul and Gotgam cultivars provide a basic foundation for more detailed study of the molecular genetics of *C. melo*.

5. Conclusions

In conclusion, differences in the expression levels of carotenoid biosynthetic genes and carotenoid content between the cultivar Ohbokggul chamoe and the native Korean cultivar Gotgam chamoe were observed. These findings will contribute to a foundation for the elucidation of carotenoid biosynthesis in *C. melo*, an important commercial crop. In addition, further investigations regarding molecular genetics and enzyme activities may help to identify key genes for improving the carotenoid accumulation in *C. melo*.

Supplementary Materials: The following are available online at http://www.mdpi.com/2304-8158/8/2/77/s1, Figure S1: Multiple alignments of the amino acid sequences of CmZDS with other ZDSs. Figure S2: Multiple alignments of the amino acid sequences of CmLCYE with other LCYEs. Figure S3: Multiple alignments of the amino acid sequences of CmCHXB with other CHXBs. Figure S4: Multiple alignments of the amino acid sequences of CmCHXE with other CHXEs. Figure S5: Multiple alignments of the amino acid sequences of CmZEP with other ZEPs.

Author Contributions: S.U.P. and H.K. designed the experiments and analyzed the data. P.A.T., J.L., C.H.P., J.K.K., Y.-H.N., and Y.B.K. performed the experiments and analyzed the data. P.A.T. and J.L. wrote the manuscript. All authors read and approved the final manuscript.

Funding: This work was financially supported by a grant from the Korea Research Institute of Bioscience and Biotechnology research initiative program.

Conflicts of Interest: The authors declare no conflict of interest.

References

1. Maynard, D.N.; Hochmuth, G.J. *Knott's Handbook for Vegetable Growers*, 5th ed.; John Wiley & Sons: Hoboken, NJ, USA, 1980; pp. 1–621.
2. Vouldoukis, I.; Lacan, D.; Kamate, C.; Coste, P.; Calenda, A.; Mazier, D.; Conti, M.; Dugas, B. Antioxidant and anti-inflammatory properties of a *Cucumis melo* LC. extract rich in superoxide dismutase activity. *J. Ethnopharmacol.* **2004**, *94*, 67–75. [CrossRef] [PubMed]
3. Ismail, H.I.; Chan, K.W.; Mariod, A.A.; Ismail, M. Phenolic content and antioxidant activity of cantaloupe (*Cucumis melo*) methanolic extracts. *Food Chem.* **2010**, *119*, 643–647. [CrossRef]
4. Choi, Y.-J.; Chun, H.; Choi, Y.; Yum, S.; Lee, S.; Kim, H.; Shin, Y.-S.; Chung, D.-S. Nutritional components content of oriental melon fruits cultivated under different greenhouse covering films. *J. Bio-Environ. Control* **2007**, *16*, 67–75.
5. Shin, Y.-S.; Lee, J.-E.; Yeon, I.-K.; Do, H.-W.; Cheung, J.-D.; Kang, C.-K.; Choi, S.-Y.; Youn, S.-J.; Cho, J.-G.; Kwoen, D.-J. Antioxidant Effects and Tyrosinase Inhibition Activity of Oriental Melon (*Cucumis melo* L. var makuwa Makino) Extracts. *J. Life Sci.* **2008**, *18*, 963–967. [CrossRef]
6. Kim, J.-H.; Suh, J.-K.; Kang, Y.-H. Anticancer effects of the extracts of oriental melon (*Cucumis melo* L. var makuwa Makino) seeds. *Korean J. Plant Res.* **2012**, *25*, 647–651. [CrossRef]
7. Britton, G. Overview of carotenoid biosynthesis. *Carotenoids* **1998**, *3*, 13–147.
8. Frank, H.A.; Cogdell, R.J. Carotenoids in photosynthesis. *Photochem. Photobiol.* **1996**, *63*, 257–264. [CrossRef] [PubMed]
9. Calzarano, F.; Osti, F.; Baranek, M.; Di Marco, S. Rainfall and temperature influence expression of foliar symptoms of grapevine leaf stripe disease (esca complex) in vineyards. *Phytopathologia Mediterranea* **2018**, *57*, 488–505.
10. Kyriacou, M.C.; Rouphael, Y. Towards a new definition of quality for fresh fruits and vegetables. *Sci. Horticult.* **2018**, *234*, 463–469. [CrossRef]
11. Havaux, M. Carotenoids as membrane stabilizers in chloroplasts. *Trends Plant Sci.* **1998**, *3*, 147–151. [CrossRef]
12. Ledford, H.K.; Niyogi, K.K. Singlet oxygen and photo-oxidative stress management in plants and algae. *Plant Cell Environ.* **2005**, *28*, 1037–1045. [CrossRef]
13. Auldridge, M.E.; McCarty, D.R.; Klee, H.J. Plant carotenoid cleavage oxygenases and their apocarotenoid products. *Curr. Opin. Plant Biol.* **2006**, *9*, 315–321. [CrossRef] [PubMed]
14. Ziegler, R.G. Vegetables, fruits, and carotenoids and the risk of cancer. *Am. J. Clin. Nutr.* **1991**, *53*, 251S–259S. [CrossRef] [PubMed]
15. Gaziano, J.M.; Hennekens, C.H. The role of beta-carotene in the prevention of cardiovascular disease. *Ann. N. Y. Acad. Sci.* **1993**, *691*, 148–155. [CrossRef] [PubMed]
16. Mayne, S.T. Beta-carotene, carotenoids, and disease prevention in humans. *FASEB J.* **1996**, *10*, 690–701. [CrossRef] [PubMed]
17. Giovannucci, E. Tomatoes, tomato-based products, lycopene, and cancer: Review of the epidemiologic literature. *J. Natl. Cancer Inst.* **1999**, *91*, 317–331. [CrossRef] [PubMed]
18. Fraser, P.D.; Bramley, P.M. The biosynthesis and nutritional uses of carotenoids. *Prog. Lipid Res.* **2004**, *43*, 228–265. [CrossRef]
19. Mactier, H.; Weaver, L. Vitamin A and preterm infants: What we know, what we don't know, and what we need to know. *Arch. Dis. Child. Fetal Neonatal Ed.* **2005**, *90*, F103–F108. [CrossRef]
20. Villamor, E.; Fawzi, W.W. Vitamin A supplementation: Implications for morbidity and mortality in children. *J. Infect. Dis.* **2000**, *182*, S122–S133. [CrossRef]

21. Thorne-Lyman, A.; Fawzi, W.W. Vitamin D during pregnancy and maternal, neonatal and infant health outcomes: A systematic review and meta-analysis. *Paediatr. Perinat. Epidemiol.* **2012**, *26*, 75–90. [CrossRef]

22. Cunningham, F., Jr.; Gantt, E. Genes and enzymes of carotenoid biosynthesis in plants. *Annu. Rev. Plant Biol.* **1998**, *49*, 557–583. [CrossRef] [PubMed]

23. Schwartz, S.H.; Tan, B.C.; Gage, D.A.; Zeevaart, J.A.; McCarty, D.R. Specific oxidative cleavage of carotenoids by VP14 of maize. *Science* **1997**, *276*, 1872–1874. [CrossRef] [PubMed]

24. Park, C.H.; Chae, S.C.; Park, S.-Y.; Kim, J.K.; Kim, Y.J.; Chung, S.O.; Arasu, M.V.; Al-Dhabi, N.A.; Park, S.U. Anthocyanin and carotenoid contents in different cultivars of chrysanthemum (*Dendranthema grandiflorum* Ramat.) flower. *Molecules* **2015**, *20*, 11090–11102. [CrossRef] [PubMed]

25. Zhu, Y.-H.; Jiang, J.-G.; Yan, Y.; Chen, X.-W. Isolation and characterization of phytoene desaturase cDNA involved in the β-carotene biosynthetic pathway in Dunaliella salina. *J. Agric. Food Chem.* **2005**, *53*, 5593–5597. [CrossRef] [PubMed]

26. Yan, P.; Gao, X.; Shen, W.; Zhou, P. Cloning and expression analysis of phytoene desaturase and ζ-carotene desaturase genes in Carica papaya. *Mol. Biol. Rep.* **2011**, *38*, 785–791. [CrossRef] [PubMed]

27. Rossmann, M.G.; Moras, D.; Olsen, K.W. Chemical and biological evolution of a nucleotide-binding protein. *Nature* **1974**, *250*, 194–199. [CrossRef]

28. Borchetia, S.; Bora, C.; Gohain, B.; Bhagawati, P.; Agarwala, N.; Bhattacharya, N.; Bharalee, R.; Bhorali, P.; Bandyopadhyay, T.; Gupta, S. Cloning and heterologous expression of a gene encoding lycopene-epsilon-cyclase, a precursor of lutein in tea (*Camellia sinensis* var assamica). *Afr. J. Biotechnol.* **2011**, *10*, 5934–5939.

29. Bouvier, F.; Keller, Y.; d'Harlingue, A.; Camara, B. Xanthophyll biosynthesis: Molecular and functional characterization of carotenoid hydroxylases from pepper fruits (*Capsicum annuum* L.). *Biochim. Biophys. Acta* **1998**, *1391*, 320–328. [CrossRef]

30. Hieber, A.D.; Bugos, R.C.; Yamamoto, H.Y. Plant lipocalins: Violaxanthin de-epoxidase and zeaxanthin epoxidase. *Biochim. Biophys. Acta* **2000**, *1482*, 84–91. [CrossRef]

31. Durocher, D.; Jackson, S.P. The FHA domain. *FEBS Lett.* **2002**, *513*, 58–66. [CrossRef]

32. Rodríguez-Villalón, A.; Gas, E.; Rodríguez-Concepción, M. Phytoene synthase activity controls the biosynthesis of carotenoids and the supply of their metabolic precursors in dark-grown Arabidopsis seedlings. *Plant J.* **2009**, *60*, 424–435. [CrossRef] [PubMed]

33. Toledo-Ortiz, G.; Huq, E.; Rodríguez-Concepción, M. Direct regulation of phytoene synthase gene expression and carotenoid biosynthesis by phytochrome-interacting factors. *Proc. Natl. Acad. Sci. USA* **2010**, *107*, 11626–11631. [CrossRef] [PubMed]

34. Fraser, P.; Kiano, J.; Truesdale, M.; Schuch, W.; Bramley, P. Phytoene synthase-2 enzyme activity in tomato does not contribute to carotenoid synthesis in ripening fruit. *Plant Mol. Biol.* **1999**, *40*, 687–698. [CrossRef] [PubMed]

© 2019 by the authors. Licensee MDPI, Basel, Switzerland. This article is an open access article distributed under the terms and conditions of the Creative Commons Attribution (CC BY) license (http://creativecommons.org/licenses/by/4.0/).

MDPI

St. Alban-Anlage 66

4052 Basel

Switzerland

Tel. +41 61 683 77 34

Fax +41 61 302 89 18

www.mdpi.com

Foods Editorial Office

E-mail: foods@mdpi.com

www.mdpi.com/journal/foods

www.ingramcontent.com/pod-product-compliance
Lightning Source LLC
LaVergne TN
LVHW071031170726
843515LV00006B/1256